CW01467833

Other Titles in This Series

148 **Vladimir I. Piterbarg,** Asymptotic methods in the theory of Gaus
147 **S. G. Gindikin and L. R. Volevich,** Mixed problems for partial di quasihomogeneous principal part, 1995
146 **L. Ya. Adrianova,** Introduction to linear systems of differential equations, 1995
145 **A. N. Andrianov and V. G. Zhuravlev,** Modular forms and Hecke operators, 1995
144 **O. V. Troshkin,** Nontraditional methods in mathematical hydrodynamics, 1995
143 **V. A. Malyshev and R. A. Minlos,** Linear infinite-particle operators, 1995
142 **N. V. Krylov,** Introduction to the theory of diffusion processes, 1995
141 **A. A. Davydov,** Qualitative theory of control systems, 1994
140 **Aizik I. Volpert, Vitaly A. Volpert, and Vladimir A. Volpert,** Traveling wave solutions of parabolic systems, 1994
139 **I. V. Skrypnik,** Methods for analysis of nonlinear elliptic boundary value problems, 1994
138 **Yu. P. Razmyslov,** Identities of algebras and their representations, 1994
137 **F. I. Karpelevich and A. Ya. Kreinin,** Heavy traffic limits for multiphase queues, 1994
136 **Masayoshi Miyanishi,** Algebraic geometry, 1994
135 **Masaru Takeuchi,** Modern spherical functions, 1994
134 **V. V. Prasolov,** Problems and theorems in linear algebra, 1994
133 **P. I. Naumkin and I. A. Shishmarev,** Nonlinear nonlocal equations in the theory of waves, 1994
132 **Hajime Urakawa,** Calculus of variations and harmonic maps, 1993
131 **V. V. Sharko,** Functions on manifolds: Algebraic and topological aspects, 1993
130 **V. V. Vershinin,** Cobordisms and spectral sequences, 1993
129 **Mitsuo Morimoto,** An introduction to Sato's hyperfunctions, 1993
128 **V. P. Orevkov,** Complexity of proofs and their transformations in axiomatic theories, 1993
127 **F. L. Zak,** Tangents and secants of algebraic varieties, 1993
126 **M. L. Agranovskiĭ,** Invariant function spaces on homogeneous manifolds of Lie groups and applications, 1993
125 **Masayoshi Nagata,** Theory of commutative fields, 1993
124 **Masahisa Adachi,** Embeddings and immersions, 1993
123 **M. A. Akivis and B. A. Rosenfeld,** Élie Cartan (1869–1951), 1993
122 **Zhang Guan-Hou,** Theory of entire and meromorphic functions: Deficient and asymptotic values and singular directions, 1993
121 **I. B. Fesenko and S. V. Vostokov,** Local fields and their extensions: A constructive approach, 1993
120 **Takeyuki Hida and Masuyuki Hitsuda,** Gaussian processes, 1993
119 **M. V. Karasev and V. P. Maslov,** Nonlinear Poisson brackets. Geometry and quantization, 1993
118 **Kenkichi Iwasawa,** Algebraic functions, 1993
117 **Boris Zilber,** Uncountably categorical theories, 1993
116 **G. M. Fel′dman,** Arithmetic of probability distributions, and characterization problems on abelian groups, 1993
115 **Nikolai V. Ivanov,** Subgroups of Teichmüller modular groups, 1992
114 **Seizô Itô,** Diffusion equations, 1992
113 **Michail Zhitomirskiĭ,** Typical singularities of differential 1-forms and Pfaffian equations, 1992
112 **S. A. Lomov,** Introduction to the general theory of singular perturbations, 1992
111 **Simon Gindikin,** Tube domains and the Cauchy problem, 1992
110 **B. V. Shabat,** Introduction to complex analysis Part II. Functions of several variables, 1992
109 **Isao Miyadera,** Nonlinear semigroups, 1992
108 **Takeo Yokonuma,** Tensor spaces and exterior algebra, 1992
107 **B. M. Makarov, M. G. Goluzina, A. A. Lodkin, and A. N. Podkorytov,** Selected problems in real analysis, 1992

(Continued in the back of this publication)

Introduction to the Theory of Diffusion Processes

TRANSLATIONS OF
MATHEMATICAL
MONOGRAPHS
VOLUME 142

N. V. Krylov

Introduction to the Theory of Diffusion Processes

American Mathematical Society

Н. В. Крылов

ВВЕДЕНИЕ В
ТЕОРИЮ ДИФФУЗИОННЫХ ПРОЦЕССОВ

Translated by Valim Khidekel and Gennady Pasechnik
from an original Russian manuscript
Translation edited by D. Louvish

The translation, editing, and keyboarding of the material for this book was done in the framework of the joint project between the AMS and Tel-Aviv University, Israel.

1991 *Mathematics Subject Classification.* Primary 58G32, 60J60, 60J65;
Secondary 35R60, 60G44, 60H15.

ABSTRACT. The theory of diffusion type random processes occupies a central position in theory of stochastic processes. The book is intended to present basic elements of this theory, at a contemporary level of mathematical rigor, in such a way that it is understandable to a reader who knows very little about probability theory, not to mention the theory of random processes. The approach used by the author is based on stochastic analysis, when diffusion processes are interpreted as solutions of Itô's stochastic integral equations. Another main tool, explained and widely used in the book, is the martingale theory. Using this approach, the author is also able to explain the connections between the theory of diffusion processes and the theory of partial differential equations. The main presentation is supplied with exercises and problems that widen the scope of the reader's understanding of the material.

The book is useful to graduate students and researchers specializing in stochastic processes and differential equations. It can be recommended as a textbook for a graduate course in diffusion processes, and also can be used by any mathematician who wants to learn about this subject.

Library of Congress Cataloging-in-Publication Data

Krylov, N. V. (Nikolaĭ Vladimirovich)
Introduction to the theory of diffusion processes/N. V. Krylov.
p. cm. — (Translations of mathematical monographs, ISSN 0065-9282; v. 142)
Includes bibliographical references and index.
ISBN 0-8218-4600-0
1. Diffusion processes. I. Title. II. Series.
QA274.75.K79 1995
519.2′33—dc20 94-37626
CIP

© Copyright 1995 by the American Mathematical Society. All rights reserved.
The American Mathematical Society retains all rights
except those granted to the United States Government.
Printed in the United States of America.

Reprinted with corrections, 1996.
∞ The paper used in this book is acid-free and falls within the guidelines
established to ensure permanence and durability.
♻ Printed on recycled paper.

Information on Copying and Reprinting can be found in the back of this volume.

10 9 8 7 6 5 4 3 2 01 00 99 98 97

Contents

Preface[1]

The theory of diffusion type random processes occupies a central place in the theory of stochastic processes. Originally being a mathematical model of Brownian motion, it gradually became a major branch of probability theory studying the large class of processes with continuous sample paths that possess "Markov property," usually interpreted as the absence of aftereffects. There are now several approaches to the study of such processes. In this book we will concern ourselves with only one approach based on stochastic analysis, which operates with such notions as stochastic differentials and stochastic integrals. The diffusion processes discussed in this book are interpreted as solutions of Itô's stochastic integral equations.

From the very beginning, the theory of diffusion processes has had links with second-order elliptic and parabolic partial differential equations, which also arise in the theory of physical diffusion and the related heat conduction processes. On intuitive level, many properties of the solutions of differential equations admit a natural stochastic interpretation. On the one hand, this often makes it possible to obtain nontrivial information about the behavior of a process using the theory of differential equations; on the other, it sometimes enables us to apply probabilistic arguments to the study of differential equations. For example, these were probabilistic methods which led Gikhman in 1951 to the first proof that the Cauchy problem for second-order degenerate parabolic equations is solvable.

Probabilistic arguments are also useful in other branches of mathematics. However, specialists in those branches, especially of older generations, are often skeptical about probabilistic proofs of analytical results. The author had often heard the claim that "whatever can be proved with probability theory can also be done without it." This essentially true statement actually challenges the value and justifiability of using probabilistic methods in other branches of mathematics. Such caution is partly justified by the formerly common style of teaching probability theory without using its basic tool, measure theory, sometimes forcing one to resort to semi-intuitive reasoning. An example of this approach is the textbook of Bernstein [2] (1946), though it is remarkable in many respects.

Nowdays the situation is quite different; measure theory having been made the foundation of probability theory by Kolmogorov in 1933 is no longer "out of bounds," and the interest in probabilistic methods has increased. This applies, in particular, to the theory of diffusion processes. The author has frequently been asked to recommend a quick introduction to the theory of diffusion processes. It turned out that this is not easy to do, since the specialized literature, such as by Itô and McKean [24] (1965),

[1]The second printing is almost identical to the first one apart from correcting quite a few mistakes and misprints found in the first edition. My students A. Zatezalo and H. Yoo found and reported to me major part of them for which I am sincerely thankful.

Stroock and Varadhan [47] (1979), Ikeda and Watanabe [20] (1989), and Durrett [11] (1984) presuppose a good command of both probability theory and the fundamentals of the theory of random processes. For this reason, they cannot be recommended to beginners. On the other hand, books treating diffusion processes as part of the theory of stochastic processes, such as Doob [8] (1953), Loève [38] (1977-78), Gikhman and Skorokhod [19] (1975), contain so much information before getting to the theory of diffusion processes that it is practically impossible for a beginner to pick out what is really necessary.

The aim of this book is to present the basic elements of the theory of diffusion processes, at the contemporary level of mathematical rigor, in a way understandable to a reader completely ignorant even of probability theory, to say nothing of the theory of stochastic processes. The only prerequisites are second-year university level mathematics and a knowledge of the existence of one-dimensional Lebesgue measure on $(-\infty, \infty)$. The level of the reader is also assumed to be high enough to allow him or her to consider the necessary set-theoretical arguments as quite natural.

All the prerequisites from the measure theory and the theory of the Lebesgue integral are assembled in the first chapter, which, though somewhat sketchy, presents all the proofs, difficult or not. Some proofs are left to the reader as exercises; this will be done in subsequent chapters as well. On the one hand, this has enabled us somewhat to reduce the volume of the book; on the other, it will enable the reader constantly to check the sufficiency of his/her command of the previous material by doing the exercises (which are often provided with hints).

Along with exercises, the book presents a large number of problems; though it is not necessary to solve them, an acquaintance with them will widen the scope of reader's knowledge. Material from problems will sometimes be used, but only in other problems. Some of the terms introduced in the book are also intended purely for reader's edification and are not used in the main part of the book. As to the definitions of essential notions, we will try to prepare the reader properly (except in Chapter I), and to give them where they begin to work. For example, convergence in probability, a basic notion of probability theory, is introduced only in Sec. III.6, after the Itô stochastic integral with variable upper limit has been defined, while the term "Bernoulli trials" never used elsewhere in this book, is defined in Sec. II.2.

We have deliberately tried to define as many terms like Bernoulli trials, Ornstein-Uhlenbeck process, Cauchy process, Brownian bridge, etc., as possible and as few "working notions" as possible. Though this certainly makes the book easier to read, it does not give the reader complete knowledge of the tools of the theory of diffusion processes, which is quite natural for an introduction to the theory. A similar merit that is at the same time a flaw is that the notions of conditional probability and conditional expectation are not even mentioned until the penultimate section, where they appear only in a very special case, in a semiheuristic explanation of certain ideas. Any reader who, after having read this book, would like to become acquainted with more profound results of the theory of diffusion processes by reading other books, should keep those features of our book in mind. We hope that he/she, having learned sufficiently many topics of the theory, will willingly and easily fill the gaps.

One consequence of our approach is that many arguments concerning the same class of objects are scattered throughout the book, but not collected at its beginning like, say, a preliminary "artillery attack." In the author's view, such an introductory chapter would have been only cumbersome and somewhat frightening and would not have helped to introduce the reader gradually to increasingly more complicated ideas.

On the other hand, our approach makes it difficult to get clear perception of the whole strength of some auxiliary notions. To correct partially this shortcoming, we point out that necessary facts from the theory of continuous martingales (including local martingales), which is widely used in the book, are proved in: Lemmas II.8.5, II.8.8, Theorems II.9.6, III.3.4, III.4.9, Lemma III.6.3, Theorems III.6.8, III.6.11, III.6.12, III.6.15, Lemma III.8.2, Sections III.9 and III.10, Theorems IV.1.5, IV.2.3, IV.3.4, Section IV.4 and Lemmas V.10.1, V.10.6.

A not very experienced reader acquainted with the main subject of the theory of diffusion processes, may wonder at the wide use of the theory of continuous martingales, where we prove all the necessary results that we need, as opposed to the absence of the notion of conditional expectation. We hope that part of the material, or the style of presentation will also interest specialists. This could apply, for example, to somewhat nonstandard construction of the Itô stochastic integral, which enable us, immediately after the definition, to compute integrals of the most interesting functions; to our investigation of solutions of the Itô stochastic equations, including the derivation of the Kolmogorov equations, using the Euler's method; and to the last chapter, which describes comparatively new methods for investigating probabilistic solutions of differential equations. Another merit of the book, as we see it, is the calculation of a large number of probability distributions and their parameters, such as often occuring in applications, distributions of random variables $\max_{t\le T}|w_t+bt|$, $\max_{t\le T}(w_t+bt)$, where w_t is a one-dimensional Wiener process, b a constant.

We have already described the content of the first chapter. The contents of the other chapters may be inferred from the table of contents and the introductions to the chapters. To refer to results in the book, we will use numbers consisting of one or more numerals. For example "Theorem III.4.9" means "Theorem 9 in Sec. 4 of Chapter III"; in Chapter III this theorem is called just "Theorem 4.9"; and in Sec. III.4, "Theorem 9."

A list of notations may be found at the end of the book; but some standard symbols and notions are not explained. For example, a domain is an arbitrary open set, $\delta^{ij}=\delta_{ij}$ is the Kronecker symbol,

$$u_{x^i}=\frac{\partial u}{\partial x^i},\quad u_{x^ix^j}=\frac{\partial^2 u}{\partial x^i\partial x^j},\quad a\wedge b=\min(a,b),$$

$$a\vee b=\max(a,b),\quad a_+=\frac{1}{2}(|a|+a),\quad a_-=\frac{1}{2}(|a|-a),$$

A^* is the matrix adjoint to A, or a column-vector if A is a row-vector. The derivative of $u(t,x)$ with respect to time t is always denoted by $\partial u/\partial t$. We will also use the somewhat extravagant convention

$$\inf\emptyset=\infty.$$

Expressions of the type $x:=y$ or $y=:x$ mean: "x is equal to y by definition"; the terms "increasing," "positive," etc. are used in the nonstrict sense, i.e., they mean "nondecreasing," "nonnegative," etc. Constants appearing in inequalities are usually not indexed; they may differ even in the same chain of inequalities. When we write $N=N(\dots)$, we are not defining a function N, but rather indicate that N depends only on the terms in the parentheses. Finally, as usual, we write a point x in a d-dimensional

Euclidean space as $(x^1, \ldots, x^d)$, considered as a column-vector.

A. D. Wentzell read the manuscript of this book, and I am sincerely grateful to him for his remarks.

N. Krylov
Moscow – Diego-Suarez
May, 1989

CHAPTER I

Elements of Measure and Integration Theory

This chapter presents results and methods of measure and integration theory that will be constantly used. We would like to accustom the reader to use more or less abstract arguments. This is absolutely necessary for understanding the subsequent chapters. A reader desiring to become not only acquainted with but also proficient in the theory of diffusion processes should study this chapter very attentively, and be ready in advance for possible slow progress, especially when he/she encounters a systematic use of set-theoretical constructions for the first time.

Sec. 3 can be postponed if one accepts the fact that the Lebesgue and Riemann integrals coincide for any Riemann-integrable Borel function. Though we will need the notion of the completion of a measure space as early as Sec. III.1, we will begin to use it systematically only in Sec. III.3, after defining the Itô stochastic integral.

1. Measurable spaces and random variables

Let Ω be a set and $\mathcal{F}$ a collection of subsets of Ω.

1. Definition. $\mathcal{F}$ is called a *σ-algebra* (in Ω), if (a) $\Omega \in \mathcal{F}$, (b) $\bigcup_n A_n \in \mathcal{F}$ for any $A_1, A_2, \dots \in \mathcal{F}$, (c) $\Omega \setminus A \in \mathcal{F}$ for any $A \in \mathcal{F}$. If $\mathcal{F}$ is a σ-algebra in Ω, the pair $(\Omega, \mathcal{F})$ is called a *measurable space*, and the elements of $\mathcal{F}$ *measurable sets* or (*random*) *events*.

In particular, this definition introduces the notion of a "random event." This and similar notions should be treated as purely formal, with no hidden meaning behind them, although we will see in Sec. II.1 that they do admit a physical interpretation, once certain real events occur in a random fashion.

Speaking of terminology, we also add that probability theory uses some figures of speech which we do not really need and shall therefore avoid. For example, a probability theoretician would generally prefer to say: "let the elementary event ω be favorable to the occurrence of the event A" (described by certain conditions), instead of: "let $\omega \in A$" (where A is the set of all $\omega \in \Omega$ such that the conditions are satisfied); "let the event A occur," instead of "take ω in A"; "if the event A occurs then the event B occurs," instead of "$A \subset B$" etc.

Let us discuss of the notion of σ-algebras.

2. Example. $\mathcal{F} = \{\emptyset, \Omega\}$, i.e., the collection consisting of only two subsets of Ω: namely Ω and the empty set, is a σ-algebra.

3. Example. The collection Σ of all subsets of Ω is obviously a σ-algebra in Ω.

Example 3 shows, in particular, that if $\mathcal{F}$ is a collection of subsets of Ω, then there always exists at least one σ-algebra in Ω that contains $\mathcal{F}$ (as a subset, naturally), for example, $\mathcal{F} \subset \Sigma$. It is also almost obvious that the intersection of any family of σ-algebras is a σ-algebra. In other words, the collection of subsets of Ω, each belonging to every σ-algebra of a given family, is a σ-algebra. Hence, one can speak of the intersection of all σ-algebras that include a given system $\mathcal{F}$ of subsets of Ω; this intersection is, obviously, the minimal σ-algebra (with respect to inclusion) that includes $\mathcal{F}$. Thus the minimal σ-algebra containing $\mathcal{F}$ exists; it is called the σ-algebra *generated* by $\mathcal{F}$, denoted by $\sigma(\mathcal{F})$.

In some spaces Ω there exist natural σ-algebras. We will say that a metric space (X, ρ), where X is a set and ρ a metric on X, is *Polish* if it is complete and separable. If the metric ρ is assumed to be fixed, we say that X is a Polish space. For example, the d-dimensional Euclidean space E_d is a Polish space. The minimal σ-algebra containing all closed balls of a Polish space X is called the *σ-algebra of Borel sets*, denoted by $\mathfrak{B}(X)$. The elements of $\mathfrak{B}(X)$ are called *Borel sets*.

It is natural to ask what does an arbitrary Borel set (or, more generally, an element of the σ-algebra generated by a system of subsets) "look like"? At first sight one might think that the very notion of a Borel σ-algebra may be useless unless this question is answered. However, this is not so, by a fortunate coincidence: even when one does not know the "structure" of an individual Borel set, working with *all* Borel sets at once turns out to be very convenient.

As an example, let X, Y be Polish spaces and f a continuous function on X with values in Y. Take some $B \in \mathfrak{B}(Y)$ and consider the question: is $\{x : f(x) \in B\}$ a Borel set in X? It turns out that the answer is positive and the easiest way to obtain it is to consider *all* $B \in \mathfrak{B}(Y)$ at once. To explain this, we will use the following fact.

4. Lemma. *Let two sets Ω, X and a function $\xi(\omega)$ defined on Ω with values in X be given. For an arbitrary set $B \subset X$ denote $\xi^{-1}(B) = \{\omega : \xi(\omega) \in B\}$ (the inverse image of B under the mapping $\xi : \Omega \to X$). Then*

(a) *ξ^{-1}, as a mapping defined on subsets of X, preserves all set-theoretical operations (i.e., the inverse image of the union of any number of sets is the union of their inverse images, and so on); in particular, if $\mathfrak{B}$ is a σ-algebra in X, then $\xi^{-1}(\mathfrak{B}) := \{\xi^{-1}(B) : B \in \mathfrak{B}\}$ is a σ-algebra in Ω;*

(b) *if $\mathcal{F}$ is a σ-algebra in Ω, then $\{B : B \subset X, \ \xi^{-1}(B) \in \mathcal{F}\}$ is a σ-algebra in X.*

This lemma can be used, for example, to built up σ-algebras from other σ-algebras. The proof is left to the reader as a simple exercise.

Returning to the continuous function $f : X \to Y$, we first note that any open set $B \subset X$ is the countable union of all closed balls in B whose radii are rational numbers and centers belong to a fixed countable dense subset of X. In particular, any open set is a Borel set. Closed sets are Borel sets, since they are complements to open (Borel) sets. In addition, if B is a closed ball in Y then, by the continuity of f, the set $f^{-1}(B)$ is closed, $f^{-1}(B) \in \mathfrak{B}(X)$ and the σ-algebra $\Sigma := \{B : B \subset Y, \ f^{-1}(B) \in \mathfrak{B}(X)\}$ (see Lemma 4(b)) contains all closed balls in Y. But $\mathfrak{B}(Y)$ is the minimal σ-algebra containing all the closed balls in Y, hence $\mathfrak{B}(Y) \subset \Sigma$, that is, for any $B \in \mathfrak{B}(Y)$ we have $B \in \Sigma$, or, in other words, $f^{-1}(B) \in \mathfrak{B}(X)$, as required.

We have shown that continuous functions are Borel functions in the sense of the following definition.

5. Definition. Let X, Y be Polish spaces, f a function on X with values in Y. Then f is called a *Borel function* if the set $\{x : f(x) \in B\}$ is a Borel set (in X) for any Borel $B \subset Y$, that is, if $f^{-1}(\mathfrak{B}(Y)) \subset \mathfrak{B}(X)$.

We will also need a more general notion.

6. Definition. Let $(\Omega, \mathcal{F})$, $(X, \mathfrak{B})$ be measurable spaces, ξ a function on Ω with values in X, that is, $\xi : \Omega \to X$. We will say that ξ is a *random element* or *measurable element* (on Ω with values in X, with respect to σ-algebras $\mathcal{F}$, $\mathfrak{B}$), if $\xi^{-1}(B) \in \mathcal{F}$ for any $B \in \mathfrak{B}$. In this case we will also say that the *function (mapping) ξ is $\mathcal{F}$-measurable*, on the assumption that the σ-algebra $\mathfrak{B}$ is fixed. If $X = E_d$, $\mathfrak{B} = \mathfrak{B}(E_d)$, then ξ is called a *random* (or *measurable*) (d-dimensional) *vector*; if $d = 1$, ξ is called a *random* (or *measurable*) *variable*. If ξ is a random element, the σ-algebra $\xi^{-1}(\mathfrak{B})$ is denoted by $\sigma(\xi)$ and is called the σ-algebra *generated* by ξ. When considering random elements, one usually omits the argument $\omega \in \Omega$.

The *indicator* $I_A(\omega)$ of a set $A \in \mathcal{F}$ (i.e., the function that equals 1 for $\omega \in A$ and 0 otherwise) is the simplest example of a random variable.

Using the almost evident formula $(f(\xi))^{-1}(B) = \xi^{-1}(f^{-1}(B))$, one can verify that a real-valued Borel function of a random element with values in a Polish space X is a random variable, that is, if ξ is a random element with values in a Polish space X, and f is a Borel function from X into E_1, then $f(\xi(\omega))$ is a random variable. This fact can be extended in an obvious manner to compositions of arbitrary measurable mappings.

Sometimes one speaks of random variables whose range is the extended real line $[-\infty, \infty]$, meaning that $\{\omega : \xi(\omega) \in B\} \in \mathcal{F}$ for any $B \in \mathfrak{B}(E_1)$ and $\{\omega : \xi(\omega) = \infty\}$, $\{\omega : \xi(\omega) = -\infty\} \in \mathcal{F}$. In other words, ξ is a random variable with values in $[-\infty, \infty]$ if and only if $\arctan \xi$ is a (finite) random variable.

Occasionally, one also has to consider functions $\xi(\omega)$ with values in $[-\infty, \infty]$ defined not for all $\omega \in \Omega$ but only on a subset $A \subset \Omega$. We will therefore adopt the convention that the symbol ξI_A denotes the function equal to ξ on A and zero on $\Omega \setminus A$ (even if ξ is defined outside A and takes infinite values).

Given a measurable space $(\Omega, \mathcal{F})$, $A \subset \Omega$ and $\xi : A \to [-\infty, \infty]$, we will say that the function ξ is *measurable on A* if ξI_A is a random variable (with values in $[-\infty, \infty]$).

There is another useful notation involving indicators and functions on Ω. We will write $I_{\xi \in B}$ for I_A, where $A = \{\xi \in B\} := \{\omega : \xi(\omega) \in B\}$. Briefly, $I_{\xi \in B}(\omega) = 1$ for ω such that $\xi(\omega) \in B$, and $I_{\xi \in B}(\omega) = 0$, if $\xi(\omega) \notin B$. For instance, if $\xi(\omega)$, $\eta(\omega)$ are real, then $I_{\xi \le \eta}(\omega) = 1$ if $\xi(\omega) \le \eta(\omega)$, and $I_{\xi \le \eta}(\omega) = 0$ if $\xi(\omega) > \eta(\omega)$.

We will write points $x \in E_d$ as $(x^1, \dots, x^d)$, relative to some orthonormal basis. In other words, we consider E_d to be the direct product of d copies of E_1. The following lemma shows how the σ-algebra of Borel sets of the direct product is related to to the Borel σ-algebras of the factors, and establishes a connection between random vectors and random variables.

7. Lemma. (a) *$\mathfrak{B}(E_1)$ is generated by the collection of all sets $[a, \infty)$, and is also generated by the collection of all sets (a, ∞), where $a \in (-\infty, \infty)$.*

(b) *Let $(\Omega, \mathcal{F})$, $(X^i, \mathfrak{B}^i)$, $i = 1, \dots, d$ be measurable spaces. Set*

$$X = X^1 \times \cdots \times X^d := \{x = (x^1, \dots, x^d) : \quad x^i \in X^i, \ i = 1, \dots, d\}.$$

Let $\mathfrak{B}$ be the minimal σ-algebra in X containing all the sets

$$\{x \in X : \quad x^i \in B^i\}, \qquad B^i \in \mathfrak{B}^i, \ i = 1, \dots, d, \tag{1}$$

that is, $\mathfrak{B} := \mathfrak{B}^1 \otimes \cdots \otimes \mathfrak{B}^d$. Then $\xi^i : \Omega \to X^i$ are random elements with respect to $\mathcal{F}$, $\mathfrak{B}^i$, $i = 1, \dots, d$ if and only if $\xi := (\xi^1, \dots, \xi^d)$ is a X-valued random element with respect to $\mathcal{F}$, $\mathfrak{B}$.

(c) *Let* (X^i, ρ_i), $i = 1, \dots, d$, *be Polish spaces,* $X = X^1 \times \cdots \times X^d$,

$$\rho(x, y) = \left[\rho_1^2(x^1, y^1) + \cdots + \rho_d^2(x^d, y^d)\right]^{1/2},$$

where $x, y \in X$. *Then* (X, ρ) *is a Polish space; in addition, if* $\mathfrak{B}^i$ *are some collections of subsets of* X^i *and* $\sigma(\mathfrak{B}^i) = \mathfrak{B}(X^i)$, *then the* σ*-algebra* $\mathfrak{B}(X)$ *is generated by the collection of all the sets of type* (1); *in particular,* $\mathfrak{B}(X) = \mathfrak{B}(X^1) \otimes \cdots \otimes \mathfrak{B}(X^d)$. *Finally,* $\xi(\omega) = (\xi^1(\omega), \dots, \xi^d(\omega))$ *is a random element on* Ω *with values in* X *(with respect to* $\mathcal{F}$, $\mathfrak{B}(X)$*) if and only if* ξ^i *are random elements on* Ω *with values in* X^i *for* $i = 1, \dots, d$.

Before proving the lemma, we set the reader a simple exercise.

8. EXERCISE. Prove that if $(\Omega, \mathcal{F})$ is a measurable space and $A, B, A_1, A_2, \dots \in \mathcal{F}$, then $\bigcap_n A_n \in \mathcal{F}$, $A \setminus B \in \mathcal{F}$.

9. PROOF OF LEMMA 7. We will prove part (a) for sets $[a, \infty)$; the proof for (a, ∞) is similar. Let $\mathfrak{A}$ denote the σ-algebra generated by all the sets $[a, \infty)$. Since $[a, \infty)$ is a closed set, we have $[a, \infty) \in \mathfrak{B}(E_1)$; and since $\mathfrak{A}$ is the minimal σ-algebra that contains all the sets $[a, \infty)$, it follows that $\mathfrak{A} \subset \mathfrak{B}(E_1)$. On the other hand,

$$[a, b) = [a, \infty) \setminus [b, \infty) \in \mathfrak{A}, \ [a, b] = \bigcap_n \Big[a, b + \frac{1}{n}\Big) \in \mathfrak{A},$$

that is, any closed ball in E_1 is an element of $\mathfrak{A}$. Hence, $\mathfrak{B}(E_1) \subset \mathfrak{A}$ and $\mathfrak{B}(E_1) = \mathfrak{A}$.

In part (b), if ξ is a random element, then for any i we have

$$\mathcal{F} \ni \xi^{-1}\{x : x^i \in B^i\} = (\xi^i)^{-1}(B^i)$$

for any $B^i \in \mathfrak{B}^i$, that is, ξ^i is a random element. It follows from the last identity that, if ξ^i are random elements, then $\xi^{-1}\{x : x^i \in \mathfrak{B}^i\} \in \mathcal{F}$ for any $B^i \in \mathfrak{B}^i$, $i = 1, \dots, d$. Consequently, the σ-algebra $\mathfrak{C}$ (see Lemma 4(b)) of all sets $B \subset X$ such that $\xi^{-1}(B) \in \mathcal{F}$ contains all the sets of type (1). Since $\mathfrak{B}$ is the minimal such σ-algebra, it follows that $\mathfrak{B} \subset \mathfrak{C}$; by definition, this means that ξ is a random element with respect to $(\mathcal{F}, \mathfrak{B})$ with values in X.

As to part (c), the fact that (X, ρ) is a Polish space we assume to be known. Further, define $\pi_i(x) = x^i$ for $x \in X$, $i = 1, \dots, d$ and let $\mathfrak{B}$ denote the minimal σ-algebra that contains all the sets of type (1). By assumption, $\pi_i^{-1}(B^i) \in \mathfrak{B}$ for any $B^i \in \mathfrak{B}^i$. By Lemma 4(b), this is also true for $B^i \in \sigma(\mathfrak{B}^i) = \mathfrak{B}(X^i)$. Therefore, $\mathfrak{B} \supset \mathfrak{B}(X^1) \otimes \cdots \otimes \mathfrak{B}(X^d)$. Since the converse is obvious ($\mathfrak{B}^i \subset \mathfrak{B}(X^i)$), it follows that

$$\mathfrak{B} = \mathfrak{B}(X^1) \otimes \cdots \otimes \mathfrak{B}(X^d).$$

In addition, the π_i's are continuous, hence they are Borel functions. In other words, $\{x : x^i \in B^i\} = \pi_i^{-1}(B^i) \in \mathfrak{B}(X)$ for any $B^i \in \mathfrak{B}(X^i)$. Hence $\mathfrak{B} \subset \mathfrak{B}(X)$.

On the other hand, it is almost obvious that if we define a metric by

$$\tilde{\rho}(x, y) = \max_{i \le d} \rho_i(x^i, y^i),$$

then the convergences in the metrics $\rho, \tilde{\rho}$ are equivalent and the sets that are closed in ρ coincide with those closed in $\tilde{\rho}$; hence, the collections of Borel sets with respect to $\rho, \tilde{\rho}$ are the same. Since any closed ball in $\tilde{\rho}$ is of the form

$$\{x \in X : x^i \in B^i, \, i = 1, \dots d\} = \bigcap_{i=1}^{d} \{x \in X : x^i \in B^i\},$$

where B^i are closed balls in (X^i, ρ_i), it follows that $\mathfrak{B}(X) \subset \mathfrak{B}(X^1) \otimes \cdots \otimes \mathfrak{B}(X^d)$ $(= \mathfrak{B})$. Together with the already proved inverse inclusion, this yields $\mathfrak{B}(X) = \mathfrak{B}(X^1) \otimes \cdots \otimes \mathfrak{B}(X^d)$. The last assertion in (c) now follows too. □

10. Exercise. Take the σ-algebra $\mathfrak{C}$ introduced in the proof. Find an example to show that in general $\mathfrak{C} \neq \mathfrak{B}$.

Lemma 7 and the definition of Borel functions imply the following corollary.

11. Corollary. *If $\xi^1, \dots, \xi^d$ are finite random variables and $f : E_d \to E_1$ is a Borel function, then $f(\xi^1, \dots, \xi^d)$ is a random variable. In particular,*

$$\xi^1 + \cdots + \xi^d, \quad \xi^1 \cdot \ldots \cdot \xi^d, \quad \max\{\xi^i, \ i = 1, \dots, d\}$$

are random variables.

In addition, $\{\omega : \xi^1(\omega) = \xi^2(\omega)\} \in \mathcal{F}$, since the set $\{(x^1, x^2) : x^1 = x^2\}$ is closed in E_2.

A fundamental property of σ-algebras is that the limit of a sequence of measurable functions is a measurable function. This convenient property distinguishes dramatically the family of measurable functions both from the family of continuous functions and from the family of Riemann-integrable functions.

12. Lemma. *Let $\xi_1, \xi_2, \dots$ be random variables on a measurable space $(\Omega, \mathcal{F})$ and let $\xi_n(\omega) \le \xi_{n+1}(\omega)$ for all $n \ge 1$, $\omega \in \Omega$. Then $\xi := \lim_{n\to\infty} \xi_n$ is a random variable.*

Proof. Passing, if necessary, from ξ_n to $\arctan \xi_n$, one can assume that the ξ_n's are uniformly bounded. Next, for any constant $a \in (-\infty, \infty)$ we have

$$\{\omega : \lim_{n\to\infty} \xi_n > a\} = \bigcup_n \{\omega : \xi_n(\omega) > a\} \in \mathcal{F}.$$

Therefore, the σ-algebra $\mathfrak{A}$ of all $B \subset E$ such that $\xi^{-1}(B) \in \mathcal{F}$ contains all the sets (a, ∞). By Lemma 7(a), we conclude that $\mathfrak{B}(E_1) \subset \mathfrak{A}$, that is, $\xi^{-1}(B) \in \mathcal{F}$ for all $B \in \mathfrak{B}(E_1)$. □

13. Theorem. *Let $\xi_1, \xi_2, \dots$ be random variables on a measurable space $(\Omega, \mathcal{F})$. Then the following are also random variables:*

$$\xi^* := \sup_n \xi_n, \quad \xi_* := \inf_n \xi_n, \quad \bar{\xi} := \varlimsup_{n\to\infty} \xi_n, \quad \underline{\xi} := \varliminf_{n\to\infty} \xi_n.$$

In addition, $A := \{\omega : \bar{\xi} = \underline{\xi}\} \in \mathcal{F}$ and

$$\xi(\omega) = \left\{ \begin{array}{ll} \lim_{n\to\infty} \xi_n, & \text{if } \bar{\xi}(\omega) = \underline{\xi}(\omega) \\ 0, & \text{if } \bar{\xi}(\omega) > \underline{\xi}(\omega) \end{array} \right\} = \bar{\xi}(\omega) I_A(\omega)$$

is also a random variable.

This theorem is an obvious corollary of Lemma 12, Corollary 11, and the fact that

$$\xi^* = \lim_{n\to\infty} \max_{k\le n} \xi_k, \ \xi_* = -\sup_n(-\xi_n), \ \underline{\xi} = \lim_{n\to\infty} \inf_{k\ge n} \xi_k.$$

In many cases is it useful to know how to approximate a random variable by simpler random variables.

14. DEFINITION. A random variable is said to be *simple* if it takes only finitely many values.

If ξ is a simple random variable, it can be represented as

$$\sum_{i=1}^{n} b_i I_{A(i)},$$

where $b_i \neq 0$, $A(i) \in \mathcal{F}$: just let $\{b_i\}$ be the set of all the nonzero values of ξ and set $A(i) = \{\omega : \xi(\omega) = b_i\}$.

Further, we set $\varkappa_n(t) = 2^{-n}[2^n t]$, where $[t]$ is the integer part of t, $n = 1, 2, \ldots$. The reader can easily verify that $\varkappa_n(t)$ is a Borel function of t, $\varkappa_n(t) \uparrow t$ as $n \to \infty$, $t - \varkappa_n(t) \leq 2^{-n}$. Hence, if ξ is a *bounded* random variable, then the functions $\varkappa_n(\xi)$ are simple random variables that converge *uniformly* to ξ. For any random variable, the functions $(\varkappa_n(\xi) \wedge n) \vee (-n)$ are simple random variables that converge to ξ for every ω.

15. EXERCISE. In the notation of part (b) of Lemma 7, is it true that $\mathfrak{B}^1 \otimes \cdots \otimes \mathfrak{B}^d = \mathfrak{B}^1 \times \cdots \times \mathfrak{B}^d$?

16. EXERCISE. Let (X, ρ) be a Polish space, $\Gamma \in \mathfrak{B}(X)$. Then the pair (Γ, ρ) is a separable metric space, in which the Borel σ-algebra $\mathfrak{B}(\Gamma)$ can be defined in the same way as before. Prove that $\mathfrak{B}(\Gamma) \subset \mathfrak{B}(X)$.

2. Probability spaces, expectations

Let us continue our brief account of prerequisites from the measure and integration theory. Let $(\Omega, \mathcal{F})$ be a measurable space; suppose that a number $\mu(A) \in [0, \infty]$ is somehow assigned to each $A \in \mathcal{F}$.

1. DEFINITION. We call μ a (positive) *measure* on $\mathcal{F}$ (or on Ω, or on $(\Omega, \mathcal{F})$) if $\mu(\emptyset) = 0$ and μ is *countably additive*, that is, for any mutually disjoint $A_1, A_2, \dots \in \mathcal{F}$,

$$\mu\Big(\bigcup_n A_n\Big) = \sum_{n=1}^{\infty} \mu(A_n).$$

A measure μ is *finite* if $\mu(\Omega) < \infty$; it is called a *probability* measure if $\mu(\Omega) = 1$. If μ is a measure on $\mathcal{F}$, the triple $(\Omega, \mathcal{F}, \mu)$ is called a *measure space*. If μ is a probability measure, $(\Omega, \mathcal{F}, \mu)$ is called a *probability space*. We say that $A \subset \Omega$ is a *null set* (with respect to $\mathcal{F}, \mu$) if there exists a set $B \in \mathcal{F}$ such that $A \subset B$ and $\mu(B) = 0$. If $C \subset \Omega$, a statement is said to be true *almost everywhere* on C (μ-*almost everywhere* on C) or *for almost all* $\omega \in \Omega$ (briefly, a.e. or μ-a.e. on C) if it is true for all $\omega \in C$ except for some null set. If $C = \Omega$, we usually omit the reference to C. In the case of a probability measure we say *almost surely* (briefly, a.s.) instead of "almost everywhere"; instead of "the measure of the set A" we say "the probability that the event A will occur," or "*the probability of the event* A."

2. EXERCISE. Let μ be a measure on $(\Omega, \mathcal{F})$. Prove that
(a) if $A_1, \ldots, A_n \in \mathcal{F}$, $A_i \cap A_j = \emptyset$ for $i \neq j$, then

$$\mu\Big(\bigcup_{i=1}^{n} A_i\Big) = \sum_{i=1}^{n} \mu(A_i).$$

(b) If $A, B, C \in \mathcal{F}$, $A \subset B$, then $\mu(B) = \mu(A) + \mu(B \setminus A) \geq \mu(A)$, $\mu(C) = \mu(C \cap A) + \mu(C \setminus A)$.

Using these assertions and the identities

$$\bigcup_{n=1}^{\infty} A_n = A_1 \cup \bigcup_{n=2}^{\infty} \Big(A_n \setminus \bigcup_{i=1}^{n-1} A_i \Big), \quad A_1 \setminus \bigcap_{n=1}^{\infty} A_n = \bigcup_{n=1}^{\infty} (A_1 \setminus A_n),$$

show that (c) if $A_1, A_2, \dots \in \mathcal{F}$, $A_n \subset A_{n+1}$ for all $n \geq 1$, then

$$\mu\Big(\bigcup_{n=1}^{\infty} A_n \Big) = \lim_{n \to \infty} \mu(A_n);$$

(d) if $A_1, A_2, \dots \in \mathcal{F}$, $A_n \supset A_{n+1}$ for $n \geq 1$ and $\mu(A_1) < \infty$, then

$$\mu\Big(\bigcap_{n=1}^{\infty} A_n \Big) = \lim_{n \to \infty} \mu(A_n).$$

(e) Prove that if f, g are real-valued functions on Ω and $f = g$ (a.e.), then there exists an $\mathcal{F}$-measurable function h, not necessarily finite, such that $h = 0$ almost everywhere and $|f - g| \leq h$ everywhere on Ω.

In the measure theory one proves the following theorem.

3. THEOREM. *There exists a unique measure ℓ on $(E_d, \mathfrak{B}(E_d))$ equal to the Euclidean volume for any closed parallelepiped with edges parallel to the coordinate axes of an orthonormal basis.*

Essentially, the reader needs only to know that this theorem is valid for $d = 1$, since in Sec. 5 we will consider products of measure spaces. The measure ℓ is called the *Lebesgue measure*. In E_d, the words "almost everywhere," without specifying the measure, always mean "ℓ-almost everywhere."

We now present what is, in a sense, the basic example of a probability space.

4. EXAMPLE. The triple consisting of the interval $[0,1] (= \Omega)$, the σ-algebra $\mathfrak{B}([0,1])$ of Borel subsets of $[0,1]$ (in the role of $\mathcal{F}$, see Exercise 1.16) and Lebesgue measure on $\mathfrak{B}([0,1])$ (in the role of μ, with $d = 1$, naturally) is a probability space.

We now define the Lebesgue integral with respect to a measure μ. We will write briefly $\mu(\xi > t) = \mu(\{\omega : \xi(\omega) > t\})$, and so on.

5. DEFINITION. Let $(\Omega, \mathcal{F}, \mu)$ be a measure space, $\xi = \xi(\omega)$ a random variable on Ω, $\xi_+ := \frac{1}{2}(|\xi| + \xi)$, $\xi_- := \frac{1}{2}(|\xi| - \xi)$. Suppose that at least one of the expressions

$$\int_0^{\infty} \mu(\xi_+ > t)\,dt, \qquad \int_0^{\infty} \mu(\xi_- > t)\,dt \tag{1}$$

is finite. Then ξ is said to be *integrable* with respect to μ, or *μ-integrable* (in the Lebesgue sense), and we write

$$\int_{\Omega} \xi(\omega)\,\mu(d\omega) = \int_0^{\infty} \mu(\xi_+ > t)\,dt - \int_0^{\infty} \mu(\xi_- > t)\,dt. \tag{2}$$

The left-hand side of (2) is called the *Lebesgue integral* of ξ with respect to μ (or μ-integral of ξ). If μ is a probability measure and ξ is integrable with respect to μ, then the left-hand side of (2) is called the *expectation* or *mean value* of ξ, denoted by $\mathbf{E}\,\xi$. The expression $\mathbf{E}\,|\xi|^p \ (=\mathbf{E}\,(|\xi|^p))$ is called the p-th *moment* of ξ.

In order to clarify Definition 5, we note that $\mu(\xi_\pm > t)$ are decreasing functions of t (see Exercise 2), and a monotone finite function on $[\varepsilon, T]$ is Riemann-integrable. The integrals in (1) are therefore understood as improper Riemann integrals (with singularities at $t = 0$ and $t = \infty$):

$$\int_0^\infty \mu(\xi_\pm > t)\,dt := \lim_{\varepsilon\downarrow 0, T\to\infty} \int_\varepsilon^T \mu(\xi_\pm > t)\,dt.$$

In addition, if, say, $\mu(\xi_+ > t) = \infty$ for some $t \in (0,\infty)$, then the first integral in (1) is put equal to infinity. That definition (2) is natural is a result of the following formulas, whose validity will be evident from what follows:

$$\int_\Omega \xi_+\,\mu(d\omega) = \int_\Omega \Big(\int_0^\infty I_{\xi_+>t}\,dt\Big)\,\mu(d\omega) = \int_0^\infty \Big(\int_\Omega I_{\xi_+>t}\,\mu(d\omega)\Big)\,dt = \int_0^\infty \mu(\xi_+ > t)\,dt,$$

$$\int_\Omega \xi\,\mu(d\omega) = \int_\Omega \xi_+\,\mu(d\omega) - \int_\Omega \xi_-\,\mu(d\omega).$$

6. EXERCISE. Throughout, ξ denotes a random variable.
(a) Prove that if $\xi \ge 0$ (a.e.), then

$$\int_\Omega \xi\,\mu(d\omega) = \int_0^\infty \mu(\xi > t)\,dt.$$

(b) Using (a) and the inequality $\mu(\xi > t) \ge \mu(\xi \ge c)$ for $t < c$ prove *Chebyshev's inequality*: if $\xi \ge 0$ (a.e.) and $c \in (0,\infty)$ is a constant, then

$$\mu(\xi \ge c) \le \frac{1}{c}\int_\Omega \xi\mu(d\omega).$$

(c) Prove that if $\xi \ge 0$ (a.e.), then $\xi = 0$ (a.e.) if and only if the integral of ξ with respect to μ equals zero. (Use the identity $(\xi > 0) = \bigcup_n (\xi > 1/n)$.)

(d) Prove that if random variables ξ and η are such that $\xi = \eta$ (a.e.) and ξ is integrable with respect to μ, then η is also integrable with respect to μ, and the integrals of ξ and η are equal. (Use the fact that $(\xi_\pm > t) \subset (\eta_\pm > t) \cup A$, where A is a null set).

(e) Prove that if ξ has a finite integral with respect to μ, then $-\infty < \xi < \infty$ (a.e.).

(f) Prove that if ξ is integrable with respect to μ and $a \in (-\infty,\infty)$ is a constant, then $a\xi$ is integrable with respect to μ and, if $a \ne 0$,

$$\int_\Omega a\xi\,\mu(d\omega) = a\int_\Omega \xi\,\mu(d\omega);$$

prove that this equality also holds for $a = 0$ if the μ-integral of ξ is finite.

We now present with proofs some more complicated properties of the Lebesgue integral.

7. THEOREM. (a) *A random variable ξ has a finite μ-integral if and only if the μ-integral of $|\xi|$ is finite. In addition, if ξ is μ-integrable, then*

$$\left|\int_\Omega \xi\,\mu(d\omega)\right| \le \int_\Omega |\xi|\,\mu(d\omega), \qquad \int_\Omega \xi\,\mu(d\omega) = \int_\Omega \xi_+\,\mu(d\omega) - \int_\Omega \xi_-\,\mu(d\omega). \tag{3}$$

(b) *If ξ, η are random variables, $\xi \le \eta$ and at least one of the random variables ξ_-, η_+ has a finite μ-integral, then both ξ and η are μ-integrable and*

$$\int_\Omega \xi\,\mu(d\omega) \le \int_\Omega \eta\,\mu(d\omega).$$

(c) *If $A(1), \dots, A(n) \in \mathcal{F}$, $c_1, \dots, c_n \in (0, \infty)$ then the random variable*

$$\xi(\omega) := \sum_{i=1}^n c_i I_{A(i)}(\omega)$$

is μ-integrable and

$$\int_\Omega \xi\,\mu(d\omega) = \sum_{i=1}^n c_i \mu(A(i)).$$

(d) (The Monotone Convergence Theorem). *If ξ_1, $\xi_2, \dots$ are random variables, $0 \le \xi_n(\omega) \le \xi_{n+1}(\omega)$ for all ω, n, then*

$$\int_\Omega \lim_{n\to\infty} \xi_n\,\mu(d\omega) = \lim_{n\to\infty} \int_\Omega \xi_n\,\mu(d\omega). \tag{4}$$

(e) *If random variables ξ, η are finite and have finite μ-integrals and $a, b \in (-\infty, \infty)$, then $a\xi + b\eta$ is μ-integrable and*

$$\int_\Omega (a\xi + b\eta)\,\mu(d\omega) = a\int_\Omega \xi\,\mu(d\omega) + b\int_\Omega \eta\,\mu(d\omega). \tag{5}$$

Equality (5) *is also valid if ξ, η are nonnegative random variables and $a > 0$, $b > 0$.*

(f) (Fatou's Lemma). *If $\xi_1, \xi_2, \dots$ are random variables and $\xi_n \ge 0$, then*

$$\int_\Omega \underline{\lim}_{n\to\infty} \xi_n\,\mu(d\omega) \le \underline{\lim}_{n\to\infty} \int_\Omega \xi_n\,\mu(d\omega).$$

(g) (The Dominated Convergence Theorem). *If ξ, η, ξ_1, $\xi_2, \dots$ are random variables, $|\xi_n| \le \eta$, η has a finite μ-integral and $\xi(\omega) = \lim_{n\to\infty} \xi_n(\omega)$, where the limit exists for all ω, then ξ is μ-integrable and*

$$\int_\Omega \lim_{n\to\infty} \xi_n\,\mu(d\omega) = \lim_{n\to\infty} \int_\Omega \xi_n\,\mu(d\omega);$$

in particular, the limit on the right exists and is finite.

PROOF. Parts (a), (b) are easily derived from the fact that for $t > 0$

$$(\xi_+ > t) \cap (\xi_- > t) = \emptyset, \qquad (\xi_+ > t) \cup (\xi_- > t) = (|\xi| > t),$$
$$\mu(\xi_+ > t) + \mu(\xi_- > t) = \mu(|\xi| > t) \qquad ((\xi_+)_- > t) = \emptyset,$$
$$\mu(\emptyset) = 0, \qquad (\xi_+ > t) \subset (\eta_+ > t),$$
$$\mu(\xi_+ > t) \le \mu(\eta_+ > t), \qquad \mu(\xi_- > t) \ge \mu(\eta_- > t).$$

To prove (c), we introduce the notation:

$$\eta = \sum_{i=1}^{n-1} c_i I_{A(i)\setminus A(n)}, \qquad \zeta = \sum_{i=1}^{n-1} c_i I_{A(i)\cap A(n)}.$$

Since $\xi = \eta$ outside $A(n)$ and $\xi = \zeta + c_n I_{A(n)}$ on $A(n)$, it follows that for $t > 0$ we have

$$(\xi > t) = (\eta > t) \cup (\zeta + c_n I_{A(n)} > t),$$

and the terms on the right are disjoint. Further, $\zeta > 0$ only on $A(n)$; therefore the set $(\zeta + c_n I_{A(n)} > t)$ equals $A(n)$ for $t < c_n$, and equals $(\zeta > t - c_n)$ for $t \ge c_n$. Hence

$$\begin{aligned}\int_\Omega \xi\, \mu(d\omega) &= \int_0^\infty \{\mu(\eta > t) + \mu(\zeta + c_n I_{A(n)} > t)\}\, dt \\ &= \int_\Omega \eta\, \mu(d\omega) + \int_0^{c_n} \mu(A(n))\, dt + \int_{c_n}^\infty \mu(\zeta > t - c_n)\, dt \\ &= \int_\Omega \eta\, \mu(d\omega) + \int_\Omega \zeta\, \mu(d\omega) + c_n \mu(A(n)).\end{aligned}$$

In addition, $\mu(A(i) \setminus A(n)) + \mu(A(i) \cap A(n)) = \mu(A(i))$, and we can proceed by induction on n. Noting that the previous arguments are also valid for $n = 1$ if we put $\eta = \zeta = 0$, we complete the proof of part (c).

To prove (d), set $\xi = \lim_{n\to\infty} \xi_n$. Since $\xi_n \le \xi_{n+1} \le \xi$, it follows by (b) that the limit on the right of (4) exists and does not exceed the left-hand side of (4). Proving the inverse inequality, one can obviously assume that the right-hand side of (4) is finite. Then, by the Chebyshev inequality, $\sup_n \mu(\xi_n > t) < \infty$ for any $t > 0$. In addition, $(\xi > t) = \bigcup_n (\xi_n > t)$, $(\xi_n > t) \subset (\xi_{n+1} > t)$. Hence it follows (see Exercise 2) that $\mu(\xi_n > t) \to \mu(\xi > t)$; in particular, $\mu(\xi > t)$ is finite for $t > 0$. The functions $\mu(\xi_n > t)$ decrease with increasing t, and since they converge to $\mu(\xi > t)$, it follows that, for any $\varepsilon > 0$, $T \in (-\infty, \infty)$, there exists n such that

$$\mu(\xi_n > t) \ge \mu(\xi > t + \varepsilon) - \varepsilon \tag{6}$$

for all $t \in [0, T]$. Indeed, otherwise there would exist points $t_n \in [0, T]$ at which the inequality inverse to (6) is true. These points, considered for some subsequence $n(k)$, would converge to a certain point $t_0 \in [0, T]$ and

$$\begin{aligned}\mu(\xi > t_0 + \varepsilon/2) &\ge \overline{\lim_{k\to\infty}}\, \mu(\xi > t_{n(k)} + \varepsilon) \ge \varepsilon + \overline{\lim_{k\to\infty}}\, \mu(\xi_{n(k)} > t_{n(k)}) \\ &\ge \varepsilon + \overline{\lim_{k\to\infty}}\, \mu(\xi_{n(k)} > t_0 + \varepsilon/2) = \varepsilon + \mu(\xi > t_0 + \varepsilon/2),\end{aligned}$$

but this is impossible, because $\mu(\xi > t)$ is finite.

By (6),

$$
\begin{aligned}
\int_{\varepsilon}^{T} \mu(\xi > t)\,dt - \varepsilon T &\le \int_{\varepsilon}^{T+\varepsilon} \mu(\xi > t)\,dt - \varepsilon T = \int_{0}^{T} \mu(\xi > t+\varepsilon)\,dt - \varepsilon T \\
&\le \int_{0}^{T} \mu(\xi_n > t)\,dt \le \int_{\Omega} \xi_n\,\mu(d\omega) \le \varliminf_{n\to\infty} \int_{\Omega} \xi_n\,\mu(d\omega).
\end{aligned}
$$

Letting first $\varepsilon \downarrow 0$ in the inequality between the extreme terms and then $T \to \infty$, we complete the proof of (d).

Now we proceed to part (e). By Exercise 6, we may assume that $a = b = 1$. If ξ and η are simple nonnegative random variables, equality (5) follows immediately from (c). To prove (5) for arbitrary $\xi, \eta \ge 0$, it now suffices to use the Monotone Convergence Theorem and the approximation of measurable functions by simple ones as described at the end of Sec. 1. Further, if $\xi \ge 0 \ge \eta$ (and ξ, η have finite integrals), then (5) remains valid. This follows easily from the above arguments, the second formula of (3), and the following relations, in which $A := (\xi + \eta \ge 0)$, $B := (\xi + \eta < 0)$:

$$
\begin{aligned}
&\xi = \xi I_A + \xi I_B, && (-\eta) = (-\eta) I_A + (-\eta) I_B, \\
&(\xi + \eta)_+ + (-\eta) I_A = \xi I_A, && (\xi + \eta)_- + \xi I_B = (-\eta) I_B.
\end{aligned}
$$

Finally, the case of arbitrary finite μ-integrable ξ, η can be reduced to the previous cases by the identity $\xi + \eta = (\xi_+ + \eta_+) - (\xi_- + \eta_-)$.

Fatou's Lemma (part (f)) follows immediately from (d) and from the relations

$$
\varliminf_{n\to\infty} \xi_n = \lim_{n\to\infty} \inf_{k\ge n} \xi_k, \quad \inf_{k\ge n} \xi_k \uparrow, \quad \inf_{k\ge n} \xi_k \le \xi_i, \; i \ge n,
$$

$$
\int_{\Omega} \inf_{k\ge n} \xi_k\,\mu(d\omega) \le \inf_{i\ge n} \int_{\Omega} \xi_i\,\mu(d\omega).
$$

We now prove part (g). Note first that the integrability of ξ follows from (b) and the inequality $|\xi| \le \eta$. In addition, $\eta + \xi_n \ge 0, \eta - \xi_n \ge 0$. Hence, by (e) and (f) we have

$$
\begin{aligned}
\int_{\Omega} \eta\,\mu(d\omega) + \int_{\Omega} \lim_{n\to\infty} \xi_n\,\mu(d\omega) &= \int_{\Omega} \lim_{n\to\infty} (\eta + \xi_n)\,\mu(d\omega) \\
&\le \varliminf_{n\to\infty} \int_{\Omega} (\eta + \xi_n)\,\mu(d\omega) = \int_{\Omega} \eta\,\mu(d\omega) + \varliminf_{n\to\infty} \int_{\Omega} \xi_n\,\mu(d\omega),
\end{aligned}
$$

$$
\int_{\Omega} \lim_{n\to\infty} \xi_n\,\mu(d\omega) \le \varliminf_{n\to\infty} \int_{\Omega} \xi_n\,\mu(d\omega),
$$

$$
\int_{\Omega} \lim_{n\to\infty} (\eta - \xi_n)\,\mu(d\omega) \le \varliminf_{n\to\infty} \int_{\Omega} (\eta - \xi_n)\,\mu(d\omega),
$$

$$
\int_{\Omega} \lim_{n\to\infty} \xi_n\,\mu(d\omega) \ge \varlimsup_{n\to\infty} \int_{\Omega} \xi_n\,\mu(d\omega).
$$

This proves (g), and with it the entire theorem. □

Some corollaries from this theorem now follow.

8. Corollary. *If $\xi_1, \xi_2, \dots$ are nonnegative random variables, then*

$$\int_\Omega \sum_{n=1}^\infty \xi_n \, \mu(d\omega) = \sum_{n=1}^\infty \int_\Omega \xi_n \, \mu(d\omega).$$

This follows immediately from the definition of the sum of a series and parts (d), (e) of Theorem 7.

If $A(1), A(2), \dots \in \mathcal{F}$, $A = \bigcup_n A(n)$, then obviously

$$I_A \le \sum_{n=1}^\infty I_{A(n)}, \quad \mu(A(n)) = \int_\Omega I_{A(n)} \, \mu(d\omega). \tag{7}$$

Hence, Corollary 8 implies

9. Corollary. *We have $\mu(A) \le \sum_n \mu(A(n))$; in particular, a countable union of sets of measure zero in $\mathcal{F}$ also has measure zero; a countable union of null sets is a null set.*

The next corollary is a consequence of Corollary 8 and Exercise 6.

10. Corollary (Borel–Cantelli Lemma). . *Let $A(1), A(2), \dots, \in \mathcal{F}$ and*

$$\sum_n \mu(A(n)) < \infty.$$

Then $\sum I_{A(n)} < \infty$ (a.e.), that is, for almost every ω there exists $k(\omega)$ such that $\omega \notin A(n)$ for all $n \ge k(\omega)$. Moreover,

$$\int_\Omega \sum_n I_{A(n)} \, \mu(d\omega) = \sum_n \mu(A(n)) < \infty.$$

The first relation in (7) is an equality if the sets $A(n)$ are mutually disjoint. Hence, by Corollary 8, we obtain

11. Corollary. *If η is a nonnegative random variable, then*

$$\nu(A) := \int_\Omega \eta I_A \, \mu(d\omega) \qquad \left(=: \int_A \eta \, \mu(d\omega) \right) \tag{8}$$

is a measure on $\mathcal{F}$.

This corollary makes it possible to construct one measure from another. Definition (8) is sometimes written in the more intuitive form

$$\nu(d\omega) = \eta(\omega) \, \mu(d\omega). \tag{9}$$

Sometimes, one needs to know to express ν-integrals in terms of μ-integrals. In view of (9), the following assertion is natural.

12. THEOREM. *Let ξ, η be nonnegative random variables. Define a measure ν by* (8). *Then*

$$\int_\Omega \xi\, \nu(d\omega) = \int_\Omega \xi\eta\, \mu(d\omega). \tag{10}$$

For simple $\xi \geq 0$, this is easily derived from (8) (and from parts (c), (e) of Theorem 7). Using approximation of ξ by simple functions and the Monotone Convergence Theorem, we can extend it to arbitrary random variables $\xi \geq 0$.

13. REMARK. Similarly to Fatou's Lemma and the Monotone Convergence Theorem, Theorem 12 can be extended to random variables with values of any sign. We will not state and prove these generalizations but leave them to the reader as a simple exercise, based on the fact that if $\xi \geq \zeta$ (or $\xi_n \geq \zeta$) then $\xi - \zeta \geq 0$ (respectively, $\xi_n - \zeta \geq 0$) and that if ζ has a finite integral, then ζ may be subtracted without harm (see the end of the proof of Theorem 7).

14. REMARK. Part (b) of Theorem 7, the Monotone Convergence Theorem, Fatou's Lemma, and the Dominated Convergence Theorem can be generalized in another direction, by stipulating that the conditions hold not for all but for almost all ω. These generalizations can be derived trivially from Theorem 7, for if the random variables involved are replaced with zeroes outside the set on which all the assumptions of (b), (d), (f), (g) of Theorem 7 hold, then these conditions will hold for all ω; by Corollary 9, these random variables will be modified only on a measurable set of measure zero so that, by Exercise 6(d), none of the integrals involved will change its value.

15. REMARK. The previous remark is true not only for random variables defined everywhere on Ω, but also for random variables defined only *almost everywhere*, that is, for random variables that are measurable on some set $A \in \mathcal{F}$ such that $\mu(\Omega \setminus A) = 0$. In such cases the integral of a random variable over Ω is defined as the integral of the variable multiplied by the indicator of A. We will first encounter integrals of such random variables in Example II.6.1. However, even now part (e) of Theorem 7 illustrates the usefulness of extending the set of integrable functions in this way; ξ, η need not be finite.

16. EXERCISE. Let $A_1, A_2, \ldots \in \mathcal{F}$, $\mu(\Omega \setminus A_n) = 0$ for all n. Prove that $\mu(\Omega \setminus \bigcap_n A_n) = 0$.

17. EXERCISE. In connection with the definitions in (8), prove that if $A \in \mathcal{F}$, $\mathcal{F}_A := \{B : B \subset A,\ B \in \mathcal{F}\}, \mu_A(B) := \mu(A \cap B)$ then $(A, \mathcal{F}_A, \mu_A)$ is a measure space and a nonnegative function η on Ω is measurable on A if and only if the mapping $\eta : A \to [0, \infty]$ is measurable with respect to $\mathcal{F}_A$. In the latter case, moreover, the μ_A-integral of η over A coincides with $\nu(A)$ from (8).

In addition to Theorem 7 (d), (f), (g), another theorem on passage to the limit under the integral sign turns out to be useful in probability theory. The proof, which is surprisingly simple, is based on the above results; we offer it as a problem (this theorem will not be used in this book).

18. PROBLEM. Prove *Scheffé's Theorem*: if $\xi, \xi_1, \xi_2, \ldots$ are random variables on a measure space $(\Omega, \mathcal{F}, \mu)$, $\xi_n \to \xi$ (a.e.), $\xi_n \geq 0$ (a.e.) and

$$\int_\Omega \xi_n \, \mu(d\omega), \longrightarrow \int_\Omega \xi \, \mu(d\omega) < \infty,$$

then

$$\int_\Omega |\xi_n - \xi| \, \mu(d\omega) \longrightarrow 0.$$

Various inequalities play an important role in applications of Lebesgue integration theory. The next theorem establishes the *Hölder inequality*; if $p = 2$, it is called the *Cauchy-Bunyakovsky inequality*. The reader should keep in mind that these inequalities are frequently used with $\eta \equiv 1$.

19. THEOREM. *Let ξ and η be nonnegative random variables on a measure space $(\Omega, \mathcal{F}, \mu)$, $p \in (1, \infty)$, $q := p(p-1)^{-1}$. Then*

$$\int_\Omega \xi\eta \, \mu(d\omega) \leq \left(\int_\Omega \xi^p \, \mu(d\omega)\right)^{\frac{1}{p}} \left(\int_\Omega \eta^q \, \mu(d\omega)\right)^{\frac{1}{q}}, \tag{11}$$

where expressions of type $0 \cdot \infty$ are put equal to zero (cf. Remark 15).

PROOF. Using standard methods of computing extremums, one easily proves that for $a, b \geq 0$ $(0 \cdot \infty := \infty)$

$$ab = \inf_{t>0}(p^{-1}t^{-p}a^p + q^{-1}t^q b^q). \tag{12}$$

Hence, $\xi\eta \leq p^{-1}t^{-p}\xi^p + q^{-1}t^q\eta^q$ (*Young's inequality*). Integrating this inequality for every fixed $t > 0$, then evaluating the infimum over all $t > 0$ and using (12), we immediately obtain (11) provided that the right-hand side of (11) does not vanish. If it does, then either $\xi = 0$ (a.e.) or $\eta = 0$ (a.e.); but then the left-hand side of (11) also vanishes. □

We have defined the Lebesgue integral for scalar functions only. The integral of a vector-valued function is defined as a vector whose coordinates are the integrals of the respective coordinates. Matrices will also be integrated in the same way, entrywise.

3. Completion of measure spaces. Relation between Riemann and Lebesgue integrals

The reader has probably observed that if $(\Omega, \mathcal{F}, \mu)$ is a measure space, null sets with respect to $\mathcal{F}, \mu$ are not necessarily elements of $\mathcal{F}$, and if ξ is a random variable and $\eta \equiv \xi$(a.e.), then η is not necessarily measurable with respect to $\mathcal{F}$. Sometimes (see Remark III.3.6 below) this situation is very inconvenient, and then it becomes necessary to extend the set of measurable functions, hence also the set of integrable functions.

1. THEOREM. *Let $\mathcal{F}^\mu$ denote the collection of all sets $A \subset \Omega$ for each of which there exists $B \in \mathcal{F}$ such that $I_A = I_B$ μ-a.e.; define $\overline{\mu}(A) = \mu(B)$ ($\overline{\mu}$ is well-defined by Exercise 2.6(d) and Theorem 2.7(c)). Then*

(a) *$\mathcal{F} \subset \mathcal{F}^\mu$, $\mathcal{F}^\mu$ is a σ-algebra;*

(b) $\overline{\mu}$ *is a measure on* $\mathcal{F}^\mu$ *such that* $\overline{\mu} = \mu$ *on* $\mathcal{F}$;
(c) A *is a null set with respect to* $\mathcal{F}$, μ *if and only if* $A \in \mathcal{F}^\mu$ *and* $\overline{\mu}(A) = 0$;
(d) *any null set* A *with respect to* $\mathcal{F}^\mu$, $\overline{\mu}$ *belongs to* $\mathcal{F}^\mu$ *and* $\overline{\mu}(A) = 0$;
(e) *a real-valued function* $\xi(\omega)$ *is* $\mathcal{F}^\mu$*-measurable if and only if for some* $\mathcal{F}$*-measurable function* η *we have* $\xi = \eta$ μ*-a.e., or, what is the same by* (c), (d), $\overline{\mu}$*-a.e.*

PROOF. (a) Obviously, $\mathcal{F} \subset \mathcal{F}^\mu$, $\Omega \in \mathcal{F}^\mu$, and if $A \in \mathcal{F}^\mu$, then $\Omega \setminus A \in \mathcal{F}^\mu$, since the indicator of $\Omega \setminus A$ is $1 - I_A$. A countable union of sets in $\mathcal{F}^\mu$ is again in $\mathcal{F}^\mu$, as is easily seen from the inequality

$$|I_{\bigcup_i A_i} - I_{\bigcup_i B_i}| \leq \sum_i |I_{A_i} - I_{B_i}| \tag{1}$$

and from Corollary 2.9. To prove (b), it suffices to point out that the equality $\overline{\mu} = \mu$ on $\mathcal{F}$ is obvious, and then use (1), Corollary 2.9, Corollary 2.8, and the fact that

$$\sum_i I_{B_i} \leq I_{\bigcup_i B_i} + \sum_{i<j} I_{B_i \cap B_j}, \qquad I_{\bigcup_i B_i} \leq \sum_i I_{B_i},$$

which in case $I_{B_i} I_{B_j} = 0$, $i < j$, almost everywhere, implies that almost everywhere

$$\sum_i I_{B_i} = I_{\bigcup_i B_i}.$$

The "only if" clause of part (c) follows from the fact that we can put $B = \emptyset$. If $I_A = I_B$ μ-a.e., $B \in \mathcal{F}$, and $\mu(B) = 0$, then A is obviously contained in the union of B and a set $C \in \mathcal{F}$ such that $\mu(C) = 0$. Hence $A \subset B \cup C$, $\mu(B \cup C) = 0$ and A is a μ-null set. This proves (c). Assertion (d) is clear since under its assumptions, $A \subset D \in \mathcal{F}^\mu$, $I_D = I_B$, μ-a.e., where $B \in \mathcal{F}$, $\mu(B) = 0$ and $I_A \leq I_B = 0$, $I_A = I_\emptyset$ μ-a.e.

To prove (e), we note that, by (c) and (d), if $\xi = 0$ (a.e.), then ξ is $\mathcal{F}^\mu$-measurable, and if $\xi = \eta$ (a.e.), where η is $\mathcal{F}$-measurable, then η, as well as $\xi - \eta$, $\xi = \eta + (\xi - \eta)$, are $\mathcal{F}^\mu$-measurable. On the other hand, the definition of $\mathcal{F}^\mu$ implies that if ξ is a simple function with respect to $\mathcal{F}^\mu$, then there exists a (simple) $\mathcal{F}$-measurable function η such that $\xi = \eta$ (a.e.). For any $\mathcal{F}^\mu$-measurable ξ, let ξ_n be a sequence of simple $\mathcal{F}^\mu$-measurable functions such that $\xi_n \to \xi$ everywhere as $n \to \infty$. Similarly, let η_n be $\mathcal{F}$-measurable functions such that $\xi_n = \eta_n$ (a.e.). Now define $\eta(\omega)$ to be the limit of η_n for those ω's for which the limit exists, and zero elsewhere. Then η is $\mathcal{F}$-measurable and since by Corollary 2.9 the set of all ω such that $\xi_n \neq \eta_n$ at least at one n is a null set, it follows that $\xi = \eta$ (a.e.), and the theorem is proved. □

2. DEFINITION. A measure space $(\Omega, \mathcal{F}, \mu)$ is said to be *complete* if every null set (with respect to $\mathcal{F}, \mu$) lies in $\mathcal{F}$. In that case we also say that the σ-algebra $\mathcal{F}$ is *complete* (with respect to $\mathcal{F}, \mu$). By Theorem 1, $(\Omega, \mathcal{F}^\mu, \overline{\mu})$ is a complete space; we will call it the *completion* of $(\Omega, \mathcal{F}, \mu)$. Since $\mathcal{F}^\mu \supset \mathcal{F}$ and $\overline{\mu} = \mu$ on $\mathcal{F}$, one usually writes μ instead of $\overline{\mu}$. The σ-algebra $\mathcal{F}^\mu$ is also called the *completion* of $\mathcal{F}$ (with respect to $\mathcal{F}$, μ).

It is important to stress that the passage to the completion of a measure space is always possible; it preserves the measurability of measurable functions and does not modify the integrals of integrable functions.

We will sometimes need completions of σ-algebras $\mathcal{G} \subset \mathcal{F}$.

3. THEOREM. *Let $(\Omega, \mathcal{F}, \mu)$ be a complete space and $\mathcal{G} \subset \mathcal{F}$ be a σ-algebra. Let $\mathcal{G}^\mu$ denote the collection of all sets $A \subset \Omega$, for each of which there exists $B \in \mathcal{G}$ such that $I_A = I_B$ (a.e.). Then*

(a) *$\mathcal{G} \subset \mathcal{G}^\mu$ and $\mathcal{G}^\mu$ is a σ-algebra, called the* completion *of $\mathcal{G}$ with respect to $\mathcal{F}, \mu$;*

(b) *all the null sets are elements of $\mathcal{G}^\mu$, and in this sense $\mathcal{G}^\mu$ is* complete *with respect to $\mathcal{F}, \mu$;*

(c) *a real-valued function ξ is $\mathcal{G}^\mu$-measurable if and only if $\xi = \eta$ (a.e.) for some $\mathcal{G}$-measurable η.*

The proof follows easily from that of Theorem 1.

The usefulness of completions of measure spaces is easily illustrated by considering the relationship between the Riemann and Lebesgue integrals: although not every Riemann-integrable function is a Borel function, nevertheless, Riemann-integrable functions are measurable with respect to the completion of the Borel σ-algebra.

Another natural reason for our appeal to Riemann integrals is as follows. The Lebesgue integral is exceptionally convenient from theoretical point of view, and it is of great importance that, once extended to the completion, it coincides with the Riemann integral for Riemann-integrable functions. We can thus use the usual rules for Riemann integration to calculate Lebesgue integrals. We give the relevant result only for the one-dimensional case and for proper Riemann integrals. It is easily extended to *absolutely* convergent improper Riemann integrals, using Theorem 2.7 and formulas of the type $f(x)I_{[-n,n]}(x) \to f(x)$, where $n \to \infty$.

4. THEOREM. *Let $f(x)$ be Riemann-integrable on* $[0, 1]$. *Then it is measurable with respect to the ℓ-completion of* $\mathfrak{B}([0,1])$ (*where ℓ is one-dimensional Lebesgue measure*), *and its Lebesgue integral over* $[0, 1]$ *exists and coincides with its Riemann integral.*

PROOF. Let $I_n = \{0 = x_{1n} < \cdots < x_{k(n)n} = 1\}$ be a sequence of nested partitions whose diameters tend to zero. Define $\overline{f}_n$ so that $\overline{f}_n(x)$ on $[x_{in}, x_{i+1,n})$ equals the supremum of f over $[x_{in}, x_{i+1,n})$ and $\overline{f}_n(1) = f(1)$; using the infimum, define $\underline{f}_n$ similarly. Then $\overline{f}_n$ and $\underline{f}_n$ are simple functions with respect to $\mathfrak{B}([0,1])$, and their Lebesgue integrals over $[0, 1]$ coincide with the upper and lower Darboux sums of f, respectively, for the partition I_n. Since f is Riemann-integrable, these sums have a common limit equal to the Riemann integral of f. Moreover, the $\overline{f}_n$'s are decreasing functions of n, the $\underline{f}_n$'s increasing functions of n, and they are uniformly bounded (f is bounded). Define

$$g(x) = \lim_{n\to\infty} \overline{f}_n(x), \qquad h(x) = \lim_{n\to\infty} \underline{f}_n(x).$$

Then g, h are Borel functions, $g \le f \le h$, and by Theorem 2.7 the Lebesgue integrals of g and h both equal the Riemann integral of f; hence they coincide. By Exercise 2.6(c), this implies that $g = h$ (a.e.), $g = f$ (a.e.), proving the theorem. □

By Theorem 4, we can use the same notation for Lebesgue integrals and Riemann integrals, e.g., instead of $\int_{[0,1]} f(x)\,\ell(dx)$ we can write $\int_0^1 f(x)\,dx$. It will always be evident from the context whether we mean this in the Riemann or Lebesgue sense.

4. Distributions of random elements, Gaussian vectors, independence

Throughout this section, $(\Omega, \mathcal{F}, P)$ is a probability space.

1. Definition. Let $(X, \mathfrak{B})$ be a measurable space, $\xi : \Omega \to X$ a random element. We define its *distribution* (on X or on $\mathfrak{B}$) to be the function $P\xi^{-1}$ that assigns to each $B \in \mathfrak{B}$ the number

$$P\xi^{-1}(B) := P(\xi^{-1}(B)) = P(\xi \in B). \tag{1}$$

By Lemma 1.4(a), $P\xi^{-1}$ is a probability measure on $(X, \mathfrak{B})$. It turns out that any probability measure on $(X, \mathfrak{B})$ is the distribution of some random element.

2. Theorem. *Let μ be a probability measure on a measurable space $(X, \mathfrak{B})$. Then there exist a probability space $(\Omega, \mathcal{F}, P)$ and a random element $\xi : \Omega \to X$ such that $P\xi^{-1} = \mu$.*

Proof. Set $(\Omega, \mathcal{F}, P) = (X, \mathfrak{B}, \mu)$, $\xi(x) = x$. Then $\{x : \xi(x) \in B\} = B$, $P\xi^{-1}(B) = \mu(B)$. □

The random element constructed in Theorem 2 is said to be *canonically given*. The following theorem establishes a relationship between the expectation of $f(\xi)$ and the distribution of ξ. This is the *theorem on change of variables* for the Lebesgue integral.

3. Theorem. *Let $(X, \mathfrak{B})$ be a measurable space, $\xi : \Omega \to X$ a random element, f a measurable mapping from $(X, \mathfrak{B})$ into $([0, \infty), \mathfrak{B}([0, \infty)))$. Then $f(\xi)$ is a nonnegative random variable and*

$$\mathbf{E} f(\xi) := \int_\Omega f(\xi(\omega))\, P(d\omega) = \int_X f(x)\, P\xi^{-1}(dx). \tag{2}$$

In particular, if $\eta : \Omega \to X$ is another random element and $P\xi^{-1} = P\eta^{-1}$, then $\mathbf{E} f(\xi) = \mathbf{E} f(\eta)$.

Proof. If $f(x) = I_B(x)$ with $B \in \mathfrak{B}$, then $f(\xi(\omega)) = I_{\xi^{-1}(B)}(\omega)$ and (2) follows directly from (1) (and from Theorem 2.7(c)). We now see by Theorem 2.7(e) that (2) holds for all simple functions f; applying the standard device of approximating measurable functions by simple functions (see the end of Sec. 1) as well as the Monotone Convergence Theorem, we see that (2) holds for all measurable $f \geq 0$. □

Note that (2) is true for functions f with values of any sign; all we need is that at least one of the terms in (2) be meaningful. To verify this, we need only to represent f as $f_+ - f_-$ and apply (2) to both f_+ and f_-.

4. Definition. Let ξ be a random d-dimensional vector on $(\Omega, \mathcal{F}, P)$ and $p(x)$ a nonnegative Borel function on E_d such that (see (2.8), (2.9)) $P\xi^{-1}(dx) = p(x)dx$, that is, for any Borel set $B \in E_d$,

$$P\{\xi \in B\} = \int_B p(x)\, \ell(dx) =: \int_B p(x)\, dx.$$

Then p is called the *density* of (the *distribution* of) ξ.

A direct corollary of Theorems 3 and 2.12 is the following

5. THEOREM. *Let ξ be a random d-dimensional vector on $(\Omega, \mathcal{F}, P)$ with density p and f a nonnegative Borel function on E_d. Then*

$$\mathbf{E} f(\xi) = \int_{E_d} f(x) p(x)\, dx.$$

One of the most important classes of distributions in probability theory is that of Gaussian, or normal distributions. To define them, we need the following lemma.

6. LEMMA. *We have*

$$\frac{1}{\sqrt{2\pi}} \int_{-\infty}^{+\infty} e^{-\frac{1}{2}x^2} dx = 1, \quad (2\pi)^{-d/2} \int_{E_d} e^{-\frac{1}{2}|x|^2} dx = 1. \tag{3}$$

PROOF. The second equality follows from the first one after we reduce the Riemann integral over E_d to repeated integrals over $dx^1, \ldots, dx^d$ and note that

$$e^{-\frac{1}{2}|x|^2} = e^{-\frac{1}{2}|x^1|^2} \ldots e^{-\frac{1}{2}|x^d|^2}. \tag{4}$$

The first equality in (3) is obtained from the following calculations:

$$\left(\int_{-\infty}^{+\infty} e^{-\frac{1}{2}x^2} dx \right)^2 = \int_{E_2} e^{-\frac{1}{2}|x|^2} dx = \int_0^{2\pi} 2 d\varphi \int_0^{\infty} e^{-\frac{1}{2}r^2} r dr$$

$$= 2\pi \int_0^{\infty} e^{-\frac{1}{2}r^2} d\frac{r^2}{2} = 2\pi. \quad \square$$

7. DEFINITION. Let ξ be a random d-dimensional vector on $(\Omega, \mathcal{F}, P)$. Then ξ is said to have the *standard normal* distribution, or to be a d-dimensional *standard normal* random vector, if the density of ξ is

$$(2\pi)^{-d/2} e^{-\frac{1}{2}|x|^2}. \tag{5}$$

It should be noted that, by Lemma 6, the integral of (5) over a Borel set $B \subset E_d$ is a *probability* measure on B. By Theorem 2, there exists a random vector ξ for which this measure is a distribution, so that there exist normally distributed vectors.

An exceptionally useful tool for studying properties of distributions is characteristic functions.

8. DEFINITION. Let ξ be a d-dimensional random vector on $(\Omega, \mathcal{F}, P)$. The function on E_d defined by the formula

$$\varphi_\xi(t) = \mathbf{E}\, e^{i(t,\xi)} := \mathbf{E} \cos(t, \xi) + i\mathbf{E} \sin(t, \xi)$$

is called the *characteristic function* of ξ (or of the distribution of ξ).

Since $|\cos x|, |\sin x| \leq 1$, it follows that $\varphi_\xi(t)$ exists for any random vector ξ. In addition, by the Dominated Convergence Theorem, $\varphi_\xi(t)$ is continuous in t and

$$|\varphi_\xi(t)| = \sup_{|z|\leq 1} \operatorname{Re} z\varphi(t) = \sup_{|z|\leq 1} \mathbf{E} \operatorname{Re} z e^{i(t,\xi)} \leq \mathbf{E} \sup_{|z|\leq 1} \operatorname{Re} z e^{i(t,\xi)} = 1.$$

9. DEFINITION. Let ξ be a d-dimensional random vector, $m \in E_d$, R a nonnegative symmetric $d \times d$ matrix. Then ξ is said to be *normally distributed with parameters* (m, R), written as

$$\xi \sim N(m, R),$$

if $\varphi_\xi(t) = \exp(i(m,t) - \frac{1}{2}(Rt, t))$. In this case one also says that ξ is a *Gaussian vector*.

10. LEMMA. *If ξ is a d-dimensional standard normal vector, then $\xi \sim N(0, I)$, where I is the $d \times d$ identity matrix, that is, $\varphi_\xi(t) = \exp\left(-\frac{1}{2}|t|^2\right)$.*

PROOF. Using Theorem 5, Definition 7 and equality (4), we see that the evaluation of $\varphi_\xi(t)$ is easily reduced to the case $d = 1$. In that case

$$\mathbf{E}\, e^{it\xi} = \frac{1}{\sqrt{2\pi}} \int_{-\infty}^{\infty} e^{itx - \frac{1}{2}x^2} dx = e^{-\frac{1}{2}t^2} \frac{1}{\sqrt{2\pi}} \int_{-\infty}^{\infty} e^{-\frac{1}{2}(x - it)^2} dx = e^{-\frac{1}{2}t^2},$$

where the last equality follows from (3) and Jordan's lemma. □

11. EXERCISE. Prove that if ξ is a standard normal vector, then $\mathbf{E}\,\xi^i = 0$, $\mathbf{E}\,\xi^i \xi^j = \delta^{ij}$ for $i, j = 1, \dots, d$.

In connection with Definition 9 and Lemma 10, it is natural to ask to what extent a distribution is determined by its characteristic function.

12. THEOREM. *Let μ, ν be finite measures on $\mathfrak{B}(E_d)$ and assume that*

$$\int_{E_d} e^{i(t,x)} \mu(dx) = \int_{E_d} e^{i(t,x)} \nu(dx) \tag{6}$$

for all $t \in E_d$. Then $\mu = \nu$ on $\mathfrak{B}(E_d)$. In particular, if $\mu(dx) = g(x)dx$, $\nu(dx) = h(x)dx$, then $g = h$ almost everywhere with respect to d-dimensional Lebesgue measure; if g and h are both continuous, then $g \equiv h$.

The proof will be postponed until the next section. Theorem 12 shows that the characteristic function contains all information about the distribution.

Note that if ξ is a d-dimensional random vector and $\xi \sim N(0, I)$, and if Q is a $k \times d$ matrix, $m \in E_k$, then for $\eta := m + Q\xi$ and $t \in E_k$ we have

$$\varphi_\eta(t) = \mathbf{E}\, e^{i(m,t)} e^{i(\xi, Q^* t)} = e^{i(m,t) - \frac{1}{2}(Rt,t)},$$

where $(Rt, t) = |Q^* t|^2$, $R = QQ^*$. This is the key to the construction of normal vectors with arbitrary parameters. Indeed, any nonnegative symmetric matrix R can be expressed as the square of some nonnegative symmetric matrix $Q := R^{1/2}$ (this is easily proved, e.g., by reducing R to diagonal form); hence there exist normally distributed vectors with arbitrary parameters.

13. EXERCISE. Using the previous arguments together with Exercise 11 and Theorems 12 and 3, show that if $\xi \sim N(m, R)$, then $m = \mathbf{E}\,\xi$,

$$R = \mathbf{E}\,(\xi - m)(\xi - m)^*, \quad \text{that is,} \quad m^i = \mathbf{E}\,\xi^i,\ R^{ij} = \mathbf{E}\,(\xi^i - m^i)(\xi^j - m^j),$$

that is, that m is the *mean value* of the vector ξ and R is its *covariance matrix*. Prove also that if $d = 1$, then $\mathbf{E}\,|\xi - m|^{2p} = c(p) R^p$ for any $p > 0$, where $c(p)$ is a finite constant depending only on p. Derive from the last equality that in the general case $\mathbf{E}\,|\xi|^p < \infty$ for any $p > 0$.

An argument similar to that after Theorem 12 easily proves the following

14. THEOREM. *Let ξ be a d-dimensional Gaussian vector, $n \in E_k$, Q a $(k \times d)$-matrix. Then $n + Q\xi$ is a k-dimensional Gaussian vector. In other words, affine image of a Gaussian vector is a Gaussian vector.*

In particular, the vector obtained by permutation of coordinates of a Gaussian vector is again Gaussian, a vector obtained by taking some of the coordinates of a Gaussian vector is again Gaussian.

It turns out that a Gaussian distribution remains Gaussian not only under affine or linear transformations, but also upon passage to a limit.

15. THEOREM. *Let $\xi(1), \xi(2), \dots$ be d-dimensional Gaussian vectors, ξ a d-dimensional random vector on $(\Omega, \mathcal{F}, P)$ and $\xi = \lim_{n\to\infty} \xi(n)$ (a.s.) (where the limit exists almost surely). Then ξ is a Gaussian vector and*

$$\mathbf{E}\,\xi = \lim_{n\to\infty} \mathbf{E}\,\xi(n), \qquad R_\xi = \lim_{n\to\infty} R_{\xi(n)}, \tag{7}$$

where $R_\xi, R_{\xi(n)}$ are the covariance matrices of $\xi, \xi(n)$.

PROOF. It follows from the Dominated Convergence Theorem that

$$\varphi_\xi(t) = \lim_{n\to\infty} \varphi_{\xi(n)}(t)$$

for all $t \in E_d$; in particular, this limit exists for all t. Hence, by Exercise 13 and Definition 9, it follows that the limits on the right in (7) exist and that ξ is a Gaussian vector. Another application of Exercise 13 allows us to identify the parameters of distribution of ξ with $\mathbf{E}\,\xi$, R_ξ. □

To discuss further properties of Gaussian vectors, we will need one more very important notion.

16. DEFINITION. Random vectors $\xi_1, \dots, \xi_k$ of dimensions $d_1, \dots, d_k$, respectively, on $(\Omega, \mathcal{F}, P)$ are said to be (jointly) *independent* if, for any Borel sets $B_j \subset E_{d_j}$, $j = 1, \dots, k$, we have

$$P\{\xi_1 \in B_1, \dots, \xi_k \in B_k\} = P\{\xi_1 \in B_1\} \cdot \ldots \cdot P\{\xi_k \in B_k\}, \tag{8}$$

where the left-hand side is, of course, understood as

$$P\left(\{\omega : \xi_1(\omega) \in B_1, \dots, \xi_k(\omega) \in B_k\}\right).$$

Given an infinite system of random vectors they are called (jointly) independent if any finite subsystem consists of jointly independent vectors.

This definition is consistent, in the sense that if vectors $\xi_1, \xi_2, \dots, \xi_k$ are independent, then any subset of them consists of independent vectors, since there is nothing to prevent us from taking some B_j in (8) equal to E_{d_j}.

17. THEOREM. *Random vectors $\xi(1), \dots, \xi(k)$ of dimensions $d_1, \dots, d_k$, respectively, defined on $(\Omega, \mathcal{F}, P)$, are independent if and only if for all $t_j \in E_{d_j}$, $j = 1, \dots, k$, we have*

$$\mathbf{E} \exp i \sum_{j=1}^{k} (t_j, \xi_j) = \prod_{j=1}^{k} \varphi_{\xi_j}(t_j). \tag{9}$$

It is more convenient to prove this theorem, like Theorem 12, in the next section. Briefly, the theorem asserts that the vectors $\xi_1, \dots, \xi_k$ are independent if and only if the characteristic function of the vector $(\xi_1^*, \dots, \xi_k^*)^*$ built up from all the coordinates of the vectors ξ_j is the product of the characteristic functions of ξ_j.

Since the characteristic functions of Gaussian vectors have a very simple form, Theorem 17 may be used to get a simple criterion for the independence of Gaussian vectors. Prior to stating it, we present a definition.

18. DEFINITION. Random vectors ξ, η on $(\Omega, \mathcal{F}, P)$, possibly of different dimensions, are said to be *uncorrelated* if $\mathbf{E}\,|\xi|^2 < \infty$, $\mathbf{E}\,|\eta|^2 < \infty$, and

$$\mathbf{E}\,\xi^i \eta^i = (\mathbf{E}\,\xi^j)(\mathbf{E}\,\eta^j) \tag{10}$$

for all possible i, j.

Note, incidentally, that $|\xi^i \eta^j| \le |\xi|^2 + |\eta|^2$, $|\xi^i| \le 1 + |\xi|^2$, so that all expressions in (10) are well defined. Furthermore, it is not hard to see that (10) is equivalent to

$$\mathbf{E}\left[(\xi^i - \mathbf{E}\,\xi^i)(\eta^j - \mathbf{E}\,\eta^j)\right] = 0.$$

19. THEOREM (on Normal Correlation). *Let $d_j \ge 1$ be integers, ξ_j d_j-dimensional random vectors on $(\Omega, \mathcal{F}, P)$, $j = 1, \dots, k$, and let the vector consisting of all the coordinates of all the ξ_j be a $d := d_1, + \dots + d_k$-dimensional Gaussian vector. Then the vectors $\xi_1, \dots, \xi_k$ are jointly independent if and only if any two of them are uncorrelated.*

PROOF. Express E_d as $E_{d_1} \times \dots \times E_{d_k}$, that is, write the points $x \in E_d$ as $(x_1, \dots, x_k)$, where $x_j \in E_{d_j}$. Then $\xi = (\xi_1, \dots, \xi_k)$ is a vector with values in E_d. If $\xi_1, \dots, \xi_k$ are pairwise uncorrelated, then (and only then) the covariance matrix R of ξ is block-diagonal, i.e., $Rt = (R_1 t_1, \dots, R_k t_k)$, where R_j is the covariance matrix of ξ_j, $t \in E_d$. Hence

$$(Rt, t) = (R_1 t_1, t_1) + \dots + (R_k t_k, t_k), \tag{11}$$

the characteristic function of ξ decomposes as a product and, by Theorem 17, the vectors $\xi_1, \dots, \xi_k$ are independent. Conversely by Theorem 17, independence of $\xi_1, \dots, \xi_k$ implies (11), so that ξ_j, ξ_r are uncorrelated for all $j \ne r \le k$. □

An elementary corollary of Theorem 17 is the following

20. THEOREM. *Let $\xi_1, \dots, \xi_k$ be independent random Gaussian vectors (possibly, of different dimensions). Then the vector $(\xi_1, \dots, \xi_k)$ built up from of all the coordinates of all the ξ_j's is also a Gaussian vector.*

21. EXERCISE. Prove that if random vectors $\xi_1, \dots, \xi_k$ are pairwise uncorrelated and c_j are constant vectors of appropriate dimensions, then $\xi_1 - c_1, \dots, \xi_k - c_k$ are pairwise uncorrelated; if the ξ_j are of the same dimension and $\mathbf{E}\,\xi_j = 0$, then

$$\mathbf{E}\,|\xi_1 + \dots + \xi_k|^2 = \mathbf{E}\,|\xi_1|^2 + \dots + \mathbf{E}\,|\xi_k|^2.$$

22. EXERCISE. Derive from (8) that, for all nonnegative Borel functions f_j on E_{d_j}, $j = 1, \dots, k$, we have

$$\mathbf{E}\,f_1(\xi_1) \cdot \ldots \cdot f_k(\xi_k) = \mathbf{E}\,f_1(\xi_1) \cdot \ldots \cdot \mathbf{E}\,f_k(\xi_k); \tag{12}$$

and if in addition $\mathbf{E}\,|\xi_j|^2 < \infty$ for $j \le k$, then ξ_j, ξ_r are uncorrelated for $j \ne r$. (Hint: first prove (12) for simple functions.) Prove that if $\xi_1, \dots, \xi_k$ are independent and f_j are Borel functions on E_{d_j}, then $f_1(\xi_1), \dots, f_k(\xi_k)$ are independent. Prove that

(12) is also true for all bounded complex-valued Borel functions f_j, provided that the functions ξ_j are independent.

23. PROBLEM. Let R be a nonsingular positive symmetric $(d \times d)$-matrix, $m \in E_d$. Define a measure on E_d with the density

$$(2\pi)^{-d/2}(\det R)^{-1/2}\exp\Big[-\frac{1}{2}(R^{-1}(x-m), x-m)\Big], \tag{13}$$

and prove, using Theorem 12, that if $\xi \sim N(m, R)$, then the function (13) is the density of ξ.

24. PROBLEM. Let ξ be a d-dimensional random vector, $\xi \sim N(0, R)$, $y \in E_d$. Prove that

$$\mathbf{E}g(\xi + Ry) = \mathbf{E}g(\xi)\exp\Big[(y, \xi) - \frac{1}{2}(Ry, y)\Big] \tag{14}$$

for any Borel function g such that at least one side of (14) is defined. (First do this for $\xi \sim N(0, I)$, using Theorem 5; then use the arguments preceding Exercise 13.)

25. PROBLEM. Under the conditions of the previous problem, prove that if g is a continuously differentiable function of x on E_d and its first derivatives by magnitude are dominated by $\exp \lambda|x|$ as $|x| \to \infty$, for some constant λ, then

$$\mathbf{E}g_{(Ry)}(\xi) = \mathbf{E}g(\xi)(y, \xi),$$

where

$$u_{(y)} := \sum_{i=1}^{d} u_{x^i} y^i. \tag{15}$$

(Either use the hints to Problem 24 and integrate by parts or differentiate (14).)

26. PROBLEM. Let ξ and η be d- and d_1-dimensional random vectors, respectively, such that the pair (ξ, η) is normally distributed and $\mathbf{E}\xi = 0$, $\mathbf{E}\eta = 0$. Given g as in the previous problem and any $y \in E_{d_1}$, prove that

$$\mathbf{E}g_{(ay)}(\xi) = \mathbf{E}g(\xi)(y, \eta),$$

where $a = (a^{ij})$, $a^{ij} = \mathbf{E}\xi^i\eta^j$. (In addition to the result of Problem 25, use the Normal Correlation Theorem).

5. Lemma on π- and λ-systems; applications. Fubini's theorem

1. DEFINITION. Let X be a set, $\mathfrak{A}$ a family of subsets of X. Then $\mathfrak{A}$ is called a *π-system* if $A_1 \cap A_2 \in \mathfrak{A}$ for any $A_1, A_2 \in \mathfrak{A}$. It is called a *$\lambda$-system* if

a) $X \in \mathfrak{A}$ and $A_2 \setminus A_1 \in \mathfrak{A}$ for any $A_1, A_2 \in \mathfrak{A}$ such that $A_1 \subset A_2$;

b) for any $A_1, A_2, \ldots \in \mathfrak{A}$ such that $A_i \cap A_j = \emptyset$ when $i \neq j$, $\bigcup_{n=1}^{\infty} A_n \in \mathfrak{A}$.

One of the basic examples of a λ-system is the family of all sets on which two given probability measures coincide. The next *lemma on π- and λ-systems* is important, in particular, when one wants to derive the coincidence of two measures on a σ-algebra from their coincidence on a π-system generating the σ-algebra. Before proving the lemma, we offer the reader an exercise.

2. EXERCISE. Prove that if $\mathfrak{A}$ is both a λ-system and a π-system, then it is a σ-algebra.

3. LEMMA. *If Λ is a λ-system and Π is a π-system and $\Pi \subset \Lambda$, then $\sigma(\Pi) \subset \Lambda$.*

PROOF. Let Λ_1 denote the smallest λ-system containing Π (Λ_1 is the intersection of all λ-systems containing Π). It suffices to prove that $\Lambda_1 \supset \sigma(\Pi)$. To do this, it suffices to prove, by Exercise 2, that Λ_1 is a π-system, that is, it contains the intersection of any two of its sets. For $B \in \Lambda_1$ let $\Lambda(B)$ denote the family of *all* $A \in \Lambda_1$ such that $A \cap B \in \Lambda_1$. Obviously, $\Lambda(B)$ is a λ-system. In addition, if $B \in \Pi$, then $\Lambda(B) \supset \Pi$ (since Π is a π-system). Consequently, if $B \in \Pi$, then by the definition of Λ_1, we have $\Lambda(B) \supset \Lambda_1$. But this means that $\Lambda(A) \supset \Pi$ for all $A \in \Lambda_1$, so that as before, $\Lambda(A) \supset \Lambda_1$ for all $A \in \Lambda_1$, that is, Λ_1 is a π-system. □

The next theorem, which follows from the lemma, will be frequently used below.

4. THEOREM. *Let X, Y be Polish spaces, y a Borel function on X with values in Y, μ a finite measure on $\mathfrak{B}(X)$, ν a finite measure on $\mathfrak{B}(Y)$, and assume that for all continuous $f \geq 0$ defined on Y we have*

$$\int_X f(y(x))\mu(dx) = \int_Y f(y)\nu(dy). \tag{1}$$

Then (1) *holds for all Borel functions $f \geq 0$, and if $X = Y$ and $y(x) \equiv x$, then $\mu = \nu$ on $\mathfrak{B}(X)$.*

PROOF. If Γ is a closed subset of Y, then

$$I_\Gamma(y) = \lim_{n\to\infty} f_n(y),$$

where $f_n(y) = (1 + n\rho(y, \Gamma))^{-1}$ are continuous functions ($\rho(y, \Gamma)$ is the distance from y to the set Γ). Using this and the Dominated Convergence Theorem, we obtain (1) for $f = I_\Gamma$. The family $\mathfrak{A}$ of all Borel sets $B \subset Y$ such that (1) holds for $f = I_B$ is a λ-system (cf. Corollary 2.11). By what we have proved, it contains the π-system of closed sets. By Lemma 3, it also contains the σ-algebra generated by the closed sets; since the latter obviously coincides with the Borel σ-algebra, it follows that (1) holds for indicators of Borel sets in Y. It is now evident that the equality is also valid for all simple Borel functions on Y; by the Monotone Convergence Theorem and the technique of approximating measurable functions by simple functions (see the end of Sec. 1) we can extend (1) to all Borel functions $f \geq 0$. Since the last assertion of the theorem is obtained from the first one by putting $f = I_B$, $B \in \mathfrak{B}(X)$, the theorem is proved. □

5. THEOREM (Fubini). *Let $(X, \mathfrak{A}, \mu)$, $(Y, \mathfrak{B}, \nu)$ be measure spaces and suppose that for some $\mathfrak{A}$-measurable function $a(x) > 0$ and a $\mathfrak{B}$-measurable function $b(y) > 0$,*

$$\int_X a(x)\mu(dx) < \infty, \int_Y b(y)\mu(dy) < \infty. \tag{2}$$

For $C \in \mathfrak{A} \otimes \mathfrak{B}$, set

(3) $$\mu \times \nu(C) = \int_X \left(\int_Y I_C(x,y)\nu(dy) \right) \mu(dx),$$

(4) $$\mu * \nu(C) = \int_Y \left(\int_X I_C(x,y)\mu(dx) \right) \nu(dy).$$

Then $\mu \times \nu$, $\mu * \nu$ are well defined, they are measures on $\mathfrak{A} \otimes \mathfrak{B}$ and $\mu \times \nu = \mu * \nu$ on $\mathfrak{A} \otimes \mathfrak{B}$. In addition, if a nonnegative $f(x,y)$ is measurable with respect to $\mathfrak{A} \times \mathfrak{B}$, then

(5) $$\begin{aligned}\int_{X\times Y} f(x,y)\mu \times \nu(dxdy) &= \int_X \left(\int_Y f(x,y)\nu(dy) \right) \mu(dx) \\ &= \int_Y \left(\int_X f(x,y)\mu(dx) \right) \nu(dy);\end{aligned}$$

where it is also asserted that all the terms in (5) are meaningful, e.g., $f(x,y)$ is $\mathfrak{A}$-measurable in x for each y and the inner integral in the last expression is $\mathfrak{B}$-measurable in y. Finally, the equalities in (5) also hold for any complex $\mathfrak{A} \otimes \mathfrak{B}$-measurable function f, if at least one of the three terms involved becomes finite when f is replaced by $|f|$.

PROOF. First let μ and ν be finite measures. We must first prove that definition (3) is meaningful, that is, $I_C(x,y)$ is measurable in y for fixed x, and the inner integral is measurable in x. By the properties of measurable functions and Corollary 2.8, we see that the set Λ of *all* C having the properties listed above is a λ-system. In addition, Λ contains the π-system Π of all the sets: $C = A \times B = \{(x,y) : x \in A\} \cap \{(x,y) : y \in B\}$, where $A \in \mathfrak{A}$, $B \in \mathfrak{B}$, since $I_C(x,y) = I_A(x)I_B(y)$. By Lemma 3, we conclude that $\Lambda \supset \sigma(\Pi) = \mathfrak{A} \otimes \mathfrak{B}$. That (4) is meaningful is proved in a similar way. That $\mu \times \nu$, $\mu * \nu$ are measures follows directly from Corollary 2.8. That they coincide on $\mathfrak{A} \otimes \mathfrak{B}$ follows by Lemma 3 from the fact they coincide on Π, which is obvious because $I_{A\times B}(x,y) = I_A(x)I_B(y)$.

The fact that the terms in (5) are meaningful and the fact that (5) holds have already been proved for $f = I_C$, $C \in \mathfrak{A} \otimes \mathfrak{B}$; for simple $\mathfrak{A} \otimes \mathfrak{B}$-measurable functions these facts are direct consequences of the linearity of integrals; they are extended to arbitrary $\mathfrak{A} \otimes \mathfrak{B}$-measurable functions $f \geq 0$ by the standard limit procedure (see the end of Sec. 1), using the Monotone Convergence Theorem.

The last assertion of the theorem is an elementary corollary of the previous assertion and the relations $f = (\mathrm{Re} f)_+ - (\mathrm{Re} f)_- + i(\mathrm{Im} f)_+ - i(\mathrm{Im} f)_-$, $|\mathrm{Re} f|$, $|\mathrm{Im} f| \leq |f|$. This proves the theorem for finite measures μ and ν.

If μ and ν are only *σ-finite* (that is, condition (2) holds with $a > 0, b > 0$), we set $\mu_1(dx) = a(x)\mu(dx)$, $\nu_1(dy) = b(y)\nu(dy)$. Then the measures μ_1, ν_1 are finite. In addition, by what we proved above and by Theorem 2.12, the right-hand sides of (3) and (4) are meaningful and equal

$$\int_{X\times Y} I_C(x,y)a^{-1}(x)b^{-1}(y)\mu_1 \times \nu_1(dxdy).$$

Hence, in this case too, the right-hand sides of (3) and (4) coincide; and this is all we need to complete the proof. □

6. DEFINITION. The measure space $(X \times Y, \mathfrak{A} \otimes \mathfrak{B}, \mu \times \nu)$ is called the *product* of $(X, \mathfrak{A}, \mu)$ and $(Y, \mathfrak{B}, \nu)$.

We have defined the product of two spaces with *σ-finite* measures. In a similar way, one can define the product of a finite number of spaces with σ-finite measures. It is easy to verify that the product is associative.

In connection with the definition of $\mathfrak{A} \otimes \mathfrak{B}$, a natural question arises: what conditions are sufficient for a function $f(x, y)$ to be $\mathfrak{A} \otimes \mathfrak{B}$-measurable? Note that if $X = E_d$, $Y = E_k$, $\mathfrak{A} = \mathfrak{B}(E_d)$, $\mathfrak{B} = \mathfrak{B}(E_k)$, then f is $\mathfrak{A} \otimes \mathfrak{B}$-measurable if and only if f is a Borel function on the Euclidean space $X \times Y$. This follows from Lemma 1.7(c), according to which $\mathfrak{B}(X \times Y) = \mathfrak{B}(X) \otimes \mathfrak{B}(Y)$. The following lemma is also useful.

7. LEMMA. *Let $(\Omega, \mathcal{F})$ be a measurable space, and let $f(\omega, x)$ be defined for $\omega \in \Omega$, $x \in E_d$, $\mathcal{F}$-measurable in ω for each x and continuous in x for each ω. Then $f(\omega, x)$ is $\mathcal{F} \otimes \mathfrak{B}(E_d)$-measurable. For $d = 1$, the requirement that f be continuous in x can be replaced with the requirement that f be continuous in x only from the right (or only from the left). For instance, the indicator $I_{\xi > t}$ is $\mathcal{F} \otimes \mathfrak{B}(E_1)$-measurable in (ω, t) if ξ is a random variable.*

PROOF. Denote $\varkappa_n(t) = 2^{-n}[2^n t] + 2^{-n}$, $\varkappa_n(x) = (\varkappa_n(x^1), \dots, \varkappa_n(x^d))$. Obviously, $\varkappa_n(t) \downarrow t$, $f(\omega, \varkappa_n(x)) \to f(\omega, x)$ as $n \to \infty$. It will thus suffice to prove that $f(\omega, \varkappa_n(x))$ is $\mathcal{F} \otimes \mathfrak{B}(E_d)$-measurable.

We have

$$f(\omega, \varkappa_n(x)) = \sum_r I_{\varkappa_n(x) = r} f(\omega, r),$$

where the summation goes over the (countable) set of all values of $\varkappa_n(x)$. Clearly, $\varkappa_n(x)$ is a Borel function, and by Fubini's Theorem the function $f(\omega, r)$ is $\mathcal{F}$-measurable for every r. By the definition of $\mathcal{F} \otimes \mathfrak{B}(E_d)$, these functions are measurable as functions of (ω, x). It remains to recall that products and sums of measurable functions are measurable. □

The last assertion of this lemma, together with Fubini's Theorem, justifies the calculations done just before Exercise 2.6. We now fulfill the promises made in the previous section.

8. PROOF OF THEOREM 4.12. Let $\psi(t)$ denote the left- and the right-hand sides of (4.6). By Fubini's Theorem and Lemma 4.10, for any fixed $y \in E_d$, $s \in (-\infty, \infty)$ we have

$$\begin{aligned} \int_{E_d} \exp\Big(-\frac{1}{2}|x - y|^2 s^2\Big)\mu(dx) &= \mathbf{E} \int_{E_d} \exp is(x - y, \xi)\mu(dx) \\ &= \mathbf{E}\, \psi(s\xi) \exp(-is(y, \xi)) \\ &= \int_{E_d} \exp\Big(-\frac{1}{2}|x - y|^2 s^2\Big)\nu(dx), \end{aligned} \tag{6}$$

where $\xi \sim N(0, I)$. Now let f be a continuous bounded function on E_d. Multiplying the outer terms of (6) by $f(y)$ and integrating with respect to y, we apply Fubini's

Theorem and Definition 4.7 to get

$$\int_{E_d} \mathbf{E} f(x+s^{-1}\xi)\mu(dx) = (2\pi)^{-d/2}\int_{E_d}\mu(dx)\Big(\int_{E_d} f(x+s^{-1}y)\exp\Big(-\frac{|y|^2}{2}\Big)dy\Big)$$

$$= (2\pi)^{-d/2}\int_{E_d}\mu(dx)\int_{E_d} f(y)\exp\Big(-\frac{1}{2}|x-y|^2s^2\Big)dy$$

$$= \int_{E_d}\mathbf{E} f(x+s^{-1}\xi)\nu(dx).$$

Letting $s \to \infty$ and using the Dominated Convergence Theorem, we obtain (1), hence $\mu = \nu$. If in addition, $\mu(dx) = gdx$, $\nu(dx) = hdx$, it follows from the equality $\mu = \nu$ that the integral of $g - h$ over any Borel set vanishes; in particular,

$$\int_{E_d} |g-h|dx = \int_{g-h\geq 0}(g-h)dx + \int_{g-h<0}(h-g)dx = 0$$

and $g = h$ (a.e.). Since the last assertion of Theorem 4.12 is obvious, this completes the proof. □

9. Proof of Theorem 4.17. Let $d = d_1+\cdots+d_k$. Expressing E_d as $E_{d_1}\times\cdots\times E_{d_k}$, define the random vector $\xi = (\xi_1,\dots,\xi_k)$ with values in E_d and let μ be its distribution on E_d. Let ν_j be the distribution of ξ_j on E_{d_j}. Then $\nu = \nu_1\times\cdots\times\nu_k$ is a measure on E_d. It follows immediately from (4.9), by Fubini's Theorem and Theorem 4.3, that (4.6) is true for all $t = (t_1,\dots,t_k)\in E_d$. Hence, by Theorem 4.12, $\mu(B) = \nu_1\times\cdots\times\nu_k(B)$ for all Borel sets $B \subset E_d$. Putting $B = B_1\times\cdots\times B_k$ and using the definitions of μ, ν_j and of direct products of measures, we obtain (4.8).

On the other hand, if (4.8) is given, then $\mu(B) = \nu(B)$ for all Borel sets B of the form $B_1\times\cdots\times B_k$, hence, also for all $B \in \mathfrak{B}(E_d)$. The required equality (4.9) now follows from Theorem 4.3 and Fubini's Theorem. □

CHAPTER II

The Wiener Process

In this section we begin our study of diffusion processes, the simplest of which is the Wiener process. We will first consider one-dimensional Wiener processes, whose existence will be proved in Sec. 2 using Fourier series. We will need only one equality from the theory of Fourier series (see below (2.4)), which can also be proved via elementary results of complex function theory. Then we will explain why it is useful to consider the Wiener process at random times and introduce the notion of a Markov time. Having done this, we will introduce in Sec. 4 multidimensional Wiener processes and in Sec. 5 prove their main property, known as the strong Markov property.

The rest of the chapter aims mainly at explaining why the notion of a martingale is natural and at preparing to define Itô's stochastic integral. A reader interested primarily in stochastic integrals can proceed to Chap. III immediately after Theorem 6.9, referring from time to time (starting from Sec. III.3) to the results from the theory of martingales collected in Lemmas 8.5, 8.8 in the present chapter.

In Secs. 8–10 we will establish important relationships between the theory of diffusion processes and the theory of differential equations. These relationships will be constantly in view from Sec. V.2 on. Section 7 can be skipped without any effect on one's understanding of the sequel.

1. Brownian motion and the Wiener process

In 1827 the British botanist R. Brown noticed that microscopic particles suspended in a liquid move in a disorderly fashion (it became clear afterwards, that this was due to collisions with the molecules of the liquid). The characteristics of this motion, such as the mean-square deviation from the initial point per unit time, depend on the temperature of the liquid, its viscosity and other physical parameters. The phenomenon, which became known as *Brownian motion* was thus interesting for physicists. Brownian motion was investigated on a physical level of rigor by Einstein, Smolukhovskiĭ and Bachelier. A rigorous mathematical model of Brownian motion was constructed in 1923 by Wiener, after whom the corresponding stochastic process is named the Wiener process.

Wiener processes are extremely interesting from mathematical point of view mainly because they can be used to describe a very wide class of continuous stochastic processes. A Wiener process may be one-dimensional or multidimensional. We will start with one-dimensional Wiener processes.

1. Definition. Let $(\Omega, \mathcal{F}, P)$ be a probability space, $T \in (0, \infty]$ a number, and suppose that for any $\omega \in \Omega$, $t \in [0, T)$ there is defined a real-valued function $w_t(\omega)$. We will say that w_t (or $w_t(\omega)$) is a (one-dimensional) *Wiener process* (on $(\Omega, \mathcal{F}, P)$) for $t \in [0, T)$, if

(a) for any integer n and any $t_1, \dots, t_n \in [0, T)$, the vector $(w_{t_1}, \dots, w_{t_n})$ is Gaussian and $\mathbf{E}\, w_t = 0$, $\mathbf{E}\, w_t w_s = t \wedge s$ for all $t, s \in [0, T)$,

(b) $w_t(\omega)$ is a continuous function of t on $[0, T)$ and $w_0(\omega) = 0$ for any $\omega \in \Omega$.

If $T < \infty$, $w_T(\omega)$ is defined and $w_t(\omega)$ is continuous on $[0, T]$, we will say that w_t is a Wiener process on the time interval $[0, T]$. For fixed ω we call $w_t(\omega)$, as a function of t, *a sample path* (or *trajectory*) of the Wiener process.

Let us elucidate why this definition is physically natural. Of course, we will merely speak about an interpretation of the mathematical notions and axioms just introduced above, an interpretation which is as indisputable from the practical point of view as the analogous arguments in favor of axioms of the Euclidean geometry. A reader who considers these explanations unconvincing can safely skip them.

The set Ω is interpreted as the set of "all events" that can occur in an experiment of observing the Brownian motion of a particle. It is assumed that this physical experiment can be reproduced "independently" under the "same conditions" as many times as desired, each time placing the Brownian particle at the same position in an infinite vessel and then beginning to observe it. $\mathcal{F}$ is interpreted as the family of all subsets A of Ω, for each of which the frequency $\nu_n(A)/n$ of the occurrence of A in n experiments tends to a limit denoted by $P(A)$ as $n \to \infty$. Here $\nu_n(A)$ is the number of occurrences of A in the first n experiments.

With this interpretation of $P(A)$ one easily sees that, for a simple random variable $\xi(\omega)$, the limit as $n \to \infty$ of the arithmetic means of the values of ξ recorded in n experiments is equal to $\mathbf{E}\,\xi$. That is why $\mathbf{E}\,\xi$ is (also) called the mean value of ξ. This interpretation extends naturally to all random variables ξ for which $\mathbf{E}\,\xi$ is meaningful. Independence of random variables ξ, η (in the sense of Definition I.4.16) is interpreted as follows: it is impossible to determine the value of ξ in an experiment, knowing *only* the exact value of η in the experiment, more accurately than when the value of η is completely unknown too. This interpretation is based on the following reasoning: $P\{\xi \in \Gamma_1, \eta \in \Gamma_2\}/P\{\eta \in \Gamma_2\}$ is the limit of

$$\frac{\frac{1}{n}\nu_n(\xi \in \Gamma_1, \eta \in \Gamma_2)}{\frac{1}{n}\nu_n(\eta \in \Gamma_2)} = \frac{\nu_n(\xi \in \Gamma_1, \eta \in \Gamma_2)}{\nu_n(\eta \in \Gamma_2)},$$

that is, the limit of the frequency of the event $\xi \in \Gamma_1$, measured only for those experiments in which $\eta \in \Gamma_2$. If ξ and η are independent, this limit must be equal to $P\{\xi \in \Gamma_1\}$, which is the limit of the frequency of the event $(\xi \in \Gamma_1)$, measured in *all* experiments.

What does this interpretation of independence give for our suspended particle? Let, for example, Ω be the set of all possible sample paths of a Brownian particle, $w_t(\omega)$ the value of its first coordinate at time t after the beginning of the experiment, where ω is the sample path obtained in a given experiment. Take $0 \le r < s < t$ and select from a sequence of experiments producing sample paths $\omega_1, \omega_2, \dots$ only those ω_i for which $w_s(\omega_i) - w_r(\omega_i) \in \Gamma_1$. Among these sample paths, how many ω_j, occur such that $w_t(\omega_j) - w_s(\omega_j) \in \Gamma_2$? Since in disjoint time intervals a Brownian particle is "driven" by different molecules of liquid, we conclude that the sample path starting from the point reached at time s will behave after s as if it had "forgotten" everything that happened beforehand. Therefore, a knowledge of the value of $w_s - w_r$ has no influence on the possibility that $w_t - w_s \in \Gamma_2$. Hence the limit of the frequency of the event $w_t - w_s \in \Gamma_2$ confined to cases in which $w_s - w_r \in \Gamma_1$ should not depend on Γ_1 when we repeat experiments infinitely, and it must be equal (if $\Gamma_1 = (-\infty, \infty)$) to

$P\{w_t - w_s \in \Gamma_2\}$. As we have already explained, this implies that $w_t - w_s$ and $w_s - w_r$ are independent in the sense of Definition I.4.16.

In addition, since the liquid is spatially homogeneous and the experimental conditions are invariant (as is obviously assumed), the distribution of the increment $w_t - w_s$ must depend not on t, s but only on there difference.

So far we have referred to only a few properties of our hypothetical experiment. Let us now turn to Definition 1, first appealing to the Central Limit Theorem of probability theory, which states, roughly speaking, that in many situations the sum of a very large number of very small and, in a sense, "equivalent" summands has an approximately normal distribution.

Here lies the motivation for having Gaussian vectors in Definition 1, since the motion of a suspended particle is caused by the overall effect of a large number of "equivalent" pushes by different molecules of the liquid. Clearly, the mean value of the particle's displacement from its initial location must be zero since there is no preferred direction, that is, $\mathbf{E}(w_t - w_0) = 0$. It is also natural to take the initial location of the particle as the origin, hence $w_0 = 0$.

To explain the assumption $\mathbf{E}\, w_t w_s = t \wedge s$, we will use the above-mentioned properties of the increments of w. Let $\varphi(t) = \mathbf{E}\,|w_{s+t} - w_s|^2$ for $t, s \geq 0$. As we have mentioned, φ is independent of s, and for $r \geq 0$ (cf. Exercise I.4.21)

$$\begin{aligned}\varphi(t+r) &= \mathbf{E}\,|w_{t+r}|^2 = \mathbf{E}\,\big|w_{t+r} - w_t + w_t - w_0\big|^2 \\ &= \mathbf{E}\,\big|w_{t+r} - w_t\big|^2 + 2\mathbf{E}\,(w_{t+r} - w_t)(w_t - w_0) + \mathbf{E}\,\big|w_t - w_0\big|^2 = \varphi(t) + \varphi(r).\end{aligned}$$

The only continuous solution of the equation $\varphi(t+r) = \varphi(t) + \varphi(r)$ is the function ct, where c is a scaling factor. Taking $c = 1$, we see that it is natural to assume that

$$\mathbf{E}\,\big|w_t - w_s\big|^2 = |t - s|. \tag{1}$$

Using the equality $\mathbf{E}\,\big|w_t - w_s\big|^2 = \mathbf{E}\,w_t^2 - 2\mathbf{E}\,w_t w_s + \mathbf{E}\,w_s^2$, one can easily show that condition (1) for all t, s, together with the equality $w_0 = 0$ (a.s.) is equivalent to the equality $\mathbf{E}\,w_t w_s = t \wedge s$ for all t, s. Finally, the origin of condition (b) in Definition 1 is obvious.

The above discussion of formula (1) and the facts that a linear transformation of a Gaussian vector yields a Gaussian vector (Theorem I.4.14), uncorrelated Gaussian variables are independent (Theorem I.4.19), and any vector with independent Gaussian coordinates is a Gaussian vector (Theorem I.4.20) immediately yield the following theorem.

2. THEOREM. *Condition* (a) *of Definition* 1 *holds if and only if, for any integer n and any $t_1 \leq \cdots \leq t_n$, $t_i \in [0, T)$, the random variables $w_{t_1}, w_{t_2} - w_{t_1}, \ldots, w_{t_n} - w_{t_{n-1}}$ are independent, $w_t - w_s \sim \mathcal{N}(0, |t - s|)$ for all $t, s \in [0, T)$, $w_0 = 0$ (a.s.).*

3. REMARK. Since $w_t - w_s \sim \mathcal{N}(0, |t - s|)$, it follows that $\xi := (w_t - w_s)|t - s|^{-1/2} \sim \mathcal{N}(0, 1)$ for $t \neq s$ and the distribution of $w_t - w_s$ has the density

$$p(x) = \frac{1}{\sqrt{2\pi|t - s|}} \exp\Big(-\frac{1}{2|t - s|}x^2\Big).$$

Indeed, by Lemma I.4.10 (and Theorem I.4.12), for Borel functions $f \geq 0$

$$\mathbf{E} f(w_t - w_s) = \mathbf{E} f(\xi|t-s|^{1/2}) = \frac{1}{2\pi}\int_{-\infty}^{\infty} f(x|t-s|^{1/2}) \exp\Big(-\frac{x^2}{2}\Big)dx.$$

If f is a continuous function then, by the rules for operating with Riemann integrals we can change variables in the last integral, getting the integral of fp. By Theorem I.5.4 this can also be done for a Borel function $f \geq 0$.

To prove the existence of the Wiener process (in the sense of Definition 1, of course) in the next section, we will need some constructions based on the following arguments. If we already knew that the Wiener process exists on $[0,\pi]$, it would be natural to represent it as a Fourier series, expanding $w_t(\omega)$ for any ω in a series in terms of, say, $\sin nt$. Of course, we want the Fourier series to converge sufficiently well and since $\sin n\pi = 0$, $w_\pi \sim \mathcal{N}(0,\pi)$, $w_\pi \neq 0$, we will expand $\widetilde{w}_t := w_t - \frac{t}{\pi} w_\pi$. When that is done the random variable

$$\int_0^\pi \widetilde{w}_t \sin nt\, dt \tag{2}$$

must be Gaussian, as a limit of integral sums, by Theorems I.4.14 and I.4.15. In addition, we must have

$$\mathbf{E}\Big(\int_0^\pi \widetilde{w}_t \sin nt\, dt\Big)\Big(\int_0^\pi \widetilde{w}_s \sin ms\, ds\Big) = \int_0^\pi\int_0^\pi \sin nt \sin ms\, \mathbf{E}\, \widetilde{w}_t \widetilde{w}_s\, dtds,$$

and the last expression is readily shown to vanish for $n \neq m$ because $\mathbf{E}\, w_t w_s = t \wedge s$. Consequently, by Theorem I.4.19 the random variables (2) must be independent. We will therefore represent w_t on $[0,\pi]$ as the sum of $\frac{t}{\pi} w_\pi$ and a Fourier sine series whose coefficients are independent Gaussian random variables.

2. Existence of the Wiener process

To carry out the program outlined at the end of Sec. 1, we must first construct a countable sequence of independent Gaussian variables. They will be derived from uniformly distributed random variables.

1. LEMMA. *There exist a probability space $(\Omega, \mathcal{F}, P)$ and independent random variables $\xi_0, \xi_1, \dots,$ defined on the space, each of which is* uniformly *distributed on* $[0,1]$, *that is, $P\{\xi_n \in B\} = \ell(B \cap [0,1])$ for any Borel set $B \subset (-\infty,\infty)$ and $n \geq 0$.*

PROOF. Let $(\Omega, \mathcal{F}, P) = \big([0,1], \mathfrak{B}([0,1]), \ell\big)$. Each point $\omega \in [0,1)$ has a binary representation

$$\omega = \sum_{r=1}^{\infty} 2^{-r} \varepsilon_r(\omega),$$

where $\varepsilon_r(\omega)$ is either 0 or 1. To define $\varepsilon_r(\omega)$ uniquely we require that an infinite number of $\varepsilon_r(\omega)$ be zero. Also we stipulate that $\varepsilon_r(1) = 1$.

2. EXERCISE. Observing that if δ_i are constants, equal to either 0 or 1, then

$$P\Big\{\varepsilon_1 = \delta_1, \dots, \varepsilon_r = \delta_r\Big\} = P\Big\{\omega \in [a, a + 2^{-r})\Big\}, \qquad a = \sum_{i=1}^{r} \delta_i 2^{-i},$$

prove that $\varepsilon_1, \varepsilon_2, \dots$ are independent random variables and $P\{\varepsilon_r = 0\} = P\{\varepsilon_r = 1\} = \frac{1}{2}$.

Now let $r\colon (n,k) \to r(n,k)$ be a one-to-one mapping of all pairs (n,k), $n = 0, 1, \dots, k = 1, 2, \dots$ into the set $\{1, 2, \dots\}$. Set $\varepsilon_{nk} = \varepsilon_{r(nk)}$. We claim that

$$\xi_n(\omega) := \sum_{k=1}^{\infty} 2^{-k} \varepsilon_{n,k}(\omega), \qquad n = 0, 1, 2, \dots \tag{1}$$

possess the desired properties.

Indeed, for different n, formula (1) involves different sets of ε_r, so it follows easily from Exercise 2 and Theorem I.4.17 that finite sums of the form (1) for different n are independent. By the Dominated Convergence Theorem and Theorem I.4.17, the ξ_n's are also independent. Moreover, if $x \in [0, 1]$, then

$$\{\omega\colon \xi_n > x\} = \bigcup_{k=1}^{\infty} \Big\{\omega\colon \varepsilon_{n1} = \varepsilon_1(x), \dots, \varepsilon_{n,k-1} = \varepsilon_{k-1}(x), \varepsilon_{nk} > \varepsilon_k(x)\Big\},$$

$$P\{\xi_n > x\} = \sum_{k=1}^{\infty} P\Big\{\varepsilon_{n1} = \varepsilon_1(x), \dots, \varepsilon_{n,k-1} = \varepsilon_{k-1}(x), \varepsilon_{nk} > \varepsilon_k(x)\Big\},$$

$$P\{\varepsilon_{ni} = \varepsilon_i(x)\} = \frac{1}{2}, \qquad P\{\varepsilon_{ni} > \varepsilon_i(x)\} = \frac{1}{2}(1 - \varepsilon_i(x)).$$

Hence, by the independence of $\varepsilon_r(\omega)$,

$$P\{\xi_n > x\} = \sum_{k=1}^{\infty} \frac{1}{2^k}(1 - \varepsilon_k(x)) = 1 - x = \ell\big((x, \infty) \cap [0, 1]\big).$$

It is clear that the outer terms are equal for $x < 0$ and $x > 1$ as well. Therefore the measures $\mu_1(B) := P\{\xi_n \in B\}$ and $\mu_2(B) = \ell(B \cap [0, 1])$ coincide for B in the π-system of all sets (x, ∞). By Lemma I.5.3, they coincide for all $B \in \mathfrak{B}(E_1)$. □

3. LEMMA. *On the probability space of the previous lemma there exists a sequence of independent random variables $\eta_0, \eta_1, \dots$ each of which has the standard normal distribution.*

PROOF. Denote

$$F(t) = \frac{1}{\sqrt{2\pi}} \int_{\infty}^{t} e^{-\frac{1}{2}u^2}\, du.$$

The function $F(t)$ is continuous on $(-\infty, \infty)$ and strictly increasing from 0 to 1 (see Lemma I.4.6), but never taking the values 0 or 1. Consequently, the inverse function $F^{-1}(x)$ is defined on $(0, 1)$; its values run from $-\infty$ to ∞. Let $\eta_n(\omega) = F^{-1}(\xi_n(\omega))$ if $\xi_n(\omega) \in (0, 1)$, $\eta_n(\omega) = 0$ otherwise. By Exercise I.4.22 the η_n's are independent random variables. In addition, $P\{\xi_n \notin (0, 1)\} = 0$ and for any $x \in E_1$

$$P\{\eta_n < x\} = P\Big\{F^{-1}(\xi_n) < x,\ \xi_n \in (0, 1)\Big\} = P\{\xi_n < F(x)\} = F(x).$$

Finally, by Definition I.4.7, if ξ is a standard normal random variable (which, as stated in Sec. I.4, exists on some, possibly other probability space), then $F(x) = P\{\xi < x\}$, and, as in the previous proof, the equality $P\{\eta_n < x\} = P\{\xi < x\}$ implies that the distribution of η_n coincides with that of ξ, that is, it is standard normal. □

To prove the existence of the Wiener process we will need one more lemma, which will be proved in Appendix A; it is a very special case of embedding theorems for Besov's classes (see Nikol'skiĭ [40] (1969), p. 279).

4. LEMMA. *Let $p \geq 1$, $\alpha > p^{-1}$ be numbers. Then there exists a constant N such that for any function $f(t)$ continuous on $[0,\pi]$ and all $t, s \in [0,\pi]$*

$$|f(t) - f(s)|^p \leq N|t-s|^{p\alpha-1} \int_0^\pi \int_0^\pi \frac{|f(x) - f(y)|^p}{|x-y|^{1+p\alpha}} \, dxdy; \tag{2}$$

with the convention that $\frac{0}{0} := 0$.

In our applications of this lemma $f(t)$ will be a function of ω too, with $f(\omega, t)$ a random variable for any t. In addition, when evaluating the expectation of the right-hand side of (2) the possibility of interchanging the order of the expectation sign and the integral with respect to $dxdy$ will be essential. This is guaranteed by Fubini's Theorem and the fact that $|f(x) - f(y)|^p$, $|x-y|^\beta$, $I_{x \neq y}$ are measurable functions of (ω, x, y) with respect to $\mathcal{F} \otimes \mathfrak{B}(E_2)$ (see Corollary I.1.11 and the definition of a product of σ-algebras in Lemma I.1.7).

We can now prove the existence of a Wiener process for $t \in [0,\pi]$. That the assumptions of the following theorem are meaningful follows from Lemma 3.

5. THEOREM. *Let $\eta_0(\omega)$, $\eta_1(\omega), \ldots$ be independent standard normal random variables on a probability space $(\Omega, \mathcal{F}, P)$. Then for some sequence of integers $N(k) \to \infty$ and for almost every ω the sequence of functions*

$$w_t^k(\omega) := \frac{1}{\sqrt{\pi}} t\eta_0(\omega) + \sqrt{\frac{2}{\pi}} \sum_{n=1}^{N(k)} \eta_n(\omega) \frac{1}{n} \sin nt \tag{3}$$

converges as $k \to \infty$ uniformly in $t \in [0,\pi]$, to a function $w_t(\omega)$, which is a Wiener process on $[0,\pi]$.

PROOF. The argments of the final part of Sec. 1 show that formula (3) is natural and that $\mathbf{E}|w_t^k - w_s^k|^2$ must tend to $|t-s|$ as $k \to \infty$. Since the last expectation is easily evaluated (see Exercise I.4.21), the following formula must be valid

$$\frac{1}{\pi}(t-s)^2 + \frac{2}{\pi} \sum_{n=1}^{\infty} \frac{1}{n^2} \big(\sin nt - \sin ns\big)^2 = |t-s|, \qquad t, s \in [0,\pi]. \tag{4}$$

To prove (4) rigorously for $t \in [0,\pi]$ let $g_t(x)$ denote the indicator of the set $x \in (-t,t)$. This function can be expanded in a Fourier cosine series on $(-\pi,\pi)$. By the Parseval identity, the left-hand side of (4) is

$$\frac{1}{2} \int_{-\pi}^{\pi} [g_t(x) - g_s(x)]^2 \, dx,$$

and since this is obviously equal to $|t-s|$, we obtain (4).

Formula (4) can be also derived by using the theory of functions of complex variable, if we observe that for $|z| = r < 1$, $\arg z = \varphi$, integration of the Taylor series of the principal branch of the logarithm gives

$$\sum_{n=1}^{\infty} \frac{1}{n^2} z^n = \sum_{n=1}^{\infty} \frac{1}{n^2} r^n - i \int_0^{\varphi} \ln\left(1 - re^{iu}\right) du,$$

$$\sum_{n=1}^{\infty} \frac{r^n}{n^2} \cos n\varphi = \sum_{n=1}^{\infty} \frac{1}{n^2} \operatorname{Re} z^n = \sum_{n=1}^{\infty} \frac{1}{n^2} r^n + \int_0^{\varphi} \operatorname{Im} \ln\left(1 - re^{iu}\right) du.$$

The functions $\operatorname{Im}\ln\left(1 - re^{iu}\right)$ are uniformly bounded and $\operatorname{Im}\ln\left(1 - re^{iu}\right) \to \operatorname{Im}\ln\left(1 - e^{iu}\right) = (u - \pi \operatorname{sgn} u)/2$ as $r \uparrow 1$, for $|u| \leq \pi$. Consequently,

$$\sum_{n=1}^{\infty} \frac{1}{n^2} \cos n\varphi = \sum_{n=1}^{\infty} \frac{1}{n^2} + \frac{1}{4}\varphi^2 - \frac{1}{2}\pi|\varphi|$$

for $|\varphi| \leq \pi$. Formula (4) can be derived from this using elementary trigonometric formulas for squares and products of sines in terms of cosines.

Now let $\alpha \in (1/4, 1/2)$ and set

$$c(n) = \frac{4}{\pi^2} \int_0^{\pi}\int_0^{\pi} \frac{1}{|t-s|^{1+4\alpha}} \left(\sum_{k=n}^{\infty} k^{-2} \left(\sin kt - \sin ks\right)^2 \right)^2 dtds.$$

It follows from (4) that

$$c(1) \leq \int_0^{\pi}\int_0^{\pi} \frac{1}{|t-s|^{4\alpha-1}} dtds < \infty.$$

By the Dominated Convergence Theorem, $c(n) \to 0$ as $n \to \infty$, and we can define $N(k)$ so that $c\left(N(k)\right) \leq 2^{-k}$. Now let $\delta_t^k = w_t^{k+1} - w_t^k$. Then by Lemma 4 for $p = 4$, $s = 0$, taking into account that $\mathbf{E}\eta_i, \eta_j = \delta^{ij}$ (and Exercise I.4.13), we get

$$\begin{aligned}
\mathbf{E} \sup_{t\in[0,\pi]} |\delta_t^k|^4 &\leq N\pi^{4\alpha-1} \int_0^{\pi}\int_0^{\pi} \frac{1}{|t-s|^{1+4\alpha}} \mathbf{E}\, |\delta_t^k - \delta_s^k|^4 \, dtds \\
&= N_1 \int_0^{\pi}\int_0^{\pi} \frac{1}{|t-s|^{1+4\alpha}} \left(\mathbf{E}\, |\delta_t^k - \delta_s^k|^2\right)^2 dtds \\
&= N_1 \int_0^{\pi}\int_0^{\pi} \frac{1}{|t-s|^{1+4\alpha}} \frac{4}{\pi^2} \left(\sum_{N(k)+1}^{N(k+1)} n^{-2} \left(\sin nt - \sin ns\right)^2 \right)^2 dtds \\
&\leq N_1 c\left(N(k)\right) \leq N_1 2^{-k},
\end{aligned} \tag{5}$$

where N_1 is independent of k. The first expression in this chain of relations is meaningful, since the supremum over $t \in [0,\pi]$ does not change if we replace it by the supremum over the countable set of all rational points in $[0, \pi]$. It is then clear that (see Theorem I.1.13)

$$\sup_{t\in[0,\pi]} \left|w_t^{k+1} - w_t^k\right|^4 = \sup_{t\in[0,\pi]} \left|\delta_t^k\right|^4$$

is a random variable. Further, by the Hölder inequality (Theorem I.2.19), we obtain from (5)

$$\sum_{k=1}^{\infty} \mathbf{E} \sup_{t\in[0,\pi]} \left|w_t^{k+1} - w_t^k\right| \le \sum_{k=1}^{\infty} \left(\mathbf{E} \sup_{t\in[0,\pi]} \left|w_t^{k+1} - w_t^k\right|^4\right)^{1/4} < \infty.$$

Consequently, by Corollary I.2.8 (and Exercise I.2.6 (e)) almost surely

$$\sum_{k=1}^{\infty} \sup_{t\in[0,\pi]} \left|w_t^{k+1} - w_t^k\right| < \infty. \tag{6}$$

Let Ω' denote the set of all ω that satisfy (6). Clearly, $\Omega' \in \mathcal{F}$ (see, however, Corollary I.1.11 and Theorem I.1.13) and for any $\omega \in \Omega'$ the sequence $w_t^k(\omega)$ converges uniformly in $t \in [0, \pi]$ by (6). Define $w_t(\omega)$ for $\omega \in \Omega'$ to be the limit of $w_t^k(\omega)$; for $\omega \notin \Omega'$, define $w_t(\omega) = 0$ on $[0, \pi]$.

It is evident that w_t is continuous in t for all ω, and since $P(\Omega') = 1$, it follows that $w_t^k \to w_t$ a.s. In addition, by Theorems I.4.14 and I.4.20 $\left(w_{t_1}^k, \dots, w_{t_n}^k\right)$ is a Gaussian vector for $t_1, \dots, t_n \in [0, \pi]$. By Theorem I.4.15 $\left(w_{t_1}, \dots, w_{t_n}\right)$ is also Gaussian and

$$\mathbf{E}\, w_t = \lim_{k\to\infty} \mathbf{E}\, w_t^k = 0,$$

(7)

$$\begin{aligned}\mathbf{E} \left|w_t - w_s\right|^2 &= \lim_{k\to\infty} \mathbf{E} \left|w_t^k - w_s^k\right|^2 \\ &= \lim_{k\to\infty} \left[\frac{1}{\pi}(t-s)^2 + \frac{2}{\pi}\sum_{n=1}^{N(k)} n^{-2}\left(\sin nt - \sin ns\right)^2\right] = |t - s|,\end{aligned}$$

where the last equality follows from (4). Finally, it is obvious that $w_0 = 0$, and together with (7) this gives $\mathbf{E}\, w_t w_s = t \wedge s$. □

We will now construct a Wiener process on $[0, \infty)$.

6. THEOREM. *There exists a probability space $(\Omega, \mathcal{F}, P)$ on which one can define a Wiener process w_t for $t \in [0, \infty)$.*

PROOF. Take a probability space $(\Omega, \mathcal{F}, P)$ on which there exist independent random variables $\zeta_1, \zeta_2, \dots$ with the standard normal distribution. Let $r(n, k)$ be the function from the proof of Lemma 1, and set

$$\left(\eta_0^{(n)}, \eta_1^{(n)}, \dots,\right) = \left(\zeta_{r(n,1)}, \zeta_{r(n,2)}, \dots,\right).$$

Using Theorem 5, construct a Wiener process $w_t^{(n)}$ on $[0, \pi]$ for each of the sequences $\left(\eta_0^{(n)}, \eta_1^{(n)}, \dots,\right)$, and define

$$w_t(\omega) = \begin{cases} w_t^{(0)}(\omega), & 0 \le t < \pi, \\ w_\pi^{(0)}(\omega) + w_{t-\pi}^{(1)}(\omega), & \pi \le t < 2\pi, \\ w_\pi^{(0)}(\omega) + w_\pi^{(1)}(\omega) + w_{t-2\pi}^{(2)}(\omega), & 2\pi \le t < 3\pi \text{ and so on.} \end{cases}$$

It can be shown as in the previous proof that, for any $s_i \in [0,\pi]$ and integers n_i, $i = 1,\dots,r$, the vector $\big(w_{s_1}^{(n_1)},\dots,w_{s_r}^{(n_r)}\big)$ is Gaussian. Hence (by Theorem I.4.14) $(w_{t_1},\dots,w_{t_r})$ is Gaussian for $t_i \geq 0$. In addition, as at the corresponding stage in the proof of Lemma 1, one can show that $w_r^{(n)}$ and $w_s^{(m)}$ are independent for $n \neq m$, $r,s \in [0,\pi]$. Consequently, for $t \geq s \geq 0$ and $p := \left[\frac{t}{\pi}\right]$, $q := \left[\frac{s}{\pi}\right]$

$$
\begin{aligned}
\mathbf{E}\, w_\pi^{(n)} w_\pi^{(m)} &= \delta^{nm}\pi, \qquad \mathbf{E}\, w_\pi^{(n)} w_{s-q\pi}^{(q)} = \delta^{nq}(s-q\pi),\\
\mathbf{E}\, w_{t-p\pi}^{(p)} w_{s-q\pi}^{(q)} &= \delta^{pq}(t-p\pi)\wedge(s-q\pi) = \delta^{pq}(s-q\pi),\\
\mathbf{E}\, w_t w_s &= \mathbf{E}\Big(\sum_{n=0}^{p-1} w_\pi^{(n)} + w_{t-p\pi}^{(p)}\Big)\Big(\sum_{m=0}^{q-1} w_\pi^{(m)} + w_{s-q\pi}^{(q)}\Big)\\
&= \sum_{m=0}^{q-1}\pi + (s-q\pi)I_{p-1\geq q} + \delta^{pq}(s-q\pi) = s.
\end{aligned}
$$

In the general case, clearly, $\mathbf{E}\, w_t w_s = t\wedge s$ for $s,t \geq 0$, and since obviously $\mathbf{E}\, w_t = 0$ and w_t is continuous in t, the theorem is proved. □

7. EXERCISE. (a) Prove that the random variables w_π and (1.2) are jointly independent for $n = 1,2,\dots$ and each of them is Gaussian.

(b) Let w_t^i be Wiener processes on $(\Omega,\mathcal{F},P)$ for $t \in [0,\infty)$, $i = 1,2,\dots$. We will say that they are *independent* if for any $n \geq 1$, $t(1), t(2),\dots \in [0,\infty)$, the vectors $(w_{t(1)}^1,\dots,w_{t(n)}^1)$, $(w_{t(1)}^2,\dots,w_{t(n)}^2),\dots$ are independent. Using (a) and a slight modification of the proof of Theorem 6, prove that if at least one Wiener process is defined on $(\Omega,\mathcal{F},P)$ for $t \in [0,\pi]$, then there exist a countable set of independent Wiener processes and a continuum of pairwise distinct (a.s.) Wiener processes on $(\Omega,\mathcal{F},P)$ for $t \in [0,\infty)$. Prove also that there exists Wiener processes w_t^1, w_t^2, $t \in [0,\infty)$ such that $w_t^1 - w_t^2$ is a linear function of $t \in [0,\pi]$, different from zero (a.s.).

8. REMARK. One can extract from the proofs of Lemmas 1, 3 and Theorems 5, 6 explicit formulas for $w_t(\omega)$ for $\omega \in [0,1]$. Admittedly, these formulas show that $w_t(\omega)$ is a rather exotic function of ω, lacking such properties as continuity, smoothness, etc. Nevertheless, it is a *Borel function*, and that is the most important point here.

9. EXERCISE. We have used the random variables $\xi_0,\xi_1,\dots$ of Lemma 1 to construct Gaussian random variables. It is also possible to construct many other random variables. For example, fix $p \in [0,1]$ and let $\eta_i(\omega) = 1$ if $\xi_i(\omega) \leq p$ and $\eta_i(\omega) = 0$ if $\xi_i(\omega) > p$. Define $S(n) = \eta_1 + \dots + \eta_n$. Prove that the η_i's form a sequence of *Bernoulli trials*, that is, $\eta_0,\eta_1,\dots$ are independent random variables, $P\{\eta_i = 1\} = p$, $P\{\eta_i = 0\} = 1 - p =: q$. Prove that for all $z \in (-\infty,\infty)$

$$
(8)\qquad \sum_{k=0}^{n} z^k P\big(S(n)=k\big) = \mathbf{E}\sum_{k=0}^{n} z^k I_{S(n)=k} = \mathbf{E}\, z^{S(n)} = \big(\mathbf{E}\, z^{\eta_0}\big)^n = (zp+q)^n
$$

and deduce that $S(n)$ has the *binomial distribution*, that is, $P\big(S(n)=k\big) = \binom{n}{k}p^k q^{n-k}$. Prove, finally, e.g. by differentiating (8) with respect to z, that $\mathbf{E}\, S(n) = np$, $\mathbf{E}\,\big(S(n)-np\big)^2 = npq$.

3. Some properties of the Wiener process

The Wiener processes have been the subject of many detailed studies and it would take a considerable amount of space merely to list all their presently known properties. We will limit ourselves in this chapter to those properties of Wiener processes that are necessary for our purposes. A reader interested in further information is referred to Itô and McKean [24] (1965), Durrett [11] (1984).

Throughout this section we consider a Wiener process w_t, $t \geq 0$, defined on a probability space $(\Omega, \mathcal{F}, P)$.

1. THEOREM (self-similarity of the Wiener process). *Let $c \neq 0$ be a constant. Then $c^{-1}w_{c^2t}$ is a Wiener process. There also exist $\Omega' \in \mathcal{F}$ such that $P(\Omega') = 1$ and the function $\widetilde{w}_t(\omega)$ defined by $\widetilde{w}_t(\omega) = tw_{1/t}(\omega)$ for $t > 0$, $\omega \in \Omega'$, $\widetilde{w}_0(\omega) = 0$ for $\omega \in \Omega'$, and $\widetilde{w}_t(\omega) = 0$ for $t \geq 0$, $\omega \notin \Omega'$, is a Wiener process.*

PROOF. The first assertion is elementary and follows immediately from Definition 1.1 (and Theorem I.4.14). To prove the second one, it suffices to verify, as follows from Definition 1.1 (and Theorem I.4.14), that $tw_{1/t} =: \widehat{w}_t \to 0$ as $t \downarrow 0$ (a.s.), that is, on a set $\Omega' \in \mathcal{F}$ such that $P(\Omega') = 1$.

Applying Lemma 2.4 for $p > 2$, $\alpha \in (1/p, 1/2)$ to the interval $[\varepsilon, \pi + \varepsilon]$, where $\varepsilon > 0$, and then letting $\varepsilon \downarrow 0$, we obtain ($\frac{0}{0} := 0$)

$$\mathbf{E} \sup_{t,s\in(0,\pi]} \frac{1}{|t-s|^{p\alpha-1}} \left|\widehat{w}_t - \widehat{w}_s\right|^p \leq N \int_0^{2\pi}\int_0^{2\pi} \frac{1}{|t-s|^{1+p\alpha}} \mathbf{E}\left|\widehat{w}_t - \widehat{w}_s\right|^p dtds$$
$$= N_1 \int_0^{2\pi}\int_0^{2\pi} |t-s|^{-1+p(\frac{1}{2}-\alpha)}\, dtds < \infty$$

(cf. the first inequality in (2.5)). Hence, for almost all ω there can be found a finite constant $\xi(\omega)$ (moreover, such that $\mathbf{E}\,\xi^p < \infty$), such that for $t, s \in (0, \pi]$

$$(1) \qquad \left|\widehat{w}_t - \widehat{w}_s\right| \leq \xi |t-s|^{\alpha - 1/p}.$$

This obviously implies that $\widehat{w}_t$ has a limit as $t \downarrow 0$ (a.s.). Since $\mathbf{E}\,\widehat{w}_t^2 = t$, this limit vanishes (a.s.) by Fatou's Lemma, and this proves the theorem. □

The proof of (1) can be clearly repeated with w_t in place of $\widehat{w}_t$, in which case we obtain a similar inequality for $|w_t - w_s|$ for all $t, s \in [0, \pi]$. In addition, any interval $[0, T]$, $T > \pi$, can be mapped linearly onto $[0, \pi]$. Hence, using the first assertion of Theorem 1 and the fact that $\alpha - 1/p$ can be taken arbitrarily close to $1/2$ we obtain the first assertion of the following theorem.

2. THEOREM. *Let $\varepsilon \in (0, 1/2)$, $T \in (0, \infty)$. Then there exists a random variable $N = N(\omega)$, depending also on ε, T, such that $\mathbf{E}\,N^p < \infty$ for all $p \in [0, \infty)$, and for any $\omega \in \Omega$*

$$\left|w_t(\omega) - w_s(\omega)\right| \leq N(\omega)|t-s|^{\frac{1}{2}-\varepsilon}$$

for all $t, s \in [0, T]$. In particular, $\left|w_t(\omega)\right| \leq N(\omega)t^{\frac{1}{2}-\varepsilon}$ for $t \in [0, \pi]$. In addition, $\left|w_t(\omega)\right| \leq N(\omega)t^{\frac{1}{2}+\varepsilon}$ for $t \geq T$.

The last assertion of this theorem is a direct corollary of the first one and the last assertion of Theorem 1, which enables us to reduce the investigation of a Wiener process at large t to its investigation at small t.

3. EXERCISE. Prove that $\exp(w_t - \frac{1}{2}t) \to 0$ (a.s.) as $t \to \infty$ and for any (nonrandom) sequence $t_k \to \infty$

$$\mathbf{E} \sup_k \exp\left(w_{t_k} - \frac{1}{2}t_k\right) = \infty. \tag{2}$$

(Hint for (2): note that $\mathbf{E}\exp(w_t - \frac{1}{2}t) = 1$.)

The result established in Theorem 2 can be improved. We shall formulate two more theorems, the first due to Lévy and the second, known as the *law of the iterated logarithm*, to Khinchin. We will prove Khinchin's Theorem in Sec. 9. Neither of these theorems will be needed in what follows.

4. THEOREM.

$$\overline{\lim_{\substack{0\le s<t\le 1\\ u=t-s\to 0}}} \frac{|w_t - w_s|}{\sqrt{2u\ln\frac{1}{u}}} = 1$$

with probability one (that is, almost surely).

5. THEOREM. *We have*

$$\overline{\lim_{t\downarrow 0}} \frac{w_t}{\sqrt{2t\ln\ln\frac{1}{t}}} = 1, \qquad \underline{\lim_{t\downarrow 0}} \frac{w_t}{\sqrt{2t\ln\ln\frac{1}{t}}} = -1$$

with probability one.

Theorem 5 shows the strong irregularity of the sample paths of the Wiener process at zero. Moreover, as easily follows from Definition 1.1, $w_{t+s} - w_s$ for $t \ge 0$ is a Wiener process for any fixed $s \ge 0$, so that the sample paths of the Wiener process behave in a similar irregular way at any time. The next result, which also shows that w_t is far from being a smooth function, will play an important (though indirect) role below.

6. THEOREM (on Quadratic Variation). *Let* $0 \le s < t < \infty$ *and let* $t(i,n) = s + (t-s)i2^{-n}$. *Then*

$$\sum_{i=0}^{2^n-1} \left(w_{t(i+1,n)} - w_{t(i,n)}\right)^2 \longrightarrow t - s \tag{3}$$

a.s. as $n \to \infty$.

PROOF. Let ξ_n denote the left-hand side of (3). Since the increments of w_t are independent (Theorem 1.2), their squares are independent and uncorrelated (Exercise I.4.22). Hence (cf. Exercise I.4.21),

$$\begin{aligned} \mathbf{E}\left|\xi_n - (t-s)\right|^2 &= \mathbf{E}\left\{\sum_{i=0}^{2^n-1}\left[\left(w_{t(i+1,n)} - w_{t(i,n)}\right)^2 - \left(t(i+1,n) - t(i,n)\right)\right]\right\}^2 \\ &= \sum_{i=0}^{2^n-1}\mathbf{E}\left[\left(w_{t(i+1,n)} - w_{t(i,n)}\right)^2 - \left(t(i+1,n) - t(i,n)\right)\right]^2. \end{aligned} \tag{4}$$

Note now that if $\eta \sim \mathcal{N}(0, \sigma^2)$, $\sigma^2 > 0$, then $\eta\sigma^{-1} \sim \mathcal{N}(0, 1)$ and

$$\mathbf{E}\left[\eta^2 - \sigma^2\right]^2 = \sigma^4 \mathbf{E}\left[(\eta\sigma^{-1})^2 - 1\right]^2 = N\sigma^4,$$

where N is independent of σ. We thus deduce from (4) that

$$\mathbf{E}\left|\xi_n - (t-s)\right|^2 = N \sum_{i=0}^{2^n-1} \left(t(i+1,n) - t(i,n)\right)^2 = N(t-s)^2 2^{-n},$$

$$\sum_{n=1}^{\infty} \mathbf{E}\left|\xi_n - (t-s)\right|^2 < \infty,$$

$$\sum_{n=1}^{\infty} \left|\xi_n - (t-s)\right|^2 < \infty \quad \text{(a.s.)}, \qquad \lim_{n\to\infty} \xi_n = t - s \quad \text{(a.s.)}. \quad \square$$

7. COROLLARY. *For almost every ω, the variation of $w_t(\omega)$ on $[0,1]$ is infinite (see also Problem* III.1.9 *below).*

Indeed, let $s = 0$, $t = 1$. Then for those ω at which the variance is finite, we have

$$\xi_n \le \max_{i \le 2^n - 1} \left|w_{t(i+1,n)} - w_{t(i,n)}\right| \operatorname*{Var}_{[0,1]} w_t \longrightarrow 0 \ne 1.$$

The fact that the variation of a sample path of the Wiener process is unbounded makes it difficult to justify the following reasoning. In real Brownian motion the variance of the particle must somehow depend on the temperature of the liquid. If the temperature is piecewise constant, remaining constant over every time interval $[t_i, t_{i+1})$ of some partition of $[0, \infty)$, then a natural one-dimensional model of a sample path of Brownian motion is the following function of t:

$$\sum_i f_i\left(w_{t\wedge t_{i+1}} - w_{t\wedge t_i}\right),$$

where the f_i's reflect the temperature-dependence of the variance of the particle. The difficulty lies in the passage from piecewise constant to continuously varying temperatures, since the above sum must converge to an integral with respect to dw_t as the intervals of constant temperature become finer. But we cannot define the integral with respect to dw_t over an individual sample path with the tools of usual analysis, since $w_t(\omega)$ has unbounded variation (a.s.).

The solution is to construct the integral with respect to dw_t not over an individual sample path but in the mean-square sense. This construction will be discussed in Sec. III.2; for the moment we proceed to the Markov property of the Wiener process.

8. DEFINITION. Let $\xi_t = \xi_t(\omega)$ be a random variable, given for any $t \ge 0$. For $s \ge 0$ let $\mathcal{F}_s^\xi$ denote the σ-algebra generated by all the events $\{\omega \colon \xi_t(\omega) \in B\}$, where t runs through $[0, s]$ and B runs through $\mathfrak{B}(E_1)$. In short,

$$\mathcal{F}_s^\xi := \sigma\Big(\Big\{\{\omega \colon \xi_t(\omega) \in B\} \colon t \in [0,s], B \in \mathfrak{B}(E_1)\Big\}\Big).$$

Similar notation is used when the ξ_t's are random vectors; $\mathcal{F}_s^\xi$ is called the σ-algebra *generated by the process* ξ_t up to time s. It is interpreted as the collection of all events that can be described in terms of values of ξ_t for $t \in [0, s]$.

Obviously, $\mathcal{F}_t^\xi \subset \mathcal{F}_s^\xi \subset \mathcal{F}$ for $0 \le t \le s$.

9. DEFINITION. Let $A, B \in \mathcal{F}$ be events. They are called *independent*, if $P(A \cap B) = P(A)P(B)$. Let $\mathfrak{A}, \mathfrak{B}$ be collections of sets, $\mathfrak{A}, \mathfrak{B} \subset \mathcal{F}$. They are said to be *independent* if $P(A \cap B) = P(A)P(B)$ for any $A \in \mathfrak{A}$, $B \in \mathfrak{B}$.

10. EXERCISE. Show that if $\mathfrak{A}, \mathfrak{B}$ are independent *σ-algebras* and ξ, η are random variables such that ξ is $\mathfrak{A}$-measurable and η is $\mathfrak{B}$-measurable, then ξ, η are independent. Show also that ξ, η are independent if and only if the σ-algebras $\sigma(\xi)$, $\sigma(\eta)$ are independent.

11. THEOREM (the Markov property of the Wiener process). *Fix $s \geq 0$ and set $\theta_s w_t := w_{t+s} - w_s$ for $t \geq 0$. Then $\theta_s w_t$ is a Wiener process for $t \geq 0$ and the σ-algebras $\mathcal{F}^w_s$ and $\mathcal{F}^{\theta_s w}_\infty$ are independent.*

PROOF. The assertion that $\theta_s w_t$ is a Wiener process follows, as already noted, from Definition 1.1. By Theorem 1.2 (and Theorem I.4.19) if $0 \leq t(1) \leq \cdots \leq t(n) = s \leq s + t(n+1) \leq \cdots \leq s + t(m)$, the vectors $\big(w_{t(1)}, w_{t(2)} - w_{t(1)}, \dots, w_{t(n)} - w_{t(n-1)}\big)$ and $\big(\theta_s w_{t(n+1)}, \dots, \theta_s w_{t(m)} - \theta_s w_{t(m-1)}\big)$ are independent. By Exercise I.4.22 and the fact that $w_{t(i)} = w_{t(1)} + \cdots + \big(w_{t(i)} - w_{t(i-1)}\big)$, it follows that the vectors $\big(w_{t(1)}, \dots, w_{t(n)}\big)$ and $\big(\theta_s w_{t(n+1)}, \dots, \theta_s w_{t(m)}\big)$ are independent. In particular, for any $\Gamma_i \in \mathfrak{B}(E_1)$, if

$$
\begin{aligned}
A &= \Big\{\omega \colon w_{t(1)} \in \Gamma_1, \dots, w_{t(n)} \in \Gamma_n\Big\}, \\
B &= \Big\{\omega \colon \theta_s w_{t(n+1)} \in \Gamma_{n+1}, \dots, \theta_s w_{t(m)} \in \Gamma_m\Big\}
\end{aligned}
\tag{5}
$$

then $P(A \cap B) = P(A)P(B)$. For fixed B, both $P(A \cap B)$ and $P(A)P(B)$ are measures with respect to A that coincide for a B from (5) on the π-system of all sets A from (5) (t_i, n, Γ_i may of course vary). Hence, by Lemma I.5.3, these measures coincide on the σ-algebra generated by the sets A, that is, on $\mathcal{F}^w_s$. This proves that $P(A \cap B) = P(A)P(B)$ for any B as in (5), $A \in \mathcal{F}^w_s$. Now, fixing A and considering the two sides of this formula as measures with respect to B, we similarly extend it to $\mathcal{F}^{\theta_s w}_\infty$. □

We will now try to apply this theorem to find the *joint* distribution of the random variables

$$
w^*_T := \max_{t \leq T} w_t, \qquad w_T,
$$

that is, the distribution of the vector (w^*_T, w_T), where $T \in (0, \infty)$ is a fixed constant. Obviously, $w^*_T \geq w_0 = 0$, $w^*_T \geq w_T$. Consequently, for $x \leq y_+$

$$
P\Big\{\max_{t \leq T} w_t \geq x,\ w_T \geq y\Big\} = P\big\{w_T \geq y\big\}. \tag{6}
$$

To analyze the situation $x \geq y_+$ we note that if $w^*_T \geq x$, then at some time τ the sample path of w_t reaches x for the first time, and this takes place before time T. In view of Theorem 11, it is plausible to assume that $\widetilde{w}_t := w_{t+\tau} - w_\tau = w_{t+\tau} - x$ is a Wiener process and does not depend on w_s, $s \leq \tau$. We will now rely on what is called the *reflection principle* (D. André). The probability that at time $t = T - \tau$ the value of $\widetilde{w}_t$ will be strictly less than $y - x$ (≤ 0) is the same as the probability that $\widetilde{w}_t$ will be strictly greater than $x - y$, because both $\widetilde{w}_t$ and $-\widetilde{w}_t$ are Wiener processes. Hence, the following derivation, using the facts that $\big\{\omega \colon w^*_T \geq x,\ w_T < y\big\} = \big\{\omega \colon \tau \leq T,\ w_T < y\big\}$ and

$2x - y \geq x$, is plausible:

$$
\begin{aligned}
P\Big\{\max_{t\leq T} w_t \geq x,\ w_T < y\Big\} &= P\Big\{\tau \leq T,\ \widetilde{w}_{T-\tau} + x < y\Big\} \\
&= P\Big\{\tau \leq T,\ \widetilde{w}_{T-\tau} + x > 2x - y\Big\} = P\Big\{\tau \leq T,\ w_T > 2x - y\Big\} \qquad (7)\\
&= P\Big\{\max_{t\leq T} w_t \geq x,\ w_T > 2x - y\Big\} = P\{w_T > 2x - y\}.
\end{aligned}
$$

However, these arguments are not rigorous, since Theorem 11 tells us that $w_{t+\tau} - w_\tau$ and w_s for $s \leq \tau$ are independent only for constant τ. Nevertheless, we will see in Sec. 5 that the conclusion is true for a wide class of τ, the so-called Markov (or stopping) times, so that equalities (7) are valid. For the moment we will take (7) for granted and go on to determine the distribution of (w_T^*, w_T).

12. THEOREM. *The distribution of the random vector*

$$\Big(\max_{t\leq T} w_t,\ w_T\Big),$$

where T is a fixed constant, $T \in (0, \infty)$, has the density

$$p(x, y) = (2\pi T)^{-1/2} 2T^{-1}(2x - y) \exp\Big\{-\frac{1}{2T}(2x - y)^2\Big\}$$

for $x \geq y_+$, $p(x, y) = 0$ for $x < y_+$. In particular, for $x \geq 0$,

$$P\Big\{\max_{t\leq T} w_t \geq x\Big\} = 2P\{w_T \geq x\} = \frac{2}{\sqrt{2\pi T}} \int_x^\infty e^{-\frac{1}{2T}u^2}\, du, \qquad (8)$$

and since $2P\{w_T \geq x\} = P\{|w_T| \geq x\}$, it follows that $\max_{t\leq T} w_t$ and $|w_T|$ have the same distribution.

PROOF. The second equality in (8) follows from Remark 1.3. That remark also implies that $P\{w_T = z\} = 0$ for any constant z. We can therefore replace the symbol ">" in the last expression in (7) by "$\geq$" and then, adding together the extreme terms in (6), (7) for $x = y \geq 0$, we immediately obtain (8). Further, it follows from (7), (8) that for $x \geq y_+$

$$
\begin{aligned}
P\Big\{w_T^* \geq x,\ w_T \geq y\Big\} &= P\{w_T^* \geq x\} - P\Big\{w_T^* \geq x,\ w_T < y\Big\} \\
&= \frac{1}{\sqrt{2\pi T}}\Big[\int_x^\infty 2e^{-\frac{1}{2T}u^2}\, du - \int_{2x-y}^\infty e^{-\frac{1}{2T}u^2}\, du\Big] \\
&= -\frac{1}{\sqrt{2\pi T}} \int_x^\infty \Big(\int_t^\infty 2e^{-\frac{1}{2T}u^2}\, du - \int_{2t-y}^\infty e^{-\frac{1}{2T}u^2}\, du\Big)_t' \, dt \\
&= \frac{1}{\sqrt{2\pi T}} \int_x^\infty 2\Big[e^{-\frac{1}{2T}u^2} - e^{-\frac{1}{2T}(2u-y)^2}\Big]\, du \\
&= \frac{1}{\sqrt{2\pi T}} \int_x^\infty du \int_y^u 2T^{-1}(2u - v)e^{-\frac{1}{2T}(2u-v)^2}\, dv = \int_{u\geq x}\int_{v\geq y} p(u, v)\, du dv.
\end{aligned}
$$

In other words, by Fubini's Theorem

$$P\{(w_T^*, w_T) \in B\} = \int_{E_2} I_B(u,v)p(u,v)\,dudv \tag{9}$$

for $B = [x,\infty) \times [y,\infty)$, $x \geq y_+$. Equality (9) is easily verified for sets B of this type with $x \leq y_+$ as well, by using (6). A standard application of the lemma on π- and λ-systems shows that (9) holds for all B in the σ-algebra generated by the sets $[x,\infty) \times [y,\infty)$, which by Lemma I.1.7 is just $\mathfrak{B}(E_2)$. □

13. COROLLARY. *Define $\tau(x)$, the first time at which the process w_t reaches the point $x \in E_1$, by the formula*

$$\tau(x) = \inf\{t \geq 0 \colon w_t = x\} \quad (\inf(\emptyset) := \infty).$$

Then $\tau(x)$ is a random variable, $P\{\tau(x) < \infty\} = 1$. Moreover, if $x \neq 0$, then the distribution of $\tau(x)$ has the density $p(v) = 0$ for $v \leq 0$,

$$p(v) = (2\pi)^{-1/2}|x|v^{-3/2}\exp\Big(-\frac{1}{2v}x^2\Big) \quad \textit{for } v > 0. \tag{10}$$

Indeed, if $x > 0$, then, as pointed out before Theorem 12, $\{\omega\colon w_t^* \geq x\} = \{\omega\colon \tau(x) \leq t\}$, and w_t^* is a random variable. Therefore, $\{\omega\colon \tau(x) \leq t\} \in \mathcal{F}$. It follows (see Lemmas I.1.4(b) and I.1.7) that $\tau(x)$ is a random variable and for $t > 0$

$$P\{\tau(x) \leq t\} = P\{w_t^* \geq x\} = 2P\Big\{w_1 \geq \frac{x}{\sqrt{t}}\Big\} = \sqrt{\frac{2}{\pi}}\int_{x/\sqrt{t}}^{\infty} e^{-\frac{1}{2}u^2}\,du$$

$$= \sqrt{\frac{2}{\pi}}\int_0^t \Big(\int_{x/\sqrt{v}}^{\infty} e^{-\frac{1}{2}u^2}\Big)'_v\,dv = \frac{1}{\sqrt{2\pi}}\int_0^t \frac{x}{v^{3/2}}e^{-\frac{1}{2v}x^2}\,dv = \int_{-\infty}^{\infty} I_{v\leq t}\,p(v)\,dv,$$

$$P\{\tau(x) < \infty\} = \lim_{n\to\infty} P\{\tau(x) \leq n\} = \sqrt{\frac{2}{\pi}}\int_0^{\infty} e^{-\frac{1}{2}u^2}\,du = 1.$$

This gives the desired result for $x > 0$. If $x < 0$, it suffices to use the fact that $(-w_t)$ is a Wiener process.

The density (10) plays an important role in the theory of probability and mathematical statistics (and also in the theory of partial differential equations). It is called the density of the *Wald distribution* or the *Wald density*.

14. REMARK. It sometimes helpful to keep in mind that for $x > 0$

$$\int_x^{\infty} e^{-\frac{1}{2}u^2}\,du \leq \int_x^{\infty} \frac{u}{x}e^{-\frac{1}{2}u^2}\,du = \frac{1}{x}e^{-\frac{1}{2}x^2},$$

$$\int_x^{\infty} e^{-\frac{1}{2}u^2}\,du \leq e^{-\frac{1}{2}x^2}\int_x^{\infty} e^{-\frac{1}{2}(u-x)^2}\,du = e^{-\frac{1}{2}x^2}\int_0^{\infty} e^{-\frac{1}{2}u^2}\,du = \sqrt{\frac{\pi}{2}}e^{-\frac{1}{2}x^2}.$$

Together with (8), this shows that for $x > 0$

$$P\Big\{\max_{t\le T} w_t \ge x\Big\} \le \Big(1\wedge \frac{2\sqrt{T}}{x\sqrt{2\pi}}\Big)e^{-\frac{1}{2T}x^2}.$$

The Markov property of the Wiener process enables us to prove following result, at first sight modest, but in fact extremely profound.

15. THEOREM (Blumenthal's zero-one law). *Let $\mathcal{F}^w_{0+} = \bigcap_{t>0}\mathcal{F}^w_t$. Then for any $A\in\mathcal{F}^w_{0+}$ we have $P(A) = 0$ or $P(A) = 1$.*

Before proving Theorem 5, we will deduce from it the relations

$$\text{(11)}\qquad \overline{\lim_{t\downarrow 0}}\frac{\pm w_t}{\sqrt{t}} = \infty \quad \text{(a.s.)},\qquad \overline{\lim_{t\to\infty}}\frac{\pm w_t}{\sqrt{t}} = \infty \quad\text{(a.s.)}.$$

The second of these relations follows easily from the first one and Theorem 1. By the same theorem, it will suffice to prove the first relation for the sign "+" only. We have

$$\xi := \overline{\lim_{t\downarrow 0}}\frac{w_t}{\sqrt{t}} = \lim_{t\downarrow 0}\sup_{s\in(0,t)}\frac{w_s}{\sqrt{s}},$$

where the supremum can be confined to rational $s\in(0,t)$. We see that this supremum is $\mathcal{F}^w_t$-measurable, it decreases with decreasing t, and the limit can be taken over the sequence $t = \frac{1}{n}$. Furthermore, ξ is $\mathcal{F}^w_t$-measurable for any $t>0$, that is, $\mathcal{F}^w_{0+}$-measurable. In particular, $P\{\xi\in B\}$ is zero or one for any Borel set B. Next use that $I_{(-\infty,n)}(x)$ is a right-continuous function of x for fixed n and $\frac{1}{\sqrt{t}}w_t\sim\mathcal{N}(0,1)$. Then

$$P\{\xi<n\} = \mathbf{E}\,I_{(-\infty,n)}(\xi) = \lim_{t\downarrow 0}P\Big\{\sup_{s\in(0,t)}\frac{1}{\sqrt{s}}w_s < n\Big\}$$

$$\le \lim_{t\downarrow 0}P\Big\{\frac{1}{\sqrt{t}}w_t<n\Big\} = \frac{1}{2\pi}\int_{-\infty}^{n} e^{-\frac{1}{2}x^2}\,dx < 1.$$

By Theorem 15, $P\{\xi<n\} = 0$, $\xi\ge n$ (a.s.) for any constant n and $\xi=\infty$ (a.s.) indeed.

Note that (11) of course follows from Theorem 5 as well. However, the proof of Theorem 5 is much more complicated than our arguments here, and in my opinion is also much less interesting from a conceptual point of view. Both Theorem 5 and the relations (11) show that, in an arbitrarily small time interval $[0,t]$, the sample path of the Wiener process passes through origin infinitely many times. In addition, $\overline{\lim}_{t\to\infty}|w_t| = \infty$ (a.s.), but for arbitrarily large T and any finite segment on $(-\infty,\infty)$ there can be found a time interval, after T, during which the sample path of w_t crosses the segment first downwards and then upwards.

16. PROOF OF THEOREM 15. Take $A\in\mathcal{F}^w_{0+}$ and let Λ be the set of all $B\in\mathcal{F}$ which are independent of A, that is $P(A\cap B) = P(A)P(B)$. Obviously, Λ is a λ-system. Furthermore, if $s>0$ and $B\in\mathcal{F}^{\theta_s w}_\infty =: \mathcal{G}_s$, then $B\in\Lambda$ by the Markov property of w (Theorem 11). Thus $\mathcal{G}_s\subset\Lambda$, $s>0$.

It is clear that $\theta_s w_t = \theta_u w_{t+s-u} - \theta_u w_{s-u}$ is $\mathcal{G}_u$-measurable for $u\le s$. Hence $\mathcal{G}_s\subset\mathcal{G}_u$. Since $\mathcal{G}_s$ are σ-algebras, it follows that $\Pi := \bigcup_{s>0}\mathcal{G}_s$ is a π-system and $\Pi\subset\Lambda$. By Lemma I.5.3 we have $\sigma(\Pi)\subset\Lambda$. Observe now that for any $s>0$, $t\ge 0$ the variable

$\theta_s w_t$ is $\mathcal{G}_s$-measurable and hence $\sigma(\Pi)$-measurable. Since $w_t = \lim_{n\to\infty} \theta_{1/n} w_t$, we conclude that w_t's are $\sigma(\Pi)$-measurable, and $\mathcal{F}^w_\infty \subset \sigma(\Pi) \subset \Lambda$. Finally, $A \in \mathcal{F}^w_{0+} \subset \mathcal{F}^w_\infty \subset \Lambda$ what implies by definition that $P(A) = P(A \cap A) = P(A)P(A)$. This certainly means that $P(A) = 0$ or 1. □

4. Multidimensional Wiener processes. Markov times

In the previous section we in fact used the strong Markov property of the one-dimensional Wiener process. To formulate and prove this property with due rigour and generality, we will need the notions of a filtration of σ-algebras and a Markov time with respect to a filtration of σ-algebras. These will be the subject of the present section.

Had we been restricting ourselves in this book to one-dimensional Wiener processes, it would have been sufficient to take the family $\{\mathcal{F}^w_t\}$ of Definition 3.8 as a filtration of σ-algebras. However, in the context of multidimensional Wiener processes, when we consider a one-dimensional Wiener process as one coordinate of a multidimensional Wiener process, it becomes natural to extend these σ-algebras and define Wiener processes with respect to such more general filtrations. The concept also arises quite naturally in the construction of stochastic integrals with respect to multidimensional Wiener processes.

1. Definition. Let $d \geq 1$ be an integer, $(\Omega, \mathcal{F}, P)$ a probability space, and $w^1_t, \dots, w^d_t$ one-dimensional Wiener processes, defined for $t \geq 0$, $i = 1, \dots, d$ on the probability space. Assume that the processes $w^1_t, \dots, w^d_t$ are *independent*, that is, $(w^1_{t(1)}, \dots, w^1_{t(n)}), \dots, (w^d_{t(1)}, \dots, w^d_{t(n)})$ are independent vectors for any $n \geq 0$, $t(1), \dots, t(n) \in [0, \infty)$. Then we will say that $w_t := (w^1_t, \dots, w^d_t)$ is a *d-dimensional Wiener process* (for $t \in [0,\infty)$) on $(\Omega, \mathcal{F}, P)$.

The existence of d-dimensional Wiener processes follows from Exercise 2.7. In this section we consider, for a fixed $d \geq 1$, a probability space $(\Omega, \mathcal{F}, P)$, on which a d-dimensional Wiener process w_t is defined for $t \geq 0$.

The following is a useful exercise in the application of Theorems I.4.14, I.4.19, and I.4.20.

2. Exercise. Prove that if $0 \leq t(1) \leq \cdots \leq t(n) < \infty$, then $w_{t(1)}$, $w_{t(2)} - w_{t(1)}$, $\dots$, $w_{t(n)} - w_{t(n-1)}$ are independent random vectors, $w_t - w_s \sim \mathcal{N}(0, |t-s|I)$, where I is $d \times d$ identity matrix, $t, s \geq 0$. Formulate and prove the converse (cf. Theorem 1.2).

3. Definition. Assume that for any $t \in [0,\infty)$ there is defined a σ-algebra $\mathcal{F}_t \subset \mathcal{F}$, and such that $\mathcal{F}_t \subset \mathcal{F}_s$ for $t \leq s$. Then we will call $\{\mathcal{F}_t\}$ an (increasing) *filtration* of σ-algebras (on the measurable space $(\Omega, \mathcal{F})$). A nonnegative function $\tau = \tau(\omega)$ defined on Ω is called a *Markov time* (with respect to the filtration $\{\mathcal{F}_t\}$) if $\{\omega\colon \tau(\omega) > t\} \in \mathcal{F}_t$ for any $t \in [0, \infty)$.

4. Exercise. Prove that a Markov time is a random variable and that constants $c \geq 0$ (independent of ω) are Markov times.

Recall that the random variables under consideration here can also take infinite values. In particular, Markov times on certain subsets of Ω may equal $+\infty$. The simplest (but quite noninteresting) example of a filtration of σ-algebras is $\mathcal{F}_t = \mathcal{F}$ for all $t \geq 0$. Any nonnegative random variable is a Markov time with respect to this filtration.

For a given filtration $\{\mathcal{F}_t\}$ the σ-algebra $\mathcal{F}_t$ can be interpreted as the collection of all random events whose occurrence or nonoccurrence in some ongoing experiment is completely determined before time t. In other words, $\mathcal{F}_t$ is the collection of all random events that *occur* up to time t. Thus a Markov time is a nonnegative random variable for which, at any time t, we know with certainty whether it exceeds t or not.

To discuss our most important example of a Markov time, we will need the following definition.

5. DEFINITION. If $\xi_t(\omega)$ is a d-dimensional random vector, defined on Ω for any $t \in [0, \infty)$, we will call ξ_t a (d-dimensional) *stochastic or random process*. For fixed ω, the function $\xi_t(\omega)$ (of t) is called a *sample path* (or *trajectory*) of ξ_t. The process ξ_t is said to be *continuous* if all its sample paths are continuous in t. Given a filtration $\{\mathcal{F}_t\}$, we will say that ξ_t is $\mathcal{F}_t$*-adapted* if ξ_t is $\mathcal{F}_t$-measurable for any $t \in [0, \infty)$.

The natural interpretation of the statement that ξ_t is $\mathcal{F}_t$-adapted is: $\xi_s(\omega)$, $s \in [0, t]$, is exactly known at time t.

6. EXAMPLE. Let ξ_t be a continuous d-dimensional $\mathcal{F}_t$-adapted process, $G \subset E_d$ an open set. Define the *first exit time of* ξ_t *from* G by the formula

$$\tau := \inf\{t \geq 0 \colon \xi_t \notin G\} \qquad (\inf \emptyset := \infty).$$

Then τ is a Markov time.

An appeal to the interpretation of $\mathcal{F}_t$ and Markov times makes this assertion obvious, since if we know the whole sample path of ξ_s for $s \in [0, t]$, we obviously know whether it has or has not reached the boundary of G by time t. A formal proof is as follows. Let G_n be open sets such that $\bar{G}_n \subset G_{n+1}$, $G = \cup_n G_n$, and ρ_t the set of all rational numbers in $[0, t]$. Then it is easy to see that for any $t \geq 0$

$$\big\{\omega \colon \tau(\omega) > t\big\} = \bigcup_n \bigcap_{r \in \rho_t} \big\{\omega \colon \xi_r(\omega) \in G_n\big\}.$$

Since the right-hand side involves countable intersections and unions of sets belonging to $\mathcal{F}_r \subset \mathcal{F}_t$, we obtain $\{\tau > t\} \in \mathcal{F}_t$, as claimed.

In applications of first exit times it is often helpful to keep in mind that *if the starting point* ξ_0 *belongs to* G *for some* ω, then $\xi_t \in G$ for this same ω and all $t < \tau(\omega)$. If also $\tau(\omega) < \infty$ (and ξ_t is a continuous process), then $\xi_\tau \in \partial G$.

With any Markov time τ we can associate a σ-algebra $\mathcal{F}_\tau$, which can be interpreted as the set of all events that occur in the time interval $[0, \tau]$.

7. DEFINITION. Let τ be a Markov time with respect to a filtration $\{\mathcal{F}_t\}$. Denote by $\mathcal{F}_\tau$ the family of all sets $A \in \mathcal{F}$ such that $A \cap \{\tau \leq t\} \in \mathcal{F}_t$ for any $t \in [0, \infty)$.

Since $\Omega \cap \{\tau \leq t\} = \Omega \setminus \{\tau > t\}$, $(\Omega \setminus A) \cap \{\tau \leq t\} = \{\tau \leq t\} \setminus (A \cap \{\tau \leq t\})$, $(\cup_n A_n) \cap \{\tau \leq t\} = \cup_n (A_n \cap \{\tau \leq t\})$, it follows that $\mathcal{F}_\tau$ is a σ-algebra.

8. EXERCISE. Prove that if τ is a nonrandom constant equal to t, then $\mathcal{F}_\tau = \mathcal{F}_t$, that is, the notation $\mathcal{F}_\tau$ is consistent with the previous notation.

In the previous section we used the fact that $w_{t+\tau(x)} - w_{\tau(x)}$ is a Wiener process. It is clear that if $\mathcal{F}_t$ is not "properly" related to the Wiener process (for example, $\mathcal{F}_t \equiv \mathcal{F}$) and τ is an arbitrary Markov time with respect to $\{\mathcal{F}_t\}$, we cannot expect $w_{t+\tau} - w_\tau$ to be a Wiener process. In this connection we will need another definition.

9. Definition. Let $\{\mathcal{F}_t\}$ be a filtration of σ-algebras, w_t a $\mathcal{F}_t$-adapted Wiener process; assume that for any $t, h \geq 0$ the *random vector* $w_{t+h} - w_t$ *and* σ*-algebra* $\mathcal{F}_t$ *are independent*, (that is, the σ-algebras $\sigma(w_{t+h} - w_t)$ and $\mathcal{F}_t$ are independent). Then we will say that w_t is a *Wiener process with respect to* $\{\mathcal{F}_t\}$, or that $(w_t, \mathcal{F}_t)$ is a *Wiener process.*

The definition requires the vector $w_{t+h} - w_t$ to be independent of $\mathcal{F}_t$ only for every fixed $h \geq 0$. In fact, however, under this assumption the whole process $(w_{t+h} - w_t, h \geq 0)$ is independent of $\mathcal{F}_t$.

10. Lemma. *Let* $(w_t, \mathcal{F}_t)$ *be a Wiener process. Fix* $s \geq 0$ *and for* $t \geq 0$ *let* $\theta_s w_t := w_{t+s} - w_s$. *Then* $\theta_s w_t$ *is a Wiener process for* $t \geq 0$, *and in particular, the distributions of* $(\theta_s w_{t(1)}, \dots, \theta_s w_{t(m)})$ *and* $(w_{t(1)}, \dots, w_{t(m)})$ *coincide for any* $m \geq 1$, $t(i) \geq 0$. *In addition,* $\mathcal{F}_s$ *and* $\mathcal{F}_\infty^{\theta_s w}$ *are independent* σ*-algebras.*

Proof. The first assertion follows immediately from the definitions (and Exercise I.4.22). To prove the second assertion, fix an event $A \in \mathcal{F}_s$ such that $P(A) > 0$ and define a new probability measure P_A on $\mathcal{F}$ by $P_A(C) = P(A \cap C)/P(A)$. Let $0 = t_0 \leq t_1 \cdots \leq t_n < \infty$, $\Gamma_i \in \mathfrak{B}(E_d)$, and

$$B_i = \left\{\omega \colon \theta_s w_{t_i} - \theta_s w_{t_{i-1}} \in \Gamma_i\right\}, \qquad \xi = \left(\theta_s w_{t_1} - \theta_s w_{t_0}, \dots, \theta_s w_{t_n} - \theta_s w_{t_{n-1}}\right).$$

Clearly,

$$\theta_s w_{t_i} - \theta_s w_{t_{i-1}} = w_{s+t_i} - w_{s+t_{i-1}}, \qquad B_i \in \mathcal{F}_{s+t_i}, \qquad A \in \mathcal{F}_{s+t_{i-1}}.$$

Consequently, it follows by Definition 9 that the events $A \cap B_1 \cap \dots B_i$ and B_{i+1} are independent with respect to P and (see Exercise 2)

$$\begin{aligned} P_A(\xi \in \Gamma_1 \times \cdots \times \Gamma_n) &= P_A(B_1 \cap \cdots \cap B_n) = P_A(B_1 \cap \cdots \cap B_{n-1}) P(B_n) \\ &= P(B_1) \cdot \ldots \cdot P(B_n) = P(\xi \in \Gamma_1 \times \cdots \times \Gamma_n). \end{aligned} \tag{1}$$

By the lemma on π- and λ-systems, it follows from the equality between the outermost terms for arbitrary Γ_i, that the distributions of ξ with respect to P and P_A coincide. The same holds for the distributions of functions of ξ. In particular,

$$\begin{aligned} &P\left\{A \bigcap \left\{\omega \colon \left(\theta_s w_{t_1}, \dots, \theta_s w_{t_n}\right) \in \Gamma_1 \times \cdots \times \Gamma_n\right\}\right\} \\ &\quad = P(A) P\left\{\left(\theta_s w_{t_1}, \dots, \theta_s w_{t_n}\right) \in \Gamma_1 \times \cdots \times \Gamma_n\right\}. \end{aligned} \tag{2}$$

We have proved that $P(A \cap B) = P(A)P(B)$ for sets $B \in \mathcal{F}_\infty^{\theta_s w}$ of a special type. This can be extended to arbitrary sets of $\mathcal{F}_\infty^{\theta_s w}$ by using Lemma I.5.3. It remains only to note that A was an arbitrary set of $\mathcal{F}_s$ such that $P(A) > 0$, and if $P(A) = 0$, then $P(A \cap B) = 0 = P(A)P(B)$ too. □

11. Remark. The passage from (1) to (2) in the proof was necessary because the family of all the sets $(\xi \in \Gamma_1 \times \cdots \times \Gamma_n)$ for all possible n and t_i, is not a π-system.

It is natural to ask whether there exists a filtration of σ-algebras with respect to which w_t is a Wiener process. It turns out that $\mathcal{F}_t^w$ is suitable, as may be deduced from the following exercise.

12. EXERCISE. Prove Theorem 3.11 for a d-dimensional Wiener process w_t. Prove that $(w_t, \mathcal{F}_t^w)$ is a d-dimensional Wiener process, $(w_t^i, \mathcal{F}_t^w)$ are one-dimensional Wiener processes for $i = 1, \dots, d$, and $\mathcal{F}_t^{w^i} \neq \mathcal{F}_t^w$ for $t > 0$, $d \geq 2$.

The properties of one-dimensional Wiener processes investigated in Sec. 3 are, of course, valid for the coordinates of multidimensional Wiener processes. Hence, the results of Sec. 3 provide us with some information concerning the multidimensional case. We limit ourselves to few remarks.

13. REMARK. The Blumenthal zero-one law, that is, Theorem 3.15, together with its proof, carry over literally to the multidimensional case.

14. REMARK. Let $(w_t, \mathcal{F}_t^w)$ be a d-dimensional Wiener process. It follows from Exercise 2, Definition I.4.9, Exercises I.4.13 and I.4.22, and Theorem 1.2 that if $a \in E_d$ is an arbitrary unit vector, then (a, w_t) is a one-dimensional Wiener process with respect to $\{\mathcal{F}_t\}$.

15. REMARK. As in Remark 1.3, it can be shown that the density of the distribution of $w_t - w_s$, for $t \neq s$, is

$$(2\pi|t-s|)^{-d/2} \exp\Big(-\frac{1}{2|t-s|}|x|^2\Big).$$

We will repeatedly use Markov times and their properties. In this connection we offer the reader an exercise that relies (only) on Definitions 3 and 7, the countability of the set of rational numbers, and the possibility to add (if necessary) one more point without violating its countability.

16. EXERCISE. Let $\{\mathcal{F}_t\}$ be a filtration of σ-algebras. Prove that if τ, σ are Markov times, then $\tau \wedge \sigma$, $\tau \vee \sigma$, $\tau + \sigma$ are also Markov times, and the events $\{\tau < \sigma\}$, $\{\tau \leq \sigma\} = \Omega \setminus \{\sigma < \tau\}$, $\{\tau = \sigma\} = \{\tau \leq \sigma\} \cap \{\sigma \leq \tau\}$ are elements of $\mathcal{F}_\tau \cap \mathcal{F}_\sigma$. In particular ($\sigma =$const), τ is a $\mathcal{F}_\tau$-measurable random variable. Finally, if $\tau \leq \sigma$, then $\mathcal{F}_\tau \subset \mathcal{F}_\sigma$. (Note, for example, that

$$\{\tau < \sigma\} \cap \{\tau \leq t\} = \bigcup_r \{\tau \leq r \wedge t\} \bigcap \{\sigma > r \wedge t\} \in \mathcal{F}_t,$$

where the union is taken over all the rational numbers r.)

17. EXERCISE. Under the assumptions of Exercise 16, let ξ_t be a d-dimensional right-continuous $\mathcal{F}_t$-adapted process. Prove that $\xi_\tau I_{\tau<\infty}$ is a $\mathcal{F}_\tau$-measurable vector. (Use the fact that $\xi_t I_{t \leq s}$ is a $\mathcal{F}_s \otimes \mathfrak{B}(E_1)$-measurable function of (ω, t) and the function $\omega \to (\omega, \tau(\omega) \wedge s)$ defines a measurable mapping of the space $(\Omega, \mathcal{F}_s)$ into $(\Omega \times E_1, \mathcal{F}_s \otimes \mathfrak{B}(E_1))$.)

5. Strong Markov property of the Wiener process

The notions of the previous section will now be used to continue our study of a d-dimensional Wiener process $(w_t, \mathcal{F}_t)$ defined for $t \geq 0$ on a probability space $(\Omega, \mathcal{F}, P)$. To formulate the strong Markov property in the most convenient form, we will need one more notion.

1. Definition. Let $C = C\big([0,\infty), E_d\big)$ denote the set of all continuous functions x_t defined for $t \in [0,\infty)$ with values in E_d. Elements of C will be written as $x., y., \dots$; the value of $x.$ at a point t will be denoted by x_t. By $\mathfrak{A}(C)$ we mean the *cylindrical* σ-algebra of subsets of C, that is, the minimal σ-algebra that contains all *cylindrical* sets, i.e., sets of the form

$$\{x.: x_t \in B\}, \tag{1}$$

when $t \in [0,\infty)$ and B ranges over $\mathfrak{B}(E_d)$.

Naturally, if $t_1, \dots, t_n \in [0,\infty)$ and $B_1, \dots, B_n \in \mathfrak{B}(E_d)$, then the set

$$\{x.: x_{t_1} \in B_1, \dots, x_{t_n} \in B_n\}$$

lies in $\mathfrak{A}(C)$ as the intersection of sets (1).

2. Exercise. Let Π be a family of subsets of E_d such that $\sigma(\Pi) = \mathfrak{B}(E_d)$ (for example, the family of all closed sets in E_d). Prove that the σ-algebra generated by all sets (1), where $t \geq 0$, $B \in \Pi$, is precisely $\mathfrak{A}(C)$.

3. Exercise. Prove that if $\xi_t = \xi_t(\omega)$ is a continuous d-dimensional stochastic process defined for $t \geq 0$, then $\xi. = \xi.(\omega)$ is a random element with values in $\big(C, \mathfrak{A}(C)\big)$ and $\mathcal{F}^{\xi}_{\infty} = \sigma(\xi.)$.

4. Exercise. Prove that if ξ_t, η_t are continuous d-dimensional stochastic processes defined for $t \geq 0$ (possibly on different probability spaces), such that $\big(\xi_{t(1)}, \dots, \xi_{t(n)}\big)$ and $\big(\eta_{t(1)}, \dots, \eta_{t(n)}\big)$ have the same distributions for any integer n and $t(1), \dots, t(n) \geq 0$, then $\xi., \eta.$ have the same distribution on $\big(C, \mathfrak{A}(C)\big)$. Deduce that all d-dimensional Wiener processes have the same distribution on $\big(C, \mathfrak{A}(C)\big)$.

The distribution of $w.$ on $\big(C, \mathfrak{A}(C)\big)$, called the *Wiener measure*, is often denoted by W.

We now have all we need to prove the strong Markov property of the Wiener process.

5. Theorem (strong Markov property of w). *Let τ be a Markov time. Define a process $\theta_\tau w_t$ by*

$$\theta_\tau w_t = (w_{t+\tau} - w_\tau) I_{\tau<\infty}, \qquad t \geq 0.$$

Then for any nonnegative $\mathcal{F}_\tau \otimes \mathfrak{A}(C)$-measurable function $F(x.) = F(\omega, x.)$ we have

$$\mathbf{E}\, I_{\tau<\infty} F(\theta_\tau w.) = \mathbf{E}\, I_{\tau<\infty} \Phi, \tag{2}$$

where

$$\Phi(\omega) := \int_\Omega F\big(\omega, w.(\omega')\big) P(d\omega') \tag{3}$$

is a $\mathcal{F}_\tau$-measurable function. In addition, if $P\{\tau < \infty\} > 0$, then the σ-algebras $\mathcal{F}_\tau$, and $\mathcal{F}^{\theta_\tau w}_{\infty}$ are independent with respect to the measure P_τ, where $P_\tau(A) := P\{A, \tau < \infty\}/P\{\tau < \infty\}$[1] *and $\theta_\tau w_t$ is a Wiener process with respect to this measure.*

---

[1] $P\{A, \tau < \infty\} := P(A \cap \{\tau < \infty\})$.

6. REMARK. In applications of this theorem it is helpful to keep the following (nonrigorous) comments in mind. The function

$$F(\theta_\tau w.) = F\left(\omega, \theta_{\tau(\omega)} w.(\omega)\right)$$

involves the randomness (that is, the argument ω) in two essentially different ways. The first is via the first argument and $\tau(\omega)$. The contribution of this randomness is completely known at time $\tau(\omega)$; that is, after time $\tau(\omega)$ we know $F(\omega, x.)$ completely as a function of $x.$ for any ω. Second, $F(\theta_\tau w.)$ involves the randomness via $\theta_\tau w.(\omega)$. The theorem asserts, in fact, that no information about what takes place in time interval $[0, \tau]$ can help us to learn anything about behavior of $\theta_\tau w.$. Note also that if $P\{\tau < \infty\} = 1$, then (see Theorem I.4.3 and Exercise 4)

$$\Phi(\omega) = \int_C F(\omega, x.) W(dx.) = \int_\Omega F\left(\omega, \theta_{\tau(\omega)} w.(\omega')\right) P(d\omega').$$

Thus, formula (2) enunciates a rule of what might be called step-by-step averaging: first we record everything that is known up to time τ and average over whatever is independent of observations before τ, as if these observations had not been made at all. Then average this intermediate result once again, taking into account that the observations in the time interval $[0, \tau]$ produce random results.

7. PROOF OF THEOREM 5. First consider the case when

$$F(\omega, x.) = I_A(\omega) \cdot g_1(x_{t_1}) \cdot \ldots \cdot g_m(x_{t_m}), \tag{4}$$

where $A \in \mathcal{F}_\tau$, g_i are bounded nonnegative continuous functions and $t_i \geq 0$. In this situation we will apply a useful method: approximating the Markov time τ by Markov times τ_n with only countable ranges of values. Namely, set $\tau_n(\omega) = 2^{-n}(k+1)$ for ω such that $k2^{-n} < \tau(\omega) \leq (k+1)2^{-n}$, where $k = -1, 0, 1, \ldots$, $n = 1, 2, \ldots$, and $\tau_n(\omega) = \infty$ if $\tau(\omega) = \infty$. It is easily seen that $\tau \leq \tau_n \leq \tau + 2^{-n}$, $\tau_n \downarrow \tau$, and for $t \geq 0$

$$\left\{\omega \colon \tau_n(\omega) > t\right\} = \left\{\omega \colon \tau(\omega) > 2^{-n}[2^n t]\right\} \in \mathcal{F}_{2^{-n}[2^n t]} \subset \mathcal{F}_t,$$

so that each τ_n is a Markov time. Moreover, for $t = (k+1)2^{-n}$ we have

$$A \cap \{\tau_n = t\} = (A \cap \{\tau \leq t\}) \backslash (A \cap \{\tau \leq t - 2^{-n}\}) \in \mathcal{F}_t. \tag{5}$$

It follows by Lemma 4.10 (and Exercises 3.10, I.4.22) that[2]

$$\begin{aligned}
\mathbf{E}\, I_{\tau<\infty} F(\theta_\tau w.) &= \lim_{n\to\infty} \mathbf{E}\, I_{A,\tau<\infty} g_1(\theta_{\tau_n} w_{t_1}) \cdot \ldots \cdot g_m(\theta_{\tau_n} w_{t_m}) \\
&= \lim_{n\to\infty} \sum_{t<\infty} \mathbf{E}\, I_{A,\tau_n = t}\, g_1(\theta_t w_{t_1}) \cdot \ldots \cdot g_m(\theta_t w_{t_m}) \\
&= \lim_{n\to\infty} \sum_{t<\infty} P(A, \tau_n = t) \mathbf{E}\, g_1(w_{t_1}) \cdot \ldots \cdot g_m(w_{t_m}) \\
&= P(A, \tau < \infty) \mathbf{E}\, g_1(w_{t_1}) \cdot \ldots \cdot g_m(w_{t_m}).
\end{aligned}$$

This proves (2) in the particular case. Now we again consider λ- and π-systems. Let Λ be the family of all sets $D \subset \Omega \times C$ whose indicators $I_D(\omega, x.) =: F$ possess the following properties: (a) $I_{\tau<\infty} F(\omega, \theta_\tau w.)$ is a random variable; (b) the integrand of (3) is a measurable function of ω' and the integral itself is a $\mathcal{F}_\tau$-measurable function of

[2] $I_{A,\tau<\infty}$ is the indicator of the event $A \cap \{\omega \colon \tau < \infty\}$.

ω; (c) formula (2) holds. It follows from the properties of measurable functions that Λ is a λ-system. In addition, by applying the already proved equality (2) to functions (4), we conclude as in the proof of Theorem I.5.4, that Λ contains the π-system of all sets $A \times \bigcap_{i \le n} \{x.: x_{t_i} \in \Gamma_i\}$, where $A \in \mathcal{F}_\tau$, the sets $\Gamma_i \subset E_d$ are closed, and $t_i \ge 0$. Hence Λ contains the σ-algebra Σ generated by these sets. For fixed $A \in \mathcal{F}_\tau$ the family of all sets $B \subset C$ such that $A \times B \in \Sigma$ is obviously a σ-algebra, which moreover contains the sets $\{x.: x_t \in \Gamma\}$, where $t \ge 0$ and Γ is a closed set. It follows (see Exercise 2) that this σ-algebra contains $\mathfrak{A}(C)$, that is $A \times B \in \Sigma$ for any $A \in \mathcal{F}_\tau$, $B \in \mathfrak{A}(C)$. By the definition of $\mathcal{F}_\tau \otimes \mathfrak{A}(C)$ this means that $\mathcal{F}_\tau \otimes \mathfrak{A}(C) \subset \Sigma \subset \Lambda$.

Thus the first part of the theorem is proved for indicators of all $\mathcal{F}_\tau \otimes \mathfrak{A}(C)$-measurable sets. It can now be extended in the standard way to all nonnegative $\mathcal{F}_\tau \otimes \mathfrak{A}(C)$-measurable functions.

It follows from (2) for $B \in \mathfrak{A}(C)$, $F = I_B(x.)$ that $P_\tau\{\theta_\tau w. \in B\} = P\{w. \in B\}$. This is obviously equivalent to the last part of the theorem.

Finally, the independence of σ-algebras $\mathcal{F}_\tau$ and $\mathcal{F}_\infty^{\theta_\tau w}$ follows from (2) for $F = I_A(\omega)I_B(x.)$, where $A \in \mathcal{F}_\tau$, $B \in \mathfrak{A}(C)$ and from the fact (see Exercise 3) that any event of $\mathcal{F}_\infty^{\theta_\tau w}$ has the form $\{\omega: \theta_\tau w. \in B\}$, where $B \in \mathfrak{A}(C)$. □

We will now use this theorem to prove (3.7). Recall that in that case $d = 1$. Note also that the event $\{\tau \le T\}$ belongs to $\mathcal{F}_\tau^w$. Indeed, τ is the first exit time of w_t from $(-\infty, x)$ and therefore (see Example 4.6) τ is a Markov time with respect to $\{\mathcal{F}_t^w\}$. Hence

$$\{\tau \le T\} \cap \{\tau \le t\} = \{\tau \le T \wedge t\} \in \mathcal{F}_{T\wedge t}^w \subset \mathcal{F}_t^w$$

for any $t \ge 0$. This means by definition that $\{\tau \le T\} \in \mathcal{F}_\tau^w$. Moreover, $g(t, x.) := x_{T-t\wedge T}$ is a continuous function of t and is $\mathfrak{A}(C)$-measurable for any t. By Lemma I.5.7, it is $\mathfrak{B}([0,\infty)) \otimes \mathfrak{A}(C)$-measurable, and since $\tau(\omega)$ and $x.$ are $\mathcal{F}_\tau^w$- and $\mathfrak{A}(C)$-measurable, respectively, we conclude that the pair $(\tau(\omega), x.)$ is $\mathcal{F}_\tau^w \otimes \mathfrak{A}(C)$-measurable (see Lemma I.1.7), and that $\tau, g(\tau, x.)$ are $\mathcal{F}_\tau^w \otimes \mathfrak{A}(C)$-measurable, as compositions of measurable mappings. In particular,

$$\Big\{(\omega, x.): \tau(\omega) \le T,\ x_{T-T\wedge\tau(\omega)} + x < y\Big\} \in \mathcal{F}_\tau^w \otimes \mathfrak{A}(C).$$

Theorem 5 applied to the indicator of this set yields a formula which seems intuitively natural:

$$P\Big\{\tau \le T,\ \widetilde{w}_{T-T\wedge\tau} + x < y\Big\} = \mathbf{E}\, I_{\tau\le T}\Phi(\tau), \tag{6}$$

where $\Phi(t) = P\{w_{T-T\wedge t} + x < y\}$. By the symmetry of the distribution of w_t, we have $\Phi(t) = P\{w_{T-T\wedge t} + x > 2x - y\}$, and by Theorem 5 the right-hand side of (6) becomes the third term in (3.7). This proves the second equality in (3.7). The remaining equalities in (3.7) are obvious.

8. Remark. In connection with the measurable space $(C, \mathfrak{A}(C))$ it is useful to keep in mind that if $D \subset E_d$ is a domain and

$$\tau(x.) := \inf\{t \ge 0: x_t \notin D\},$$

then $\tau(x.)$ is $\mathfrak{A}(C)$-measurable (see Exercise 4.4) and moreover, $\tau(x.)$ is a Markov time with respect to the filtration of σ-algebras $\mathfrak{A}_s(C)$, where $\mathfrak{A}_s(C)$ is the minimal σ-algebra that includes all the sets (1) for $t \le s$, $B \in \mathfrak{B}(E_d)$. This assertion was actually proved in Example 4.6. Incidentally, it follows from that example (and the definition of a distribution) that, under the assumption of Exercise 4 the distributions of the first exit times $\tau(\xi.)$, $\tau(\eta.)$ of the processes ξ_t, η_t from D coincide.

9. EXERCISE. There are other formulas connected with the Wiener process expressing different rules of step-by-step averaging. For example, Theorem 2.5 implies that $\eta_0 = w_\pi/\sqrt{\pi}$, and that $w_t - t\eta_0/\sqrt{\pi} = w_t - tw_\pi/\pi$ for $t \in [0, \pi]$ is independent of η_0, that is, of w_π. Using the Normal Correlation Theorem, prove that for any $d \geq 1$, $T \in (0, \infty)$, $t(1), \dots, t(n) \leq T$ the random vectors w_T and $(\widetilde{w}_{t(1)}, \dots, \widetilde{w}_{t(n)})$ are independent, where $\widetilde{w}_t := w_t - tw_T/T$. Deduce from this that for any nonnegative $\mathfrak{A}_T(C) \otimes \mathfrak{B}(E_d)$-measurable function $F(x., y)$

$$\mathbf{E}\,F(\widetilde{w}., w_T) = \mathbf{E}\,\Phi(w_T),$$

where $\Phi(y) = \mathbf{E}\,F(\widetilde{w}., y)$. Using Remark 4.15, rewrite this result as

$$\mathbf{E}\,F(\widetilde{w}., w_T) = (2\pi T)^{-d/2} \int_{E_d} \mathbf{E}\,F(\widetilde{w}., y) \exp\Big(-\frac{1}{2T}|y|^2\Big)\,dy.$$

6. Martingale properties of the Wiener process

Theorem 5.5 has great many applications. We have already used it to prove that our heuristic derivation of equalities (3.7) is in fact rigorous. We will now use the theorem to obtain further properties of the Wiener process related, actually, to the theory of martingales. It is convenient to present these results immediately after Theorem 5.5, although the notion of martingale will be introduced only in Sec. 8.

Let $(w_t, \mathcal{F}_t)$ be a given d-dimensional Wiener process defined on a probability space $(\Omega, \mathcal{F}, P)$ for $t \geq 0$.

1. LEMMA. *Let g be a Borel function on E_d and $S \in (0, \infty)$ a number. For $t \in [0, S]$, $x \in E_d$, set $u(t, x) = \mathbf{E}\,g(x + w_{S-t})$. Assume that either u is well defined and finite for $t \in [0, S]$, $x \in E_d$, or $g \geq 0$. Then for any $t \in [0, S]$, $x \in E_d$ and any Markov time $\tau \leq S - t$*

$$\mathbf{E}\,u(t + \tau, x + w_\tau) = u(t, x). \tag{1}$$

PROOF. As usual, since $g = g_+ - g_-$, we need only consider the case $g \geq 0$. Set $F(\omega, x.) = g(x + w_\tau + x_{S-t-\tau})$. It follows from the arguments after the proof of Theorem 5.5 that F satisfies the assumptions of that theorem. Moreover, $F(\omega, \theta_\tau w.) = g(x + w_{S-t})$, and the left-hand side of formula (5.2) equals $u(t, x)$. Finally, the function Φ defined in (5.3) is the value of $\mathbf{E}\,g(y + w_{S-t-s}) = u(t + s, y)$ at the point $y = x + w_\tau$, $s = \tau$. In other words, $\Phi = u(t + \tau, x + w_\tau)$ and for the case under consideration (1) is simply (5.2). □

2. REMARK. Formula (1) can be rewritten for nonrandom τ in another form. To do this define the operator T_t, $t \geq 0$, mapping any nonnegative Borel function g on E_d into the function $T_t g(x) := \mathbf{E}\,g(x + w_t)$. It is easily seen that if g is bounded and continuous, then $T_t g$ is a continuous function of (t, x) and therefore a Borel function. Using the lemma on π- and λ-systems, we immediately conclude that $T_t g$ is a Borel function whenever g is a Borel function and $g \geq 0$. In particular, the expression $T_s T_t g$ is meaningful and for $s \leq S - t$, $\tau = s$ we can write (1) as $T_s T_{S-t-s} = T_{S-t}$. Since

S, t, s are arbitrary, this means that the T_t form an *operator semigroup*, that is, for any $s, t \geq 0$

$$T_s T_t = T_{s+t}. \tag{2}$$

3. EXERCISE. By Remark 4.15 (and Theorem I.4.5), for $t > 0$,

$$T_t g(x) = \mathbf{E} g(x + w_t) = (2\pi t)^{-d/2} \int_{E_d} g(y) \exp\Big(-\frac{|x-y|^2}{2t} \Big) dy. \tag{3}$$

Prove equality (2) analytically, using (3), Lemma I.4.6, Fubini's Theorem, and the formula

$$\frac{|x-y|^2}{s} + \frac{|y-z|^2}{t} = \frac{s+t}{st} \Big| y - \Big(\frac{x}{s} + \frac{z}{t} \Big) \frac{st}{s+t} \Big|^2 + \frac{1}{s+t} |x-z|^2. \tag{4}$$

By Lemma 1, for (constant) $s, t \geq 0$ such that $s + t \leq S$ the function u satisfies the equation

$$u(t, x) = \mathbf{E} u(t+s, x + w_s). \tag{5}$$

In addition, by Remark 2 the function u of Lemma 1 is a Borel function. Hence the following is a generalization of Lemma 1.

4. THEOREM. *Let $T \in (0, \infty]$ and let $u(t, x)$ be a Borel finite or Borel nonnegative function defined for $t \in [0, \infty)$, $x \in E_d$. Suppose that (5) holds for all $s, t \in [0, \infty)$, $x \in E_d$, such that $t + s \leq T$. Then, for any bounded Markov time τ and any $t \in [0, \infty)$, $x \in E_d$ such that $t + \tau \leq T$, equality (1) holds, that is, equality (5) remains valid upon substitution of any bounded Markov time for s. Moreover, if $t, S \in [0, \infty)$, $x \in E_d$, τ, σ are bounded Markov times, $t + \tau \leq t + \sigma \leq S \leq T$, η is a $\mathcal{F}_\tau$-measurable random variable and the expression $\mathbf{E}\eta u(S, x + w_{S-t})$ is well defined, then*

$$\mathbf{E}\eta u(t+\tau, x + w_\tau) = \mathbf{E}\eta u(t + \sigma, x + w_\sigma), \tag{6}$$

and both sides are well defined.

PROOF. The first part of the theorem follows from Lemma 1, since there exists a finite $S \leq T$ such that $t + \tau \leq S$, and if $g(x) := u(S, x)$, $s = S - t$ we infer from (5) that $u(t, x) = \mathbf{E} g(x + w_{S-t})$.

To prove the second part we note that Theorem 5.5 is valid not only for nonnegative F but also for any F such that the left-hand side of (5.2) is meaningful. This is easily proved, using the formula $F = F_+ - F_-$. Let us apply these arguments to $F(\omega, x.) = \eta u(S, x + w_\tau + x_{S-t-\tau})$. Then we immediately deduce from (5.2) and (5) that the left-hand side of (6) is meaningful and equal to $\mathbf{E}\eta u(S, x + w_{S-t})$. Since $\mathcal{F}_\tau \subset \mathcal{F}_\sigma$, it follows that η is $\mathcal{F}_\sigma$-measurable. Then the previous arguments can be applied to σ as well as to τ, and the right-hand side of (6) also coincides with $\mathbf{E}\eta u(S, x + w_{S-t})$. □

5. EXERCISE. Using Remark 4.15 and Lemma I.4.6, prove that for any $\lambda \in E_d$, $t \geq 0$

$$\mathbf{E}\, e^{(\lambda, w_t)} = e^{\frac{1}{2}|\lambda|^2 t}.$$

Then, considering the function $u(x) = \exp\left[(\lambda, x) - \frac{1}{2}|\lambda|^2 t\right]$, deduce from Theorem 4 that for any bounded Markov time τ

$$\mathbf{E}\, e^{(\lambda, w_\tau) - \frac{1}{2}|\lambda|^2 \tau} = 1. \tag{7}$$

6. EXERCISE. Let $d = 1$, $x > 0$ and let $\tau(x)$ be as in Corollary 3.13. Observing that

$$\exp\left[\lambda w_{t\wedge\tau(x)} - \frac{1}{2}\lambda^2 (t \wedge \tau(x))\right]$$
$$= I_{\tau(x)\geq t} \exp\left[\lambda w_t - \frac{1}{2}\lambda^2 t\right] + I_{\tau(x)<t} \exp\left[\lambda x - \frac{1}{2}\lambda^2 \tau(x)\right],$$

where $\lambda w_t \leq \lambda x$ for $\tau(x) \geq t$, $\lambda \geq 0$, substitute $\lambda > 0$ and $\tau(x) \wedge t$ for τ in (7), let $t \to \infty$ and conclude, using the Dominated Convergence Theorem, that for $\nu > 0$

$$\mathbf{E}\, e^{-\nu\tau(x)} = e^{-x\sqrt{2\nu}}. \tag{8}$$

Derive this result once again from Corollary 3.13.

7. EXERCISE. For $d = 1$, $b \in (-\infty, \infty)$, $x > 0$ define

$$\tau(x) = \inf\left\{t \geq 0 \colon w_t + bt \geq x\right\},$$

that is, τ is the first exit time of the process $w_t + bt$ from $(-\infty, x)$.

Substituting $\tau(x) \wedge t$ into (7), noting that $w_s + bs = x$ if $s = \tau(x) < \infty$, and letting $t \to \infty$, prove the following generalization of (8) for $\nu > 0$,

$$\mathbf{E}\, e^{-\nu\tau(x)} = e^{-\lambda x}, \tag{9}$$

where $\lambda = \sqrt{b^2 + 2\nu} - b$. Letting $\nu \downarrow 0$ deduce that $P\{\tau(x) < \infty\} = 1$ for $b \geq 0$ and $P\{\tau(x) < \infty\} = \exp\left(-2|b|x\right)$ for $b < 0$.

8. EXERCISE. Fix $\varepsilon \in (0, \infty)$ and define

$$h(t, x) = (t + \varepsilon)^{-d/2} \exp\left(\frac{1}{2(t+\varepsilon)}|x|^2\right). \tag{10}$$

Using Remark 4.15, Lemma I.4.6, and a formula similar to (4), deduce from Theorem 4 that for any $t \in [0, \infty)$, $x \in E_d$ and bounded Markov times τ

$$h(t, x) = \mathbf{E}\, h(t + \tau, x + w_\tau).$$

We will now deduce from Theorem 4 one more result, which will be the basis for our construction of the Itô stochastic integral in Sec. III.2.

9. THEOREM. *Let τ, σ be bounded Markov times with respect to $\{\mathcal{F}_t\}$. Then for $i, j = 1, \ldots, d$*

$$\mathbf{E}\, w_\tau^i = 0, \qquad \mathbf{E}\, w_\tau^i w_\sigma^j = \delta^{ij}\mathbf{E}\,(\tau \wedge \sigma). \tag{11}$$

PROOF. It follows immediately from the definition of the Wiener process that the functions x^i, $x^i x^j - \delta^{ij} t$ satisfy (5). Hence, assertion (11) for $\tau = \sigma$ is a consequence of the first part of Theorem 4 (with $t = 0$, $x = 0$). Therefore, it remains only to prove that the left-hand side of the second equality in (11) is unchanged after the substitution of $\underline{\gamma} := \tau \wedge \sigma$ for τ and σ.

We use the second part of Theorem 4 and the fact that, as we already know, $\mathbf{E}\left|w_\tau^i\right|^2 = \mathbf{E}\,\tau < \infty$, $\mathbf{E}\left|w_\sigma^i\right|^2 < \infty$, $\mathbf{E}\left|w_\tau\right|\cdot\left|w_\sigma\right| < \infty$. Then we obtain, for $u(t,x) = x^j$, $\bar{\gamma} = \tau \vee \sigma$,

$$\mathbf{E}\, w_\tau^i w_\sigma^j I_{\tau<\sigma} = \mathbf{E}\, w_{\underline{\gamma}}^i I_{\underline{\gamma}<\sigma} w_{\bar{\gamma}}^j = \mathbf{E}\, w_{\underline{\gamma}}^i I_{\underline{\gamma}<\sigma} w_{\underline{\gamma}}^j = \mathbf{E}\, w_{\underline{\gamma}}^i w_{\underline{\gamma}}^j I_{\tau<\sigma}.$$

The equality between the outermost terms obviously remains valid if we replace τ by σ and vice versa, or if we substitute the set $\{\tau = \sigma\}$ for $\{\tau < \sigma\}$. Summing up all these relations, we obtain the desired assertion. □

This theorem and the equality $|a - b| = a + b - 2(a \wedge b)$ yield the following result.

10. COROLLARY. $\mathbf{E}\left|w_\tau - w_\sigma\right|^2 = \mathbf{E}\,|\tau - \sigma| d$.

Theorem 9, like Theorem 4 can be used for actual calculations.

11 EXAMPLE. Let $d = 1$, $p, q \in (0, \infty)$,

$$\tau = \inf\left\{t \geq 0 : w_t \notin (-p, q)\right\},$$

that is, τ is the first exit time of w_t from the interval $(-p, q)$. Clearly, $|w_{t\wedge\tau}| \leq p \vee q$. Hence,

$$\mathbf{E}\,\tau = \lim_{t\to\infty} \mathbf{E}\, t \wedge \tau = \lim_{t\to\infty} \mathbf{E}\, w_{t\wedge\tau}^2 \leq (p \vee q)^2 < \infty. \tag{12}$$

This gives $\mathbf{E}\,\tau < \infty$, and consequently $\tau < \infty$ (a.s.) and by the Dominated Convergence Theorem we can extend (12) (see Remark I.2.15):

$$\mathbf{E}\,\tau = \mathbf{E}\, w_\tau^2 = \mathbf{E}\, w_\tau^2 I_{w_\tau=-p} + \mathbf{E}\, w_\tau^2 I_{w_\tau=q} = p^2 P\{w_\tau = -p\} + q^2 P\{w_\tau = q\}. \tag{13}$$

In addition,

$$-pP\{w_\tau = -p\} + qP\{w_\tau = q\} = \mathbf{E}\, w_\tau = \lim_{t\to\infty} \mathbf{E}\, w_{t\wedge\tau} = 0,$$
$$P\{w_\tau = -p\} + P\{w_\tau = q\} = 1.$$

The last relations, together with (13), give

$$P\{w_\tau = -p\} = \frac{q}{p+q}, \qquad P\{w_\tau = q\} = \frac{p}{p+q}, \qquad \mathbf{E}\,\tau = pq.$$

12. EXERCISE. Let $d \geq 1$, $R > 0$, and let τ be the first exit time of w_t from the ball $\{|x| < R\}$. Prove that $\mathbf{E}\,\tau = R^2 d^{-1}$.

13. PROBLEM. Let $T = \infty$, $u \geq 0$. Under the assumptions of Theorem 4, let γ, τ be arbitrary finite Markov times, $\gamma \leq \tau$, and let equality (1) *hold* for $t = 0$, $x = 0$ and $u(0,0) < \infty$. Using Theorem 4 and Scheffé's Theorem, prove that

$$\lim_{t\to\infty} \mathbf{E}\, I_{\tau>t}\big|u(t, w_t) - u(\tau, w_\tau)\big| = 0.$$

Deduce from this and Theorem 4 that

$$\lim_{t\to\infty} \mathbf{E}\, u(t, w_t) I_{\gamma>t} = \lim_{t\to\infty} \mathbf{E}\, u(\tau, w_\tau) I_{\gamma>t} = 0, \qquad u(0,0) = \mathbf{E}\, u(\gamma, w_\gamma).$$

We conclude this section with an application of Lemma 1 to determine the distribution of $\max_{t\leq T}(w_t + bt)$. We will return to this distribution again in Exercise 9.4 and in Exercise IV.3.8, which is related in a natural way to Problem IV.3.10.

14. EXERCISE. Let $d = 1$, $c > 0$, $T \in (0,\infty)$, $b \in E_1$. Define $g(x) = 1$ for $x < c - bT$, $g(c - bT) = 0$ and $g(x) = -e^{-2b(x+bT-c)}$ for $x > c - bT$; also, let $u(t,x) = \mathbf{E}\, g(x + w_{T-t})$ for $(t,x) \in [0,T] \times E_1$, $\tau = \inf\{t \geq 0 \colon w_t + bt \geq c\}$. Using the equality

$$2b\big(x + b(T-t)\big) + \frac{1}{2(T-t)}x^2 = \frac{1}{2(T-t)}\big(x + 2b(T-t)\big)^2,$$

prove by direct calculation that $u(t, c-bt) = 0$ for $t \in [0,T]$, and therefore $u(\tau, w_\tau) = 0$ for $\tau \leq T$, since in the latter case $w_\tau + b\tau = c$. Observing that $w_T + bT < c$ and $u(T, w_T) = g(w_T) = 1$ for $\tau > T$, conclude via Lemma 1 that

$$\begin{aligned} P\big\{\max_{t\leq T}(w_t + bt) < c\big\} &= P\{\tau > T\} \\ &= \mathbf{E}\, u(\tau, w_\tau) I_{\tau\leq T} + \mathbf{E}\, u(T, w_T) I_{\tau>T} = \mathbf{E}\, u\big(\tau\wedge T, w_{\tau\wedge T}\big) = u(0,0) \\ &= \frac{1}{\sqrt{2\pi T}} \int_{-\infty}^{c-bT} e^{-\frac{1}{2T}x^2}\, dx - \frac{1}{\sqrt{2\pi T}} \int_{c-bT}^{\infty} e^{-\frac{1}{2T}x^2 - 2b(x+bT-c)}\, dx. \end{aligned} \tag{14}$$

15. REMARK. Using (14), one can prove (9) again, and with even weaker restrictions on the parameters (see also Problem 9.12).

7. Burkholder-Davis-Gundy inequalities and Wald identities for the Wiener process

Let $(w_t, \mathcal{F}_t)$ be a one-dimensional Wiener process and τ a Markov time with respect to $\{\mathcal{F}_t\}$. We begin the present section by proving the *Burkholder-Davis-Gundy inequalities*

$$\mathbf{E}\sup_{t\leq\tau}|w_t|^p \leq N\,\mathbf{E}\,\tau^{p/2}, \qquad \mathbf{E}\,\tau^{p/2} \leq N\,\mathbf{E}\sup_{t\leq\tau}|w_t|^p, \tag{1}$$

where $p \in (0,\infty)$, $N = N(p) < \infty$ are constants, and the suprema are of course evaluated on the assumption that $t \neq \infty$. Inequalities of this kind play an important role in the theory of martingales, they are known as the Burkholder-Davis-Gundy inequalities. We recall that, as mentioned in the introduction to this chapter, the results of this section will not be used later on.

The proof of inequalities (1) will be divided into three lemmas.

1. LEMMA. *Let $\xi, \eta \geq 0$ be random variables, $f(\delta)$ a nonnegative function for $\delta \geq 0$, $f(\delta) \to 0$ as $\delta \to 0$, and suppose that for all $\delta \in (0,1]$, $\lambda > 0$,*

$$P\Big\{\xi \geq 2\lambda,\ \eta \leq \delta\lambda\Big\} \leq f(\delta)P\{\xi \geq \lambda\}. \tag{2}$$

Then for any $p \in (0,\infty)$ there exists a constant $N < \infty$, depending only on p and f, and such that $\mathbf{E}\,\xi^p \leq N\,\mathbf{E}\,\eta^p$.

PROOF. By Fubini's Theorem (and Lemma I.5.7)

$$\mathbf{E}\,\xi^p = p\mathbf{E}\int_0^\infty \lambda^{p-1} I_{\lambda\leq\xi}\,d\lambda = p\int_0^\infty \lambda^{p-1}P\{\xi \geq \lambda\}\,d\lambda = 2^p p\int_0^\infty \lambda^{p-1}P\{\xi \geq 2\lambda\}\,d\lambda.$$

Replacing $P\{\xi \geq 2\lambda\}$ by

$$P\Big\{\xi \geq 2\lambda,\ \eta \leq \delta\lambda\Big\} + P\Big\{\xi \geq 2\lambda,\ \eta > \delta\lambda\Big\} \leq f(\delta)P\{\xi \geq \lambda\} + P\{\eta > \delta\lambda\},$$

we obtain

$$\mathbf{E}\,\xi^p \leq 2^p f(\delta)\mathbf{E}\,\xi^p + \Big(\frac{2}{\delta}\Big)^p\mathbf{E}\,\eta^p. \tag{3}$$

Choosing δ so that $2^p f(\delta) \leq 2^{-1}$ and combining the like terms in (3), we arrive at the desired equality. It must be noted, however, that gathering like terms is possible only provided $\mathbf{E}\,\xi^p < \infty$. This difficulty is easily overcome if one notices that (2) is also true for $\xi \wedge n$ in place of ξ, where n is any positive constant. Consequently, the finite expression $\mathbf{E}(\xi \wedge n)^p$ admits an estimate and it remains only to let $n \to \infty$ and use the Monotone Convergence Theorem. □

In the next lemma we will prove (2) for $\xi = |w|^*_\tau$, $\eta = \tau^{1/2}$, where

$$|w|^*_\tau = \sup_{t\leq\tau}|w_t|.$$

2. LEMMA. *For any $\lambda > 0$, $\delta > 0$*

$$P\Big\{\sup_{t\leq\tau} w_t \geq 2\lambda,\ \tau^{1/2} \leq \delta\lambda\Big\} \leq \delta^2 P\Big\{\sup_{t\leq\tau} w_t \geq \lambda\Big\}, \tag{4}$$

$$P\Big\{\sup_{t\leq\tau}(-w_t) \geq 2\lambda,\ \tau^{1/2} \leq \delta\lambda\Big\} \leq \delta^2 P\Big\{\sup_{t\leq\tau}(-w_t) \geq \lambda\Big\}, \tag{5}$$

$$P\Big\{|w|^*_\tau \geq 2\lambda,\ \tau^{1/2} \leq \delta\lambda\Big\} \leq 2\delta^2 P\Big\{|w|^*_\tau \geq \lambda\Big\}. \tag{6}$$

In particular, the first inequality in (1) is true.

PROOF. Since

$$\Big\{\omega\colon |w|^*_\tau \geq 2\lambda\Big\} = \Big\{\omega\colon \sup_{t\leq\tau} w_t \geq 2\lambda\Big\}\bigcup\Big\{\omega\colon \sup_{t\leq\tau}(-w_t) \geq 2\lambda\Big\},$$

it follows that the left-hand side of (6) is at most the sum of the left-hand sides of (4), (5), and the coefficients of δ^2 in (4), (5) are at most $P\{|w|^*_\tau \geq \lambda\}$. Hence (6) follows from (4), (5).

We now recall the notation $\tau(x)$ from Corollary 3.13. Then for $\mu := (\delta\lambda)^2$ the left-hand side of (4) is equal to

$$P\Big\{\tau(2\lambda) \le \tau \le \mu\Big\} \le P\Big\{\tau(\lambda) \le \tau,\ \tau(2\lambda) \le \mu\Big\}$$
$$= P\Big\{\tau(\lambda) \le \tau \wedge \mu,\ \max_{t \le \mu - \tau(\lambda)} \theta_{\tau(\lambda)} w_t \ge \lambda\Big\}$$
$$\le P\Big\{\tau(\lambda) \le \tau \wedge \mu,\ \max_{t \le \mu} \theta_{\tau(\lambda)} w_t \ge \lambda\Big\}.$$

In this expression $\{\omega\colon \tau(\lambda) \le \tau \wedge \mu\} \in \mathcal{F}^w_{\tau(\lambda)}$ (Exercise 4.16), and we conclude by the strong Markov property and Theorem 3.12 that the last probability is

$$P\Big\{\tau(\lambda) \le \tau \wedge \mu\Big\} P\Big\{\max_{t \le \mu} w_t \ge \lambda\Big\} \le P\Big\{\sup_{t \le \tau} w_t \ge \lambda\Big\} P\Big\{|w_\mu| \ge \lambda\Big\}.$$

By the Chebyshev inequality,

$$P\Big\{|w_\mu| \ge \lambda\Big\} = P\Big\{|w_\mu|^2 \ge \lambda^2\Big\} \le \lambda^{-2} \mathbf{E}\, w_\mu^2 = \delta^2,$$

and this proves (4). The proof of (5) is similar. □

3. Lemma. *For any $\lambda > 0$, $\delta > 0$*

$$P\Big\{\tau^{1/2} \ge 2\lambda,\ |w|^*_\tau \le \delta\lambda\Big\} \le \delta\sqrt{\frac{8}{3\pi}}\, P\{\tau^{1/2} \ge \lambda\}, \tag{7}$$

In particular, the second inequality in (1) *is true.*

Proof. For $\mu := \lambda^2$, the left-hand side of (7) is clearly at most

$$P\Big\{\tau \ge \lambda^2,\ |w|_\mu \le \delta\lambda,\ \max_{t \le 3\mu} |\theta_\mu w_t + w_\mu| \le \delta\lambda\Big\}$$
$$\le P\Big\{\tau \ge \lambda^2,\ \max_{t \le 3\mu} |\theta_\mu w_t| \le 2\delta\lambda\Big\}.$$

Since $\{\tau \ge \lambda^2\} \in \mathcal{F}^w_\mu$, it follows by the Markov property that the last probability is

$$P\{\tau \ge \lambda^2\} P\Big\{\max_{t \le 3\mu} |w_t| \le 2\delta\lambda\Big\}$$
$$\le P\{\tau \ge \lambda^2\} P\Big\{|w_{3\mu}| \le 2\delta\lambda\Big\} = P\{\tau \ge \lambda^2\} \cdot \sqrt{\frac{2}{\pi}} \int_0^{2\delta/\sqrt{3}} e^{-x^2/2}\, dx.$$

It remains to note that $\exp(-x^2/2) \le 1$. □

4. Remark. Using Theorem 3.2 rather than the strong Markov property, one could prove a rougher estimate than the first inequality of (1). Indeed, for $\varepsilon \in (0, 1/2)$ and any $r \ge 0$, $q > 1$, $p = q(q-1)^{-1}$,

$$\mathbf{E} \sup_{t \le \tau} |w_t|^r \le \mathbf{E} \sup_{t < \infty} |w_t|^r \big(t^{\frac12+\varepsilon} + t^{\frac12-\varepsilon}\big)^{-r} \big(\tau^{\frac12+\varepsilon} + \tau^{\frac12-\varepsilon}\big)^r$$
$$\le \Big(\mathbf{E} \sup_{t < \infty} |w_t|^{rp} \big(t^{\frac12+\varepsilon} + t^{\frac12-\varepsilon}\big)^{-rp}\Big)^{1/p} \Big(\mathbf{E}\big(\tau^{\frac12+\varepsilon} + \tau^{\frac12-\varepsilon}\big)^{rq}\Big)^{1/q},$$

and the first factor here is finite by Theorem 3.2.

5. EXERCISE. Using formulas (3.11), prove that for $r > 0$ the supremum of

$$\mathbf{E}\,|w_\tau|^r(\tau+1)^{-r/2}$$

over all finite (a.s.) Markov times τ is infinite.

We now proceed to the *Wald identities*, which in fact generalize Theorem 6.9 for $d = 1$.

6. THEOREM. *Let* $\mathbf{E}\,\tau^{1/2} < \infty$. *Then* $\mathbf{E}\,w_\tau = 0$. *If* σ *is also a Markov time and* $\mathbf{E}\,\tau \vee \sigma < \infty$, *then* $\mathbf{E}\,w_\sigma w_\tau = \mathbf{E}\,\sigma \wedge \tau$.

To prove this theorem, it suffices to apply Theorem 6.9 with $\tau \wedge t$, $\sigma \wedge t$ in place of τ, σ, then use the Dominated Convergence Theorem and the fact that for $\gamma = \tau \vee \sigma$ by (1) we have

$$\mathbf{E}\,\sup_{t\le\tau}|w_t| \le N\,\mathbf{E}\,\tau^{1/2}, \qquad \mathbf{E}\,\sup_{t\le\gamma}|w_t|^2 \le N\,\mathbf{E}\,\gamma.$$

7. EXERCISE. Prove that for the $\tau = \tau(x)$ introduced in Corollary 3.13

$$\mathbf{E}\,\tau^{1/2} = \infty, \qquad \mathbf{E}\,w_\tau = x \neq 0, \qquad \mathbf{E}\,\tau = \infty, \qquad \mathbf{E}\,w_\tau^2 = x^2 \neq \mathbf{E}\,\tau.$$

8. The Wiener process and the heat equation. Martingales

The Wiener process was originally a model of Brownian motion, which is caused by the thermal motion of molecules. The physical processes connected with heat propagation can also be described by the heat equation. It is therefore not surprising that there is a connection between the Wiener process and the heat equation. For example, it is proved in the theory of differential equations that under quite broad assumptions on the function g the right-hand side of formula (6.3) satisfies the following *heat equation* for $t > 0$, $x \in E_d$:

$$\frac{\partial u}{\partial t} = \frac{1}{2}\Delta u,$$

where Δ is the *Laplace operator*, defined as

$$\Delta u(x) = u_{x^1x^1}(x) + \cdots + u_{x^dx^d}(x). \tag{1}$$

To present a precise and convenient formulation of the result we need, consider a d-dimensional Wiener process $(w_t, \mathcal{F}_t)$ defined for $t \ge 0$ on some probability space. Fix $T \in (0,\infty)$ and a vector $b \in E_d$.

1. THEOREM. *Fix* $K, \varepsilon \in (0,\infty)$, $p \in [1,\infty)$ *and define the function* h *by formula* (6.10). *Assume that* $g(x)$ *is a continuous function on* E_d, $|g(x)|^p \le Kh(T,x)$ *for all* $x \in E_d$. *Set* $u(t,x) = \mathbf{E}\,g(x + w_{T-t})$ *for* $t \in [0,T]$, $x \in E_d$. *Then* u *is continuous in* $\bar{Q}$, *where* $Q = [0,T) \times E_d$, *infinitely differentiable in* Q, *such that*

$$\frac{\partial u}{\partial t} + \frac{1}{2}\Delta u = 0 \tag{2}$$

in Q, *and* $|u|^p \le Kh$ *in* $\bar{Q}$. *If* g *is also continuously differentiable and* $|g_x(x)|^p \le Kh(T,x)$ *on* E_d, *then* u *is differentiable with respect to* x *in* $\bar{Q}$, u_x *is continuous in* $\bar{Q}$, $u_x(t,x) = \mathbf{E}\,g_x(x + w_{T-t})$, *and* $|u_x|^p \le Kh$ *in* $\bar{Q}$.

By formula (6.3), this is a purely analytical result. For completeness we will prove it in Appendix B. In this section we will demonstrate the connection between the Wiener process and an equation more general than (2), concentrating mainly on assertions that are converse to Theorem 1. We well need the following lemma.

2. LEMMA. *Let $f(t,x)$ be a Borel function, defined for $t \geq 0$, $x \in E_d$. Then*
(a) $F(t,s,x,x.) := f(t+s, x+x_s)$, considered for $t,s \geq 0$, $x \in E_d$, $x. \in C$, is a measurable function of $(t,s,x,x.)$,
(b) it is a measurable function of $(x,x.)$ for any fixed $t,s \geq 0$,
(c) if $f \geq 0$, or if the integral

$$\int_0^r f(s+t, x+x_s)ds \tag{3}$$

is meaningful for any $t,r \geq 0$, $x \in E_d$, $x. \in C$, then this integral is a measurable function of $(r,t,x,x.)$.

PROOF. Let $Y = [0,\infty) \times [0,\infty) \times E_d \times C$, $\mathfrak{L} = \mathfrak{B}([0,\infty)) \otimes \mathfrak{B}([0,\infty)) \otimes \mathfrak{B}(E_d) \otimes \mathfrak{A}(C)$. It follows from the definition of product of σ-algebras that the functions $f_1(y) = t$, $f_2(y) = s$, $f_3(y) = x$, $f_4(y) = x.$, where $y = (t,s,x,x.)$, are $\mathfrak{L}$-measurable on Y. By Lemma I.1.7, the pair (f_2, f_4), for example, is $\mathfrak{L}$-measurable. In addition, we have already seen that $f_5(s,x.) := x_s$, which is a continuous function of s and a $\mathfrak{A}$-measurable function of $x.$, is measurable as a function on $[0,\infty) \times C$ (Lemma I.5.7). The composite function $f_5(f_2, f_4)$ is obviously $\mathfrak{L}$-measurable. In other words, x_s is a $\mathfrak{L}$-measurable function of $(t,s,x,x.)$. Therefore (see Corollary I.1.11), $t+s$, $x+x_s$, the pair $(t+s, x+x_s)$ and, finally, F, as a composition of measurable mappings, are $\mathfrak{L}$-measurable. We have proved (a). For $f \geq 0$ assertions (b), (c) can be proved in the same way as the parallel parts of Fubini's Theorem. To do so we need only rewrite the integral (3) as

$$\int_0^\infty I_{s\leq r} f(t+s, x+x_s)\,ds$$

and notice that $I_{s\leq r}$, $I_{s\leq r}F$ are also measurable as functions of $(r,t,s,x,x.)$. In the general case we apply, as usual, the formula $f = f_+ - f_-$. □

In view of (6.3), the following result, in the particular case $\tau = T - t$ is well known in the theory of differential equations.

3. LEMMA. *Let $S \in [0,T)$ and let $u(t,x)$ be a real-valued function, continuous in $[S,T] \times E_d$ and equal to zero for $|x| \geq K$, where K is a constant. Assume that the derivatives $\partial u/\partial t$, u_x, u_{xx} exist and are continuous and bounded in $(S,T) \times E_d$. Denote*

$$-f(t,x) = \frac{\partial u}{\partial t}(t,x) + \frac{1}{2}\Delta u(t,x) + u_{(b)}(x), \tag{4}$$

where the last summand is understood in the sense of (I.4.15). Then, for any $(t,x) \in [S,T] \times E_d$ and any Markov time $\tau \leq T - t$, for $\xi_s := (t+s, x+w_s+bs)$

$$u(t,x) = \mathbf{E}\Big[\int_0^\tau f(\xi_s)\,ds + u(\xi_\tau)\Big]. \tag{5}$$

PROOF. By considering the function $v(t,x) = u(t, x+bt)$ in place of u, we easily reduce the assertion of the theorem to the case $b = 0$, to which we henceforth confine ourselves. Denote by $w(t,x)$ the right-hand side of (5) for $\tau = T - t$, and first prove that $u = w$. The right-hand side of (5) is meaningful by Lemma 2, and it is continuous in x by the Dominated Convergence Theorem, so w is also continuous in x. The continuity of u, w and Theorem I.4.12 imply that to prove equality $u = w$ it suffices to show that the Fourier transforms of u and w with respect to x are equal. We introduce the following notation

$$\tilde{g}(\lambda) = \int_{E_d} e^{i(\lambda,x)} g(x)\,dx.$$

Clearly, if $\int |g|dx < \infty$, then $\tilde{g}$ is defined. Moreover, by Fubini's Theorem (which is applicable in view of the measurability properties established in Lemma 2), the boundedness of u and f, and the fact that each of them has compact support, we obtain

$$\int_{E_d} |w(t,x)|dx \le \int_0^{T-t} ds \int_{E_d} |f(t+s,x)|dx + \int_{E_d} |u(T,x)|dx < \infty.$$

Applying Fubini's Theorem once more (and Exercise 4.2), we obtain

$$\begin{aligned}\widetilde{w}(t,\lambda) &= \mathbf{E}\Big[\int_0^{T-t}\int_{E_d} f(t+s,x)e^{i(\lambda,x-w_s)}\,dxds + \tilde{u}(T,\lambda)e^{-i(\lambda,w_{T-t})}\Big] \\ &= \int_0^{T-t} \tilde{f}(t+s,\lambda)e^{-\frac{1}{2}|\lambda|^2 s}\,ds + \tilde{u}(T,\lambda)e^{-\frac{1}{2}|\lambda|^2(T-t)}.\end{aligned} \tag{6}$$

Now we find $\tilde{f}$ using (4) (with $b = 0$). Following the rules for differentiating an integral with respect to a parameter and integration by parts (for the Riemann integral of continuous functions with compact support), we obtain

$$\int_{E_d} \frac{\partial}{\partial t}u(t,x)e^{i(\lambda,x)}\,dx = \frac{\partial}{\partial t}\tilde{u}(t,\lambda),$$

$$\int_{E_d} u_{x^kx^k}(t,x)e^{i(\lambda,x)}\,dx = -i\lambda^k\int_{E_d} u_{x^k}(t,x)e^{i(\lambda,x)}\,dx = -|\lambda^k|^2\tilde{u}(t,\lambda),$$

$$\tilde{f}(t,\lambda) = -\frac{\partial}{\partial t}\tilde{u}(t,\lambda) + \frac{1}{2}|\lambda|^2\tilde{u}(t,\lambda).$$

Substituting the last equality into (6) and integrating by parts, we get $\widetilde{w} = \tilde{u}$, and therefore $w = u$ on $[S,T] \times E_d$.

Now define

$$F(t,x,x.) = \int_0^{T-t} f(t+s,x+x_s)\,ds + u(T,x+x_{T-t}).$$

We have proved the equality $u(t,x) = \mathbf{E}\, F(t,x,w.)$. This gives, in particular,

$$(7)\qquad \begin{aligned} u(t,x) &= \mathbf{E}\int_0^{\tau} f(\xi_s)\,ds + \mathbf{E}\Big[\int_{\tau}^{T-t} f(\xi_s)\,ds + u(\xi_{T-t})\Big] \\ &= \mathbf{E}\int_0^{\tau} f(\xi_s)\,ds + \mathbf{E}\, F(t+\tau, x+w_\tau, \theta_\tau w.). \end{aligned}$$

Note also that the mappings

$$(\omega, x.) \longrightarrow \tau(\omega), \qquad (\omega, x.) \longrightarrow w_{\tau(\omega)}(\omega), \qquad (\omega, x.) \longrightarrow x.,$$

are $\mathcal{F}_\tau \otimes \mathfrak{A}(C)$-measurable (see Exercises 4.16, 4.17). Consequently, Lemma I.1.7, implies that the mapping $(\omega, x.) \longrightarrow (\tau, w_\tau, x.)$ is $\mathcal{F}_\tau \otimes \mathfrak{A}(C)$-measurable, and an appeal to Lemma 2 shows that $F(t+\tau, \xi_\tau, x.)$ is $\mathcal{F}_\tau \otimes \mathfrak{A}(C)$-measurable. We may thus use the strong Markov property when evaluating the last term in (7) which, together with the equality $u(t,x) = \mathbf{E}\, F(t,x,w.)$, has been proved for *all* $(t,x) \in [S,T] \times E_d$, immediately brings us to (5). □

If we regard (4) as an *equation* for u and assume that $u(T,\cdot)$ is known, then formula (5) for $\tau = T-t$ enables us to calculate $u(t,x)$ and is therefore a probabilistic representation of the solution of the heat equation in the strip $[S,T] \times E_d$. A similar representation can be derived for other domains. To make our future arguments shorter we introduce several extremely useful notions.

4. DEFINITION. Let η_t be a real-valued continuous $\mathcal{F}_t$-adapted process, defined for $t \geq 0$, such that for any bounded Markov time τ (with respect to $\{\mathcal{F}_t\}$) the expectation $\mathbf{E}\,|\eta_\tau|$ is finite. Then we will say that η_t is a *martingale* (with respect to $\{\mathcal{F}_t\}$ or that $(\eta_t, \mathcal{F}_t)$ is a martingale) if $\mathbf{E}\,\eta_\tau = \mathbf{E}\,\eta_0$ for any bounded Markov time τ. We will say that η_t is a *submartingale* (*supermartingale*) with respect to $\{\mathcal{F}_t\}$, or that $(\eta_t, \mathcal{F}_t)$ is a submartingale (supermartingale), if, for any bounded Markov times τ, σ, such that $\tau \leq \sigma$, we have $\mathbf{E}\,\eta_\tau \leq \mathbf{E}\,\eta_\sigma$ ($\mathbf{E}\,\eta_\tau \geq \mathbf{E}\,\eta_\sigma$).

We met many examples of martingales in Sec. 6, also seeing there some of their uses. In terms of Definition 4, Lemma 3 simply states that the process

$$\int_0^{r\wedge(T-t)} f(\xi_s)\,ds + u(\xi_{r\wedge(T-t)})$$

is a martingale (in r). A process η_t is obviously a submartingale if and only if $(-\eta_t)$ is a supermartingale; it is a martingale if and only if it is simultaneously a supermartingale and a submartingale. Thus, properties of martingales and supermartingales may be derived from those of submartingales; we will bear this in mind constantly when referring to the following lemma.

5. LEMMA. *Let η_t, η_t^1, $\eta_t^2, \ldots$ be submartingales, τ, σ Markov times, $\tau \leq \sigma \leq T$, and ζ a nonnegative bounded $\mathcal{F}_\tau$-measurable variable. Then*

(a) *We have*

$$(8)\qquad \mathbf{E}\,\zeta\eta_\tau \leq \mathbf{E}\,\zeta\eta_\sigma, \qquad \mathbf{E}\,\sup_i \eta_\tau^i \leq \mathbf{E}\,\sup_i \eta_\sigma^i;$$

in particular, $(\eta_t)_+ = \max(\eta_t, 0)$ *is a submartingale, and if* η_t *is a martingale then* $|\eta_t| = \max(\eta_t, -\eta_t)$ *is a submartingale.*

(b) *If* $\psi(x)$ *is a nondecreasing convex function defined on* $(-\infty, \infty)$, *then* Jensen's inequality *is valid*:

$$\mathbf{E}\,\psi(\eta_\tau) \le \mathbf{E}\,\psi(\eta_\sigma). \tag{9}$$

Moreover, if η_t *is a martingale, this inequality is true under the sole assumption that* $\psi(x)$ *is a convex function.*

(c) *Let* c_t *be a real-valued process that is* $\mathcal{F}_t$*-adapted, continuous in* t *and bounded in* (ω, t) *on* $\Omega \times [0, S]$ *for any* $S \in (0, \infty)$. *Let* η_t *be a martingale. Then the process*

$$\varkappa_t := \int_0^t \eta_s c_s e^{-\varphi_s}\, ds + \eta_t e^{-\varphi_t},$$

where $\varphi_t := \int_0^t c_s\, ds$, *is also a martingale. Moreover,*

$$\lambda_t := -\int_0^t \eta_s c_s\, ds + \eta_t \int_0^t c_s\, ds$$

is a martingale.

PROOF. (a) Using the Dominated Convergence Theorem one easily reduces the proof of the first inequality in (8) to the case $\zeta = I_A$, where $A \in \mathcal{F}_\tau$. In that case set $\gamma(\omega) = \tau(\omega)$ for $\omega \in A$, $\gamma(\omega) = \sigma(\omega)$ for $\omega \notin A$. Since $A \in \mathcal{F}_\tau \subset \mathcal{F}_\sigma$, $\Omega \setminus A \in \mathcal{F}_\sigma$ and

$$\{\omega : \gamma \le t\} = \Big(A \bigcap \{\omega : \tau \le t\}\Big) \bigcup \Big((\Omega \setminus A) \bigcap \{\omega : \sigma \le t\}\Big) \in \mathcal{F}_t,$$

it follows that γ is a Markov time, $\gamma \le \sigma$. Therefore, $\mathbf{E}\,\eta_\gamma \le \mathbf{E}\,\eta_\sigma$, which is *equivalent* to the first inequality in (8).

Next, $\eta_t^1 \le \sup_i \eta_t^i = \lim_{n\to\infty} \max_{i\le n} \eta_t^i$. Hence, by the Monotone Convergence Theorem it suffices to prove the second inequality in (8) for finitely many η^i. In this case the events

$$A_i := \Big\{\omega : \max_{j \le i-1} \eta_\tau^j < \eta_\tau^i = \max_{j \le n} \eta_\tau^j\Big\} \qquad (\max_{j\le 0} := -\infty)$$

belong to $\mathcal{F}_\tau$ (see Exercise 4.17), are disjoint, and satisfy $\cup_i A_i = \Omega$. Consequently,

$$\mathbf{E}\max_{i\le n} \eta_\tau^i = \sum_{i\le n} \mathbf{E}\, I_{A_i} \eta_\tau^i \le \sum_{i\le n} \mathbf{E}\, I_{A_i} \eta_\sigma^i \le \sum_{i\le n} \mathbf{E}\, I_{A_i} \max_{j\le n} \eta_\sigma^j = \mathbf{E}\max_{j\le n} \eta_\sigma^j.$$

Part (b) follows immediately from (a) if one notes that $\psi(x) = \sup_i (a_i x + b_i)$, where $a_i x + b_i$ are straight lines supporting the graph of ψ at all points $(x, \psi(x))$ with rational x.

To prove (c), we first note that φ_s can be represented as a limit of integral sums, and it is seen that φ_s is $\mathcal{F}_s$-measurable. In addition, if we replace c_s in the definition of λ_t by $c_s \exp(-\varphi_s)$, we obtain a process related to $\varkappa_t$ in an obvious way. This shows that we need only consider λ_t.

Note also that by part (b) $\mathbf{E}\,|\eta_s| \le \mathbf{E}\,|\eta_S|$ if $s \le S$, whence $\mathbf{E}\,|\eta_s|$ is bounded on $[0, S]$. Hence we may apply Fubini's Theorem in the following calculations where we use part (a) as well:

$$\mathbf{E}\,|\lambda_\tau| \le \mathbf{E}\,|\eta_\tau| \int_0^\tau |c_s|\,ds + \int_0^T \mathbf{E}\,|\eta_s c_s|\,ds < \infty,$$

$$\mathbf{E}\,\lambda_\tau = \int_0^T \mathbf{E}\,(\eta_\tau - \eta_s) c_s I_{\tau > s}\,ds = \int_0^T \mathbf{E}\,c_s I_{\tau>s} (\eta_{\tau \vee s} - \eta_s)\,ds = 0 = \mathbf{E}\,\lambda_0. \quad \square$$

It may be useful to draw the reader's attention to a surprising result: If $\mathbf{E}\,\eta_\tau \le \mathbf{E}\,\eta_\sigma$ for all bounded Markov times $\tau \le \sigma$, then inequality (9) holds for a wide class of functions ψ.

6. THEOREM. *Fix $S \in [0, T)$ and let $G \subset (-\infty, T) \times E_d$ be a domain. Assume $Q := \big([S, T) \times E_d\big) \cap G \ne \emptyset$, let Γ be the boundary of the set $\{x\colon (T, x) \in \partial Q\}$, and assume that the d-dimensional Lebesgue measure of Γ is zero. Let $u(t, x)$ be a real-valued Borel function defined in $\bar{Q}$, continuous in $\bar{Q} \setminus \{(T, x)\colon x \in \Gamma\}$ and such that*

$$|u(t, x)| \le K(t + \varepsilon)^{-d/2} \exp \frac{|x - bt|^2}{2(t + \varepsilon)} =: h(t, x),$$

in Q, where $K, \varepsilon \in (0, \infty)$ are some constants.

Assume that $\partial u/\partial t$, u_x, u_{xx} exist and are continuous in Q. Let c be a continuous real-valued bounded function, defined in Q. Fix $(t, x) \in Q$ and denote

$$(10)\qquad \begin{gathered} \xi_s = \big(t + s, x + w_s + bs\big), \qquad c_s = c(\xi_s), \qquad \varphi_s = \int_0^s c_r\,dr, \\ \tau_Q = \inf\{s \ge 0\colon \xi_s \notin Q\} \quad (= \tau_G), \\ -g = \frac{\partial u}{\partial t} + \frac{1}{2}\Delta u + u_{(b)} - cu. \end{gathered}$$

Finally, let $|g| \le h$ in Q. Then, for any Markov time $\tau \le \tau_Q$,

$$(11)\qquad u(t, x) = \mathbf{E}\bigg[\int_0^\tau g(\xi_s) e^{-\varphi_s}\,ds + u(\xi_\tau) e^{-\varphi_\tau}\bigg].$$

PROOF. Take bounded domains $G(n)$ such that $\bar{G}(n) \subset G(n+1)$, $G = \cup_n G(n)$, and construct infinitely differentiable functions $\zeta_n(s, y)$ such that $\zeta_n = 1$ on $G(n)$, $\zeta_n = 0$ outside $G(n+1)$. Define $u_n(s, y) = u(s, y)\zeta_n(s, y)$ for $(s, y) \in Q$, $u_n(s, y) = 0$ for $(s, y) \notin Q$, $s \ge S$, and

$$Q(n) = \big([S, T - 1/n) \times E_d\big) \cap G(n).$$

Let us consider only n such that $(t,x) \in Q(n)$. Applying Lemma 3 to u_n and noting that $u_n(\xi_s) = u(\xi_s)$ for $s \le \tau_{Q(n)} =: \tau(n)$, we see that the process

$$\eta_s = \int_0^{s\wedge\tau(n)} f(\xi_r)\,dr + u\big(\xi_{s\wedge\tau(n)}\big)$$

is a martingale. By Lemma 5 (c) with $\zeta_n\big(\xi_{s\wedge\tau(n)}\big)c_{s\wedge\tau(n)}$ for c_s, the process $\varkappa_{s\wedge\tau(n)}$ is also a martingale. The latter can be rewritten, after integration by parts, and turns out to be equal to

$$\int_0^{s\wedge\tau(n)} (cu+f)(\xi_r)e^{-\varphi_r}\,dr + u\big(\xi_{s\wedge\tau(n)}\big)e^{-\varphi_{s\wedge\tau(n)}}.$$

Since $g = cu+f$, this actually means that (11) is valid upon substitution of $\tau\wedge\tau(n)$ for τ. Now let $n\to\infty$, noting that $\tau \le T-t$, $\tau(n)\to\tau_Q$, $\tau\wedge\tau(n)\to\tau\wedge\tau_Q=\tau$. In addition,

$$u\big(\xi_{\tau\wedge\tau(n)}\big) \longrightarrow u(\xi_\tau), \tag{12}$$

provided that $\xi_\tau \notin \{(T,x)\colon x\in\Gamma\}$. Since $\ell(\Gamma)=0$ and the distribution of w_{T-t} has a density, we have

$$\begin{aligned} P\Big\{\xi_\tau \in \{(T,x)\colon x\in\Gamma\}\Big\} &= P\Big\{\tau = T-t,\ x+w_{T-t}+b(T-t)\in\Gamma\Big\} \\ &\le P\Big\{x+w_{T-t}+b(T-t)\in\Gamma\Big\} = 0. \end{aligned}$$

Thus relation (12) is valid almost surely. Using Exercise 6.8 and Lemma 5 (c), we see that

$$\mathbf{E}\int_0^{T-t} |g(\xi_s)|\,ds \le \int_0^{T-t} \mathbf{E}\,h(\xi_s)\,ds = \int_0^{T-t} h(t,x)\,ds = h(t,x)(T-t) < \infty,$$

$$h(t,x) = \mathbf{E}\Bigg[\int_0^{\tau\wedge\tau(n)} ch(\xi_s)e^{-\varphi_s}\,ds + h\big(\xi_{\tau\wedge\tau(n)}\big)e^{-\varphi_{\tau\wedge\tau(n)}}\Bigg],$$

and the formula obtained from this by substituting τ for $\tau\wedge\tau(n)$ is also valid. Hence, by the Dominated Convergence Theorem and Fatou's Lemma $(u+h\ge 0)$

$$\begin{aligned} u(t,x)+h(t,x) &= \mathbf{E}\int_0^{\tau}(g+ch)(\xi_s)e^{-\varphi_s}\,ds + \lim_{n\to\infty}\mathbf{E}\,e^{-\varphi_{\tau\wedge\tau(n)}}(u+h)\big(\xi_{\tau\wedge\tau(n)}\big) \\ &\ge \mathbf{E}\Bigg[\int_0^{\tau} g(\xi_s)e^{-\varphi_s}\,ds + u(\xi_\tau)e^{-\varphi_\tau}\Bigg] + \mathbf{E}\Bigg[\int_0^{\tau} ch(\xi_s)e^{-\varphi_s}\,ds + h(\xi_\tau)e^{-\varphi_\tau}\Bigg]. \end{aligned}$$

Since the last expectation is precisely $h(t,x)$, we have obtained (11) with an inequality sign. The reverse inequality is obtained by replacing u with $(-u)$. □

7. REMARK. It follows from the arguments about (12) that the assumption $\ell(\Gamma) = 0$ is not necessary if u is continuous in $\bar{Q}$.

Formula (11) for $\tau = \tau_Q$, $g = 0$ is sometimes referred to as the *Feynman-Kac formula.* It is clear from (11) that $u(t,x)$ is uniquely determined by g and the values of u on the set of all points $(s,y) \in \partial Q$, for each of which there exists $\delta \in (0,s]$ and a continuous function x_t with values in E_d, defined on $[s-\delta, s]$, such that $(t, x_t) \in Q$ for $t \in [s-\delta, s]$, $x_s = y$. This set is called the *parabolic boundary of* Q. Thus (11) implies a uniqueness theorem for the solution of (10) as an equation in u with given values on the parabolic boundary of Q.

We end this section with yet another useful property of martingales.

8. LEMMA. *Let η_t, ζ_t be martingales such that for any bounded Markov times τ we have $\mathbf{E}\eta_\tau^2 < \infty$, $\mathbf{E}\zeta_\tau^2 < \infty$. Then for any bounded Markov times τ, σ*

$$\mathbf{E}\,\eta_\tau\zeta_\sigma = \mathbf{E}\,\eta_{\tau\wedge\sigma}\zeta_{\tau\wedge\sigma}. \tag{13}$$

PROOF. The proof essentially duplicates the corresponding part of the proof of Theorem 6.9, using Lemma 5 (a) instead of Theorem 6.4. Set $\underline{\gamma} = \tau \wedge \sigma$, $\eta_t^n = (n \wedge \eta_t) \vee (-n)$. By the Dominated Convergence Theorem and the Cauchy-Bunyakovskiĭ inequality, $\mathbf{E}\left(\eta_\tau - \eta_\tau^n\right)^2 \to 0$, $\mathbf{E}\left|(\eta_\tau - \eta_\tau^n)\zeta_\sigma\right| \to 0$. In addition, $\eta_{\tau\wedge\sigma}$, $I_{\underline{\gamma}<\sigma}$ are $\mathcal{F}_{\underline{\gamma}}$-measurable (see Exercises II.4.16, II.4.17). Hence, for $\overline{\gamma} = \tau \vee \sigma$ by Lemma 5 (a)

$$\mathbf{E}\,\eta_\tau I_{\tau<\sigma}\zeta_\sigma = \mathbf{E}\,\eta_{\underline{\gamma}} I_{\underline{\gamma}<\sigma}\zeta_{\overline{\gamma}} = \lim_{n\to\infty} \mathbf{E}\,\eta_{\underline{\gamma}}^n I_{\underline{\gamma}<\sigma}\zeta_{\overline{\gamma}} = \lim_{n\to\infty} \mathbf{E}\,\eta_{\underline{\gamma}}^n I_{\underline{\gamma}<\sigma}\zeta_{\underline{\gamma}} = \mathbf{E}\,\eta_{\underline{\gamma}}\zeta_{\underline{\gamma}} I_{\tau<\sigma}.$$

If we replace $I_{\tau<\sigma}$ by $I_{\tau=\sigma}$, the equality between the outermost terms becomes evident. If we replace $I_{\tau<\sigma}$ by $I_{\sigma<\tau}$, we get the corresponding equality in the same way. The three equalities, summed up, give (13). □

9. Some applications of Theorem 8.6

At the end of the previous section we mentioned the possibility of applying results like Theorem 8.6 to prove uniqueness theorems for solutions of differential equations. In the same field, it can also be used to prove *Liouville's Theorem.*

1. THEOREM. *Let u be a real-valued bounded function defined in $[0,\infty) \times E_d$ and having continuous derivatives $\partial u/\partial t$, u_x, u_{xx} such that $\partial u/\partial t + \frac{1}{2}\Delta u = 0$ in $(0,\infty) \times E_d$. Then u is a constant in $[0,\infty) \times E_d$.*

PROOF. Let w_t be a d-dimensional Wiener process. Define a Wiener process $\widetilde{w}_t$ such that $\widetilde{w}_t = t w_{1/t}$ for $t > 0$ (a.s.) (see Theorem 3.1) and set

$$\xi = \varliminf_{t\to\infty} u(t, w_t).$$

Since $\xi = \eta$ (a.s.) where

$$\eta = \varliminf_{s\downarrow 0} u\Big(\frac{1}{s}, \frac{1}{s}\widetilde{w}_s\Big),$$

and η is $\mathcal{F}_{0+}^{\widetilde{w}}$-measurable, it follows by Remark 4.13 that ξ and η equal some constant c (a.s.).

For $\varepsilon \in (-\infty, \infty)$, define

$$\tau_\varepsilon = \inf\Big\{t \geq 0 \colon u(t, w_t) \leq c + \varepsilon\Big\}.$$

Then τ_ε is a Markov time and $P\{\tau_\varepsilon < \infty\} = 1$ for $\varepsilon > 0$ by the foregoing. By Theorem 8.6 with $S = 0$, $G = (-\infty, T) \times E_d$ we obtain

(1) $$u(0,0) = \mathbf{E}\, u\big(T \wedge \tau_\varepsilon, w_{T\wedge\tau_\varepsilon}\big).$$

For $T \to \infty$, $\varepsilon > 0$ this implies by the Dominated Convergence Theorem that

$$u(0,0) = \mathbf{E}\, u\big(\tau_\varepsilon, w_{\tau_\varepsilon}\big) \leq c + \varepsilon.$$

Since $\varepsilon > 0$ is arbitrary, we obtain $u(0,0) \leq c$. Considering the lim sup of $u(t, w_t)$ as $t \to \infty$, we see that it is at most $u(0,0)$ (a.s.). This shows that $\lim_{t\to\infty} u(t, w_t)$ exists (a.s.), and $u(0,0) = c$. Next, it follows from (1) that for $\varepsilon < 0$

$$u(0,0) = \mathbf{E}\,(c + \varepsilon) I_{\tau_\varepsilon < T} + \mathbf{E}\, u(T, w_T) I_{\tau_\varepsilon \geq T}.$$

Letting $T \to \infty$, we obtain

$$u(0,0) = (c + \varepsilon)\, P\{\tau_\varepsilon < \infty\} + c\, P\{\tau_\varepsilon = \infty\} = u(0,0) + \varepsilon\, P\{\tau_\varepsilon < \infty\}.$$

Thus $\tau_\varepsilon = \infty$ (a.s.), and since $\varepsilon < 0$ is arbitrary, it follows that $u(t, w_t) \geq u(0,0)$ for all t (a.s.). Taking $(-u)$ in place of u, we obtain $u(0,0) - u(t, w_t)$ for all t (a.s.). Finally, for $t > 0$

$$0 = \mathbf{E}\,\big|u(0,0) - u(t, w_t)\big| = (2\pi t)^{-d/2} \int_{E_d} \big|u(0,0) - u(t,x)\big| e^{-\frac{1}{2t}|x|^2}\, dx,$$

and this implies that $u(t,x) = u(0,0)$ for almost all x. Since u is continuous in $[0,\infty) \times E_d$, we have $u \equiv u(0,0)$. □

The boundedness of u is essential, as can be demonstrated for $d = 1$ by the function $u(t,x) = \exp(x - t/2)$. Theorem 1 is also discussed in Exercise IV.4.4.

Theorem 8.6 can be used to study properties of the Wiener processes. Here are some examples.

2. THEOREM. *Let $\varepsilon > 0$, $T \in (0,\infty)$, and let w_t be a d-dimensional Wiener process. Then*

(2) $$P\Big\{\max_{t\leq T} |w_t + bt| \leq \varepsilon\Big\} \geq \exp\Big(-\frac{T\pi^2 d^2}{8\varepsilon^2} - \varepsilon|b| - \frac{1}{2}|b|^2 T\Big) > 0.$$

PROOF. Take $D = \big(-\frac{\varepsilon}{\nu}, \frac{\varepsilon}{\nu}\big)^d$, where $\nu = \sqrt{d}$, $G = (-\infty, T) \times D$, $Q = [0,T) \times D$,

$$u(t,x) = \Big(\prod_{i=1}^{d} \cos\frac{\pi\nu}{2\varepsilon} x^i\Big) \exp\Big(\frac{\pi^2 d^2}{8\varepsilon^2} t - (b,x) + \frac{1}{2}|b|^2 t\Big).$$

It is readily verified that

$$\frac{\partial u}{\partial t} + \frac{1}{2}\Delta u + u_{(b)} = 0$$

in Q, $u(t,x) = 0$ for $x \in \partial D$. Hence, by Theorem 8.6 for $(t,x) = (0,0)$

$$1 = u(0,0) = \mathbf{E}\,u(\xi_{\tau_Q}) = \mathbf{E}\,u(T, w_T + bT) I_{\tau_Q = T}. \tag{3}$$

In addition, on the set $\{\tau_Q = T\}$ we have $|w_T + bT| \le \varepsilon$,

$$u(T, w_T + bT) \le \exp\left(\frac{1}{8\varepsilon^2} T\pi^2 d^2 + \varepsilon|b| + \frac{1}{2}|b|^2 T\right).$$

Finally, it is obvious that $\{\tau_Q = T\} \subset \{\max_{t \le T} |w_t + bt| \le \varepsilon\}$. Combining this result with (3) we immediately obtain (2). □

3. Problem. Let $\varepsilon > 0$, $d = 1$. Use the Fourier method to solve the equation

$$\frac{\partial u}{\partial t} + \frac{1}{2} u_{xx} + b u_x = 0$$

in $(0,T) \times (-\varepsilon, \varepsilon)$ with boundary values $u(T,x) = 1$, $u(t, \pm\varepsilon) = 0$ for $t \in (0,T)$, and prove that if $\lambda_k = \pi(2k+1)/2\varepsilon$, then

$$P\Big\{\max_{t \le T} |w_t + bt| < \varepsilon\Big\} = \frac{e^{b\varepsilon} + e^{-b\varepsilon}}{\varepsilon} \sum_{k=0}^{\infty} \frac{(-1)^k \lambda_k}{b^2 + \lambda_k^2} e^{-(\lambda_k^2 + b^2)T/2}. \tag{4}$$

Deduce from this that for $b = 0$, $T \to \infty$ the left-hand side is equivalent to $4\pi^{-1} \exp(-\pi^2 T / 8\varepsilon^2)$.

4. Exercise. Let $d = 1$, $\varepsilon > 0$. Solve the equation $\frac{\partial u}{\partial t} + u_{(b)} + \frac{1}{2} u_{xx} = 0$ in $[0,T) \times (-\infty, \varepsilon)$ with boundary values $u(T,x) = 1$ on $(-\infty, \varepsilon]$, $u(t,\varepsilon) = 0$ on $[0,T)$ (the solution is easily constructed by using, for example, the function u of Exercise 6.14). Using this result and a formula similar to (3), find another proof of (6.14):

$$P\Big\{\max_{t \le T} (w_t + bt) < \varepsilon\Big\} = \frac{1}{\sqrt{2\pi T}} \int_{-\infty}^{\varepsilon} e^{by - \frac{1}{2} b^2 T} \left(e^{-\frac{1}{2T} y^2} - e^{-\frac{1}{2T}(2\varepsilon - y)^2}\right) dy \tag{5}$$

(see also Exercise IV.3.8 and Problem IV.3.10).

5. Problem. Theorem 2 shows that with a positive probability a multidimensional Wiener process on the time interval $[0,T]$ is located in the ε-tube about the straight line $(-b)t$. Using the Markov property, prove that the process w_t, $t \in [0,T]$, is located with a positive probability in the ε-tube about any fixed polygonal line, and, in general, about any fixed continuous path x_t, $t \in [0,T]$ that starts at the origin. By the zero-one law, deduce that for any $x_\cdot \in C$ with $x_0 = 0$

$$\varliminf_{n \to \infty} \max_{t \le T} \left|\sqrt{n}\, w_{t/n} - x_t\right| = 0 \quad \text{(a.s.)}.$$

For $d = 2$ this result says, literally speaking, that over any positive time interval $[0,T]$ a Wiener process writes out one's name infinitely often (written out continuously with initial point at the origin).

We will use Theorem 8.6 to prove one more result that will be helpful in Sec. III.2.

6. THEOREM. *Let $T, K, \varepsilon \in (0, \infty)$, let h be the function defined by (6.10), and $g(x)$ a function continuous on E_d, $|g(x)| \le Kh(T,x)$ on E_d. Let $(w_t, \mathcal{F}_t)$ be a d-dimensional Wiener process defined on some probability space for $t \ge 0$; set $u(t,x) = \mathbf{E}\, g(x + w_{T-t})$ for $t \in [0,T]$, $x \in E_d$. Then the process $u(t \wedge T, w_{t\wedge T})$ is a martingale. If in addition the first derivatives of g are continuous and $|g_x(x)| \le Kh(T,x)$ on E_d, then for any $\lambda \in E_d$ the process*

$$(\lambda, w_{t\wedge T})u(t \wedge T, w_{t\wedge T}) - \int_0^{t\wedge T} (\lambda, u_x(s, w_s))\, ds$$

is also a martingale. If in addition $|g(x)|^2$, $|g_x(x)|^2 \le Kh(T,x)$ on E_d, then the process

$$u^2(t \wedge T, w_{t\wedge T}) - \int_0^{t\wedge T} \left|u_x(s, w_s)\right|^2 ds$$

is also a martingale.

PROOF. The first part of the theorem is in fact a consequence not of Theorem 8.6 but of Lemma 6.1, which establishes the martingale property of the process $u(t \wedge T, w_{t\wedge T})$, and of Theorem 8.1, which guarantees its continuity. Moreover, by Theorem 8.1 in $[0,T) \times E_d$

$$\frac{\partial(u^2)}{\partial t} + \frac{1}{2}\Delta(u^2) = |u_x|^2, \qquad \frac{\partial v}{\partial t} + \frac{1}{2}\Delta v = (\lambda, u_x),$$

where $v = (\lambda, x)u$. Hence, by Theorem 8.6, to prove the remaining assertions it suffices to use the additional information given by Theorem 8.1 regarding u_x when g_x exists, and to apply the almost obvious inequality $|v| \le K|\lambda| \cdot |x| h(t,x) \le N\hat{h}(t,x)$, which holds in $[0,T] \times E_d$, where N is a constant and $\hat{h}$ is defined by substituting $\varepsilon/2$ for ε in (6.10). □

Finally, let us use Theorem 8.6 to derive Khinchin's law of the iterated logarithm. As already mentioned in Sec. 3, this result will not be used in the main body of this book.

7. THEOREM. *Let w_t be a one-dimensional Wiener process for $t \ge 0$ and $\alpha(t)$ a strictly positive continuously differentiable function defined for $t > 0$. Denote*

$$h(t,x) = \frac{1}{\sqrt{t}} e^{\frac{1}{2t}x^2}, \qquad \beta(t) = h\big(t, \alpha(t)\big), \qquad \gamma(t) = \big(\beta^{-1}(t)\big)'$$

and assume that $\beta^{-1}(t)$, $t\alpha^{-1}(t)$ increase for large t and

$$\frac{\alpha(t)}{\sqrt{t}} \longrightarrow \infty \tag{6}$$

as $t \to \infty$. Then

$$\begin{aligned} &\int_2^\infty \frac{\alpha(t)}{t}\gamma(t)\, dt < \infty \Longrightarrow \varlimsup_{t\to\infty} \frac{w_t}{\alpha(t)} \le 1 \quad (a.s.), \\ &\int_2^\infty \frac{\alpha(t)}{t}\gamma(t)\, dt = \infty \Longrightarrow \varlimsup_{t\to\infty} \frac{w_t}{\alpha(t)} \ge 1 \quad (a.s.). \end{aligned} \tag{7}$$

In addition, if we take $\alpha(t) = (1+\varepsilon)\sqrt{2t\ln\ln t}$ *for* $\varepsilon \in (0,1)$ *and large* t, *then the integral in* (7) *converges; if we take* $\alpha(t) = (1-\varepsilon)\sqrt{2t\ln\ln t}$, *the integral diverges and*

$$1+\varepsilon \geq \varlimsup_{t\to\infty} \frac{w_t}{\sqrt{2t\ln\ln t}} \geq 1-\varepsilon \quad (a.s.), \qquad \varlimsup_{t\to\infty} \frac{w_t}{\sqrt{2t\ln\ln t}} = 1 \quad (a.s.).$$

PROOF. The last parts of the theorem are based on fairly elementary calculations, which are left to the reader. Moreover, as in the proof of Liouville's Theorem, it is easy to see that the upper limit in (7) is a constant (a.s.). Note also that $(-w_t)$ is a Wiener process together with w_t, and their distributions on C coincide. In particular, the distributions of $F(w.)$ and $F(-w.)$ coincide, where $F(x.)$ is the $\mathfrak{A}(C)$-measurable function defined by

$$F(x.) = \varlimsup_{t\to\infty} \frac{x_t}{\alpha(t)} = \lim_{n\to\infty} \sup_{r\geq n} \frac{x_r}{\alpha(r)},$$

where $n = 1, 2, \dots$, and r runs through the set of rational numbers. Consequently, the quantities

$$\varlimsup_{t\to\infty} \frac{w_t}{\alpha(t)}, \qquad \varlimsup_{t\to\infty} \frac{-w_t}{\alpha(t)}, \qquad \varlimsup_{t\to\infty} \frac{|w_t|}{\alpha(t)} = \Big(\varlimsup_{t\to\infty} \frac{w_t}{\alpha(t)}\Big) \vee \varlimsup_{t\to\infty} \frac{-w_t}{\alpha(t)},$$

which are equal a.s. to constants, coincide a.s., and it remains only to prove that

$$\int_2^\infty \frac{\alpha(t)}{t}\gamma(t)\,dt < \infty \Longrightarrow \varlimsup_{t\to\infty} \frac{|w_t|}{\alpha(t)} \leq 1 \quad \text{(a.s.)}. \tag{8}$$

$$\int_2^\infty \frac{\alpha(t)}{t}\gamma(t)\,dt = \infty \Longrightarrow \varlimsup_{t\to\infty} \frac{|w_t|}{\alpha(t)} \geq 1 \quad \text{(a.s.)}. \tag{9}$$

The main role in the following arguments is played by formula (11), which can be proved in several ways (cf. Problem 13). The most natural proof is the following. Let $u(t,x)$ be an infinitely differentiable strictly positive function on $[0,\infty) \times E_d$ such that $u(t,x) = u(t,-x)$, define $\hat\beta(t) := u\big(t,\alpha(t)\big)$ and set

$$v(t,x) = u(t,x)/\hat\beta(t),$$

$$f(t,x) = \frac{\partial v}{\partial t}(t,x) + \frac{1}{2}\Delta v(t,x), \qquad \tau(s) = \inf\big\{t \geq s : |w_t| \geq \alpha(t)\big\}.$$

We claim that if $0 < s < T < \infty$, then

$$\begin{aligned} &P\{s < \tau(s) \leq T\} + \mathbf{E}\,v(T, w_T) I_{\tau(s)>T} = \frac{1}{\sqrt{2\pi s}} \int_{-\alpha(s)}^{\alpha(s)} v(s,x) e^{-\frac{1}{2s}x^2}\,dx \\ &+ \int_s^T \frac{1}{\sqrt{2\pi r}} \int_{-\alpha(r)}^{\alpha(r)} P\Big\{\sup_{[s,r]} \Big|w_p - \frac{p}{r}w_r + \frac{p}{r}x\Big|\alpha^{-1}(p) < 1\Big\} f(r,x) e^{-\frac{1}{2r}x^2}\,dx\,dr. \end{aligned} \tag{10}$$

Assuming that this assertion is true, let us derive the theorem from it. For a fixed $s > 0$, modify $h(t,x)$ for $t \leq s$ so that it can be taken instead of u. This is convenient, since for $v = \beta^{-1}h$ we have $f = \gamma h$, so that the exponential functions inside the integrals in (10) drop out and

$$\mathbf{E}\, v(T, w_T) I_{\tau(s)>T} \leq \mathbf{E}\,\beta^{-1} h(T, w_T) I_{|w_T|<\alpha(T)}$$
$$= \frac{1}{\sqrt{2\pi T}} \int_{-\alpha(T)}^{\alpha(T)} \beta^{-1}(T) h(T,x) e^{-\frac{1}{2T}x^2}\, dx = \frac{\alpha(T)}{T\beta(T)}\sqrt{\frac{2}{\pi}}.$$

This expression tends to zero as $T \to \infty$, by (6). Performing the substitution and letting $T \to \infty$, we obtain

$$P\{s < \tau(s) < \infty\} = \frac{\alpha(s)}{s\beta(s)}\sqrt{\frac{2}{\pi}}$$
$$(11) \qquad + \frac{1}{\sqrt{2\pi}} \int_s^\infty \frac{\alpha(r)}{r}\gamma(r) \int_{-1}^{1} P\Big\{ \sup_{[s,r]} \Big| w_p - \frac{p}{r} w_r + \frac{\alpha(r)}{r} x p \Big| \alpha^{-1}(p) < 1 \Big\}\, dx dr.$$

If the integral in (8) converges, then we see, substituting 1 for the last probability in (11), that the left-hand side of (11) vanishes as $s \to \infty$. In addition, for any $\varepsilon > 0$, $s > 0$ the set of all ω for which the upper limit in (8) exceeds $1 + \varepsilon$ is obviously contained in $\{s < \tau(s) < \infty\} \cup \{|w_s| \geq \alpha(s)\}$. Hence,

$$P\Big\{ \overline{\lim_{t\to\infty}} \frac{|w_t|}{\alpha(t)} \geq 1 + \varepsilon \Big\} \leq \lim_{s\to\infty} P\big\{|w_s| \geq \alpha(s)\big\} \leq \lim_{s\to\infty} P\Big\{|w_1| \geq \frac{\alpha(s)}{\sqrt{s}}\Big\} = 0,$$

and this shows that the first relation in (8) yields the second one.

To prove (9) we analyze (11) more carefully. Denote

$$A(s,r,x) = \{\omega : \sup_{[s,r]} \Big| w_p - \frac{p}{r} w_r + \frac{\alpha(r)}{r} x p \Big| \alpha^{-1}(p) < 1\}, \quad p(s,r,x) = P(A(s,r,x)).$$

Take $\varepsilon \in (0,1)$ and choose $s > 0$ so that $p\alpha^{-1}(p)$ increases as a function of p for $p \geq s$ and $\gamma(r) \geq 0$ for $r \geq s$. Then for $|x| \leq \varepsilon$, $p \in [s,r]$

$$\frac{\alpha(r)}{r}|x|\frac{p}{\alpha(p)} \leq |x| \leq \varepsilon, \qquad \frac{p}{r}|w_r|\alpha^{-1}(p) \leq |w_r|\alpha^{-1}(r).$$

It follows that for $|x| \leq \varepsilon$

$$(12) \qquad \Big\{|w_r|\alpha^{-1}(r) \leq \varepsilon\Big\} \bigcap \Big\{\sup_{p\geq s} |w_p|\alpha^{-1}(p) < 1 - 2\varepsilon\Big\} \subset A(s,r,x).$$

Therefore, the lower limit of $p(s,r,x)$ as $r \to \infty$ is larger than the lower limit of the probability of the left event in (12), and since

$$P\Big\{|w_r|\alpha^{-1}(r) \leq \varepsilon\Big\} = P\Big\{|w_1|\sqrt{r}\alpha^{-1}(r) \leq \varepsilon\Big\} \longrightarrow 1,$$

it follows that for $|x| \leq \varepsilon$

$$(13) \qquad \underline{\lim}_{r\to\infty}\, p(s,r,x) \geq P\Big\{\sup_{p\geq s} |w_p|\alpha^{-1}(p) < 1 - 2\varepsilon\Big\}.$$

By Fatou's Lemma the lower limit of the integral with respect to x in (11) is larger than the right-hand side in (13) multiplied by 2ε. But since the left-hand side of (11) is at most one, it follows from condition (9) that the probability in (13) vanishes, the upper limit in (9) is greater $1-2\varepsilon$ (a.s.), and as ε is arbitrary, it is greater than or equal to one (a.s.).

Thus, it remains only to prove (10). For any fixed T this expression involves values of u, v and their derivatives on some compact set only. Therefore, fixing T and multiplying u by a suitable cutoff function of x, we may assume that $u(t,x)=0$ for $|x| \geq K$, $t \in [0,T]$ (although $\hat{\beta} > 0$ for $t \in [0,T]$). That being the case, we see by Theorem 8.6 (or Lemma 8.3) that the process

$$\eta_t := v\big(t \wedge T, w_{t\wedge T}\big) - \int_0^{t\wedge T} f(r, w_r)\,dr$$

is a martingale. In addition, $\tau(s)$ is the first exit time of the process (t, w_t) from the domain $\big((-1,s)\times E_d\big) \cup \big\{(t,x)\colon t>0, |x|<\alpha(t)\big\}$. Consequently, $\tau(s)$ is a Markov time and by Lemma 8.5 (a)

$$\begin{aligned} \mathbf{E}\,\eta_{\tau(s)\wedge T} I_{\tau(s)>s} &= \mathbf{E}\,\eta_s I_{\tau(s)>s}, \\ \mathbf{E}\,v\big(\tau(s)\wedge T, w_{\tau(s)\wedge T}\big) I_{\tau(s)>s} &= \mathbf{E}\,v(s, w_s) I_{\tau(s)>s} \\ &\quad + \mathbf{E}\,I_{\tau(s)>s} \int_s^{\tau(s)\wedge T} f(r, w_r)\,dr. \end{aligned} \tag{14}$$

The left-hand side of the last equality coincides with that of (10), since $v(t,w_t)=1$ for $s<t=\tau(s)<\infty$, and the first term on the right coincides with the first summand on the right of (10), since $\{\tau(s)>s\} = \big\{|w_s|<\alpha(s)\big\}$. It remains to note that, by Fubini's Theorem, the last expression in (14) equals

$$\int_s^T \mathbf{E}\,I_{\tau(s)>r} f(r,w_r)\,dr = \int_s^T \mathbf{E}\,I_{\sup_{[s,r]} |w_p|\alpha^{-1}(p)<1} f(r,w_r)\,dr;$$

and after this use the equality $w_p = (w_p - pw_r/r) + pw_r/r$ and Exercise 5.9. □

8. PROBLEM. Slightly modifying the arguments following formula (11), prove that under the assumptions of Theorem 7

$$\int_2^\infty \frac{\alpha(t)}{t}\gamma(t)\,dt < \infty \Longrightarrow \varlimsup_{t\to\infty} \big(w_t - \alpha(t)\big) \leq 0 \quad \text{(a.s.)}. \tag{15}$$

9. PROBLEM. Prove that for any continuous function α the upper limit in (15) is either $+\infty$ (a.s.) or $-\infty$ (a.s.).

10. PROBLEM. Let $\alpha(t)$ be a continuous function such that $\alpha(t)>0$ for $t\geq 2$ and the upper limit in (15) is strictly positive. For example, $\alpha=(1-\varepsilon)\tilde{\alpha}$, where $\tilde{\alpha}(t)$ satisfies the assumptions of Theorem 7, and the integral in (15) corresponding to $\tilde{\alpha}$ diverges. Let u be any function which is continuous and bounded in the closure of the domain $Q=\big\{(t,x)\colon t>2,\ x<\alpha(t)\big\}$ and satisfies equation (8.2) in Q. Prove that u vanishes in Q if it vanishes on $\big\{(t,x)\colon t>2,\ x=\alpha(t)\big\}$. (It turns out that this assertion is false if α satisfies the assumptions of Theorem 7 and the integral in (15)

converges. A counterexample is given by the function $u(t,x) = P\{(t+s, x+w_s) \in Q$ for all $s > 0\}$.)

11. PROBLEM. Let w_t be a one-dimensional Wiener process defined on a probability space $(\Omega, \mathcal{F}, P)$ and $\xi_0 \sim \mathcal{N}(0,1)$ a random variable defined on the same space, such that ξ_0 and $\mathcal{F}^w_\infty$ are independent (cf. Exercise 2.7). Consider the *Langevin equation*

$$\xi_t = \xi_0 + \sqrt{2}w_t - \int_0^t \xi_s\, ds, \quad t \geq 0,$$

which is solvable for every ω; the resulting ξ_t is knows as the *Ornstein-Uhlenbeck process*. Prove that the vector $(\xi_{t(1)}, \dots, \xi_{t(n)})$ has a normal distribution for any $t(1), \dots, t(n) \in [0, \infty)$. Deduce, using Exercise 5.4, that the processes $\xi_\cdot$, $\eta_\cdot$ have the same distributions in C, where

$$\eta_t := e^{-t} w_{e^{2t}}, \quad t \geq 0.$$

Prove also that

$$\overline{\lim_{t\to\infty}} \frac{\xi_t}{\sqrt{2\ln t}} = 1 \quad \text{(a.s.)}.$$

12. PROBLEM. Note that it follows from (5) for $b \geq 0$ that the left-hand side is at most $\exp\left(\frac{1}{2}b^2 T + \varepsilon b\right)$ multiplied by a number that tends to zero as $T \to \infty$. Repeating the arguments of Exercise 6.7, deduce that formula (6.9) remains valid for $x > 0$, $b \geq 0$, $\nu > -b^2/2$.

13. PROBLEM. Derive the second equality in (14) for $v \equiv h\beta^{-1}$ from part (c) of Lemma 8.5 concerning the process λ_t and from Exercise 6.8.

This problem shows that we do not need to apply Theorem 8.6 if we do not want to explain why the equalities in (14) are natural for $v \equiv h\beta^{-1}$.

10. The Wiener process and the Laplace operator

In this section we will continue to study the relationship between the Wiener process and partial differential equations. While in Sec. 8 we were dealing with parabolic equations, we now proceed to simplest second-order elliptic equations with the Laplace operator, which acts on a smooth function u as in (8.1).

Assume, as in Sec. 8, that $(w_t, \mathcal{F}_t)$ is a d-dimensional Wiener process on some probability space $(\Omega, \mathcal{F}, P)$ and $b \in E_d$ a vector. Fix a bounded domain $D \subset E_d$.

1. THEOREM. *Let $u(x)$ be a given real-valued function, continuous in $\bar{D}$ and having continuous derivatives u_x, u_{xx} in D. Let $c \geq 0$ be a continuous bounded function defined in D. Fix $x \in D$ and denote $\xi_t = x + w_t + bt$, $c_t = c(\xi_t)$, $\varphi_t = \int_0^t c_s\, ds$, $\tau_D = \inf\{t \geq 0 \colon \xi_t \notin D\}$, and*

$$(1) \qquad -g = \frac{1}{2}\Delta u + u_{(b)} - cu.$$

Finally, assume that g is bounded in D. Then $\mathbf{E}\tau_D < \infty$ and for any Markov time $\tau \leq \tau_D$

$$(2) \qquad u(x) = \mathbf{E}\left[\int_0^\tau g(\xi_t)e^{-\varphi_t}\, dt + u(\xi_\tau)e^{-\varphi_\tau}\right].$$

PROOF. It is readily verified (cf. Remark 4.14) that if U is an orthogonal $(d \times d)$-matrix, then $\widetilde{w}_t := Uw_t$ is a d-dimensional Wiener process together with w_t. Choose U so that $Ub = \big(|b|, 0, \dots, 0\big)$. Since τ_D is the first exit time of $Ux + \widetilde{w}_t + Ubt$ from UD and UD is a bounded domain, it follows that UD has a bounded projection on the first coordinate axis, and τ_D does not exceed the first exit time $\tilde{\tau}$ of $\widetilde{w}_t^1 + |b|t$ from $(-R, R)$, where $R \in (0, \infty)$ is a constant. If $|b| = 0$, then (see Example 6.11) $\mathbf{E}\tau_D \le \mathbf{E}\tilde{\tau} = R^2 < \infty$. If $|b| \ne 0$, then for any $T \in (0, \infty)$

$$R \ge \mathbf{E}\Big[\widetilde{w}^1_{T\wedge\tilde{\tau}} + |b|(T \wedge \tilde{\tau})\Big] = \mathbf{E}\,\widetilde{w}^1_{T\wedge\tilde{\tau}} + |b|\mathbf{E}\,T \wedge \tilde{\tau} = |b|\mathbf{E}\,T \wedge \tilde{\tau}. \tag{3}$$

Letting $T \to \infty$, we again see that $\mathbf{E}\tilde{\tau} < \infty$. Thus $\mathbf{E}\tau_D < \infty$.

To prove (2) we note that if we add the equality $0 = \partial u/\partial t$ to (1) and take $S = t = 0$, $G = (-\infty, T) \times D$ in Theorem 8.6, we infer from that theorem and Remark 8.7 that formula (2) holds for $\tau \le T \wedge \tau_D$, that is, it remains valid upon substitution of $T \wedge \tau$ for τ. Since u, g are bounded, $\mathbf{E}\tau_D < \infty$, $\mathbf{E}\tau < \infty$, $\tau < \infty$ (a.s), it follows that after this substitution it suffices to let $T \to \infty$ and use the Dominated Convergence Theorem. □

This theorem, like Theorem 8.6, can be used to study properties of the Wiener process.

2. EXERCISE. Let $d = 1$, $b \ne 0$, $p, q \in (0, \infty)$, and let τ be the first exit time of $w_t + bt$ from $(-p, q)$. Solving the equation $\frac{1}{2}u_{xx} + bu_x = 0$ in $(-p, q)$ with boundary values $u(-p) = 0$, $u(q) = 1$, prove that

$$u(0) = P\{w_\tau + b\tau = q\} = \frac{e^{2bp} - 1}{e^{2bp} - e^{-2bq}}.$$

Prove also (using, for example, the equalities in (3)) that

$$b\mathbf{E}\,\tau = \frac{q\big(e^{2bp} - 1\big) + p\big(e^{-2bq} - 1\big)}{e^{2bp} - e^{-2bq}}. \tag{4}$$

Letting $p \to \infty$ and noticing that under these conditions $\tau \to \gamma$, where γ is the first exit time of $w_t + bt$ from $(-\infty, q)$, prove that $P\{\gamma < \infty\} = e^{2bq}$ if $b < 0$ and $P\{\gamma < \infty\} = 1$ if $b > 0$ (cf. Exercise 6.7).

3. EXERCISE. Drop the restriction on b in the previous exercise. By solving the equation $\frac{1}{2}u_{xx} + bu_x - vu = 0$ in $(-p, q)$ with boundary values $u(-p) = u(q) = 1$, show that for $v > 0$

$$\mathbf{E}\,e^{-v\tau} = \Big[e^{-bp}\sinh\mu q + e^{bq}\sinh\mu p\Big] \cdot \big[\sinh\mu(p+q)\big]^{-1}, \tag{5}$$

where $\mu = \sqrt{b^2 + 2v}$. Letting $p \to \infty$ deduce again (6.9).

It is natural to compare these results with (9.4), (9.5), since, for example, the left-hand side of (9.4) is simply $P\{\tau > T\}$ for $p = q = \varepsilon$ and

$$\mathbf{E}\,e^{-v\tau} = 1 - v\mathbf{E}\int_0^\infty I_{T<\tau}e^{-vT}\,dT = 1 - v\int_0^\infty P\{T < \tau\}e^{-vT}\,dT.$$

Differentiating (5) with respect to v for $v = 0$, $p = q = \varepsilon$, $b \ne 0$, prove that $\mathbf{E}\,\tau = \varepsilon b^{-1}\tanh b\varepsilon$, which agrees with (4) for $p = q = \varepsilon$.

We present one more example of application of Theorem 1.

4. THEOREM. *Let $|x_0| > \varepsilon > 0$ and let τ be the first hitting time of the ball $\bar{S}_\varepsilon(x_0)$ by the process w_t, where $S_\varepsilon(x_0) = \{x \in E_d\colon |x - x_0| < \varepsilon\}$. Then $P\{\tau < \infty\} = \varepsilon^{d-2}|x_0|^{2-d}$ for $d > 2$ and $P\{\tau < \infty\} = 1$ for $d \leq 2$.*

PROOF. For $R > |x_0|$ we let τ_R denote the first exit time of w_t from $S_R(x_0)$ and define

$$u_R(x) = \left(|x - x_0|^{2-d} - R^{2-d}\right)\left(\varepsilon^{2-d} - R^{2-d}\right)^{-1}, \quad d > 2,$$
$$u_R(x) = \left(\ln|x - x_0| - \ln R\right)\left(\ln\varepsilon - \ln R\right)^{-1}, \quad d = 2,$$
$$u_R(x) = \left(|x - x_0| - R\right)\left(\varepsilon - R\right)^{-1}, \quad d = 1.$$

It is readily verified that $\Delta u_R = 0$ in $S_R(x_0) \setminus S_\varepsilon(x_0)$, $u_R = 1$ on $\partial S_\varepsilon(x_0)$ and $u_R = 0$ on $\partial S_R(x_0)$. Hence, by Theorem 1 (cf. (6.13))

$$u_R(0) = \mathbf{E}\, u_R(w_{\tau\wedge\tau_R}) = 1 \cdot P\{\tau < \tau_R\} + 0 \cdot P\{\tau_R < \tau\}. \tag{6}$$

Since the events $\{\tau < \tau_R\}$ expand with increasing R and their union is $\{\tau < \infty\}$, we easily obtain the desired result by letting $R \to \infty$ in (6). □

This theorem shows that, with probability one, one- and two-dimensional Wiener processes reach ε-neighborhood of any point x_0 (note that in Sec. 3 we have already deduced this result in one-dimensional case from (3.11)). The strong Markov property implies that, having reached the ε-neighborhood of x_0, the process w_t will return to the ε-neighborhood of the origin with probability one. This property of Wiener processes is called *recurrence*. Thus, if $d \leq 2$, the Wiener process is recurrent. In contrary to the situation in one-dimensional case, however, two-dimensional Wiener process will reach a fixed point x_0 with probability zero, as follows immediately from (6) when $\varepsilon \downarrow 0$. This assertion is obviously true for $d \geq 3$ as well. It turns out that the Wiener process for $d \geq 3$ is not only nonrecurrent but tends to infinity as $t \to \infty$.

5. THEOREM. *If $d \geq 3$, then $|w_t| \to \infty$ as $t \to \infty$ (a.s.).*

PROOF. Letting $tw_{1/t} = \widetilde{w}_t$ and redefining $\widetilde{w}_t$ in such a way that $\widetilde{w}_t$ becomes a Wiener process (see Theorem 3.1), as in Theorem 9.1 one finds that for some constant c

$$\varliminf_{t\to\infty} |w_t| = c \quad \text{(a.s.)}. \tag{7}$$

It will obviously suffice to prove that $c = \infty$. Suppose that $c < \infty$. Let τ_R be the first exit time of w_t from $S_R = S_R(0)$. Since τ_R is finite (a.s.), it follows from (7) that

$$P\Big\{\inf_{t\geq 0} |\theta_{\tau_R} w_t + w_{\tau_R}| < c + 1\Big\} = 1. \tag{8}$$

By the strong Markov property, the left-hand side of (8) is equal to $\mathbf{E}F(w_{\tau_R})$, where for $|x| = R > c + 1$, by Theorem 4,

$$F(x) := P\Big\{\inf_{t\geq 0} |w_t + x| < c + 1\Big\} \leq (c+1)^{d-2} R^{2-d}.$$

Thus, for $R > c + 1$ the left-hand side of (8) does not exceed $(c+1)^{d-2}R^{2-d} < 1$, and this contradicts (8). □

It is proved in the theory of differential equations, for certain domains D, that if g, φ are continuously differentiable in $\bar{D}$, then there exists a continuous function u in $\bar{D}$ that is equal to φ on ∂D, has two continuous derivatives in D, and satisfies the equation $\Delta u = -2g$ in D. Besides, for $x \in D$

$$u(x) = \int_D g(y)G(x,y)\,dy + \int_{\partial D} \varphi(y)\,\pi(x,dy), \tag{9}$$

where $G(x,y)$ is the *Green's function*, $\pi(x,\Gamma)$ is the *harmonic* (with respect to x) *measure* (with respect to Γ), defined on Borel subsets of ∂D.

What is the probabilistic meaning of G and π? Take $b = 0$ in Theorem 1, that is, $\xi_t = x + w_t$. Then, in addition to (9), we get another representation of u:

$$u(x) = \mathbf{E}\int_0^{\tau_D} g(x + w_t)\,dt + \mathbf{E}\,\varphi(x + w_{\tau_D}).$$

For Borel sets $\Gamma \subset E_d$ and fixed $x \in D$, define

$$G(x,\Gamma) = \mathbf{E}\int_0^{\tau_D} I_\Gamma(x + w_t)\,dt \;\left(= \mathbf{E}\int_0^{\tau_D} I_{\Gamma\cap D}(x + w_t)\,dt \right).$$

The function $G(x,\Gamma)$, as a function of Γ, is called the *Green's measure* (for the point x, domain D and process $x + w_t$). It must be noted that for any Borel function $f(y)$, the function $\eta(\omega,t) := f(x + w_t)$ is measurable with respect to (ω,t) and the integral with respect to t from 0 to τ_D can be replaced by an integral from 0 to ∞ if we multiply the integrand by the indicator of the set $\{(\omega,t)\colon t < \tau_D\}$. The indicator in question is also measurable with respect to (ω,t), being $\mathcal{F}$-measurable with respect to ω and left continuous in t. It follows from this and Fubini's Theorem that G is well defined (this also follows from Lemma 8.2). It is clear that G is a measure with respect to Γ, and since $\mathbf{E}\,\tau_D < \infty$, it follows that G is a finite measure.

It is clear that for fixed ω the function $I_\Gamma(x + w_t(\omega))$ is the indicator of the set $\{t\colon x + w_t(\omega) \in \Gamma\}$. The Lebesgue integral of this function with respect to t from 0 to τ_D is (by definition) the time spent by the sample path $x + w_t(\omega)$ in the set Γ until its first exit from D occurs. Therefore $G(x,\Gamma)$ is the mean time spent in Γ until τ_D.

It turns out that for any bounded Borel function g

$$\mathbf{E}\int_0^{\tau_D} g(x + w_t)\,dt = \int_D g(y)\,G(x,dy).$$

Indeed, for $g = I_\Gamma$ this equality holds by definition; for a simple Borel function g it is obvious, and it remains to recall that any bounded Borel function can be uniformly approximated by simple functions.

Thus it follows from the aforesaid (for $\varphi \equiv 0$) that

$$\int_D g(y)G(x,y)\,dy = \int_D g(y)\,G(x,dy)$$

for all functions g that are continuously differentiable in $\bar{D}$. But any function g continuous in $\bar{D}$ can be uniformly approximated by such functions (for example, polynomials). Hence, by Theorem I.5.4,

$$G(x,y)\,dy = G(x,dy).$$

This clarifies the probabilistic meaning of $G(x,y)$.

6. EXERCISE. Prove that if $d = 1$, $R > 0$, $D = (-R,R)$, then $G(x,y) = \frac{1}{R}\min\big((R+x)(R-y),(R+y)(R-x)\big)$. Deduce from this that the mean time spent by a one-dimensional Wiener process in the interval $(-\varepsilon,\varepsilon)$ until the first exit from the interval $(-R,R)$ is $2\varepsilon R - \varepsilon^2$ for $R \geq \varepsilon > 0$.

Similarly, we can define *the exit distribution of* $x + w_t$ from D as

$$\tilde{\pi}(x,\Gamma) = P\{x + w_{\tau_D} \in \Gamma\}.$$

Now, taking $g \equiv 0$, we see that $\tilde{\pi} = \pi$, that is, the exit distribution is a harmonic measure.

Sometimes a probabilistic reasoning can be used to determine $G(x,y)$ and $\pi(x,\Gamma)$.

7. PROBLEM. Let $d \geq 2$, $D = \{x\colon x^d < 0\}$. For $x \in D$ and a fixed bounded Borel function φ on E_d define

$$u(x) = \mathbf{E}\,\varphi\big(x + w_{\tau(x)}\big),$$

where τ is the first exit time of $x + w_t$ from D, that is,

$$\tau(x) = \inf\{t \geq 0\colon x^d + w_t^d \geq 0\}.$$

For $y \in E_{d-1}$, define $\varphi(y) = \varphi(y^1,\dots,y^{d-1},0)$ and let $\tilde{x} = (x^1,\dots,x^{d-1})$. Using Corollary 3.13 and the fact that $\tau(x)$ is determined by $w_\cdot^d$ and that $(w_\cdot^1,\dots,w_\cdot^{d-1})$ is independent of $w_\cdot^d$, show that the following derivation is rigorous:

$$\begin{aligned} u(x) &= \mathbf{E}\Big[\mathbf{E}\,\varphi\big(x^1 + w_t^1,\dots,x^{d-1} + w_t^{d-1},0\big)\big|_{t=\tau(x)}\Big] \\ &= \mathbf{E}\,\big(2\pi\tau(x)\big)^{\frac{1-d}{2}} \int_{E_{d-1}} \varphi(y)\exp\Big(-\frac{1}{2\tau(x)}|\tilde{x}-y|^2\Big)dy \\ &= (2\pi)^{-\frac{d}{2}} \int_{E_{d-1}} |x^d|\varphi(y)\int_0^\infty t^{-\frac{d+2}{2}}\exp\Big(-\frac{1}{2t}\big(|\tilde{x}-y|^2 + |x^d|^2\big)\Big)dt\,dy \qquad (10) \\ &= \int_{E_{d-1}} \varphi(y)K(x,y)\,dy, \end{aligned}$$

where $K(x,y)$ is the *density of the* $(d-1)$*-dimensional Cauchy distribution* (or the *Poisson kernel*), defined by

$$K(x,y) = c|x^d|\big(|\tilde{x}-y|^2 + |x^d|^2\big)^{-\frac{d}{2}}, \qquad c = (2\pi)^{-\frac{d}{2}}\int_0^\infty t^{-\frac{d+2}{2}}e^{-\frac{1}{2t}}\,dt.$$

Prove also that the last expression in (10) is infinitely differentiable and satisfies the equation $\Delta u = 0$ in D. Show, moreover, that if $x^d \uparrow 0$ and φ is continuous at $\tilde{x}$, then $u(x) \to \varphi(\tilde{x})$.

8. Problem. In the notation of the previous problem, take $x = (0, \dots, 0, -t)$, $t \geq 0$, and set $\sigma(t) = \tau(x)$. Define the $(d-1)$-dimensional *Cauchy process* ξ_t by

$$\xi_t = \left(w^1_{\sigma(t)}, \dots, w^{d-1}_{\sigma(t)}\right).$$

Prove that ξ_t has independent increments, each of its coordinates is a one-dimensional Cauchy process, but $\xi^1_t, \dots, \xi^{d-1}_t$ are not independent. Explain how a knowledge of ξ^1_t imposes restrictions on ξ^2_t (assume, for example, that $\xi^1_t(\omega)$ makes a jump at time t_0).

9. Problem. For the process ξ_t of the previous problem prove that the characteristic function of the sum $\xi_1 + \xi_1$ $(= 2\xi_1)$ is the product of the characteristic functions of its terms. Explain why this does not contradict Theorem I.4.17.

10. Problem. Let u be a function with compact support, infinitely differentiable on E_d. Define $\sqrt{-\Delta} u$ as the inverse Fourier transform of $|\lambda|\tilde{u}(\lambda)$, where $\tilde{u}(\lambda)$ is the Fourier transform of u. Prove that if $\nu > 0$ and

$$\sqrt{-\Delta} u + \nu u = g,$$

then

$$u(x) = \mathbf{E} \int_0^\infty e^{-\nu t} g(x + \xi_t)\, dt$$

on E_d, where ξ_t is a d-dimensional Cauchy process. Compare this result with Theorem 1.

CHAPTER III

Itô's Stochastic Integral

There are several ways to define Itô's stochastic integral. Historically the first definition of the stochastic integral of a nonrandom function is due to Paley, Wiener and Zygmund [41] (1933). Its slight generalization, the stochastic integral of a nonrandom function with respect to a random orthogonal measure (Sec. 1), brings us to one of the possible constructions of Itô's stochastic integral (Sec. 2). The definitions of Secs. 1, 2 turn out to be rather convenient, in the sense that in Sec. 2 they quickly yield various results of the theory of stochastic integrals, such as Itô's formula (in a particular case) and Clark's representation of any functional of a Wiener process as an Itô's stochastic integral. These definitions are also convenient because of their generality. For example, they require almost no modification if one wants to treat stochastic integrals with respect to arbitrary (even discontinuous) martingales. Itô's classical definition (see Theorem 11.11), despite having certain advantages, does not possess this property. This may create considerable, and for that matter by no means justified, difficulties in the transition from stochastic integrals with respect to Wiener processes to stochastic integrals with respect to more general martingales.

In Sec. 2 we define Itô's stochastic integral over the halfline $(0, \infty)$. Then, in Sec. 3, we consider stochastic integrals with variable upper limit, which, as turns out in Sec. 4, can be defined under fewer assumptions on the integrand. All the subsequent material is closely connected with the notion of quadratic variation or quadratic characteristic of local continuous martingales. This will be defined in Sec. 5, first for stochastic integrals with respect to Wiener processes. It will then be used in Sec. 10 to introduce stochastic integrals with variable upper limit against multidimensional admissible continuous local martingales, in particular, against stochastic integrals with variable upper limit with respect to a Wiener process. This tool proves very useful later, in investigating various properties of diffusion processes. The definition adopted in Sec. 10 is very similar to the definition in Sec. 1 and is very close to that of Motoo and Watanabe [39] (1965). Thus, the reader will meet three constructions of the stochastic integral: a generalization of the method of Paley, Wiener and Zygmund, which gradually gives rise to stochastic integrals with variable upper limit; Itô's construction; and a more subtle construction of Motoo and Watanabe.

An essential part of this chapter is devoted to the derivation of "Itô's formula", which forms the basis for stochastic calculus. In Sec. 8 we present a technically simple derivation; however we feel that it explains badly the essence of the subject. Another derivation, presented in Secs. 9, 10, is better in this respect and closer to Itô's original method.

Although we will be repeatedly returning to the same questions, there will be no duplications. Each topic will be analyzed each time anew, in a different conceptual context and with more advanced aims.

A reader encountering the theory of stochastic integrals for the first time, does not need more than a general idea of the material in Secs. 9–11. The results of these sections are not necessary for an understanding of basic facts from the theory of Itô's stochastic equations (and Sec. 11 will not be used subsequently at all). Nevertheless, those sections are necessary for a more thorough understanding of subsequent chapters.

1. Integral with respect to a Random Orthogonal Measure

We begin with some results from the theory of $\mathcal{L}_p$ spaces, $p \in [1, \infty)$. Let X be a set, Π *some* family of subsets of X, $\mathfrak{A}$ a σ-algebra in X, and μ a measure on $(X, \mathfrak{A})$. Suppose that $\Pi \subset \mathfrak{A}$ and $\Pi_0 := \{\Delta \in \Pi : \mu(\Delta) < \infty\} \neq \emptyset$. Let $S(\Pi) = S(\Pi, \mu)$ denote the set of all *step functions*, that is, functions

$$\sum_{i=1}^{n} c_i I_{\Delta(i)}(x),$$

where $c_i \in (-\infty, \infty)$, $\Delta(i) \in \Pi_0$, $n < \infty$ is an integer. Let $\mathcal{L}_p(\Pi, \mu)$ denote the set of all $\mathfrak{A}^\mu$-measurable functions f on X with values in the extended line $[-\infty, \infty]$ for each of which there exists a sequence $f_n \in S(\Pi)$ such that

$$\int_X |f - f_n|^p \mu(dx) \longrightarrow 0 \tag{1}$$

as $n \to \infty$.

A sequence $f_n \in S(\Pi)$ satisfying (1) is called a *defining sequence* for f. Note that since $|t|^p$ is convex, we have $|a + b|^p \leq 2^{p-1}|a|^p + 2^{p-1}|b|^p$, $|f|^p \leq 2^{p-1}|f_n|^p + 2^{p-1}|f - f_n|^p$ and therefore, if $f \in \mathcal{L}_p(\Pi, \mu)$, then

$$\|f\|_p := \left(\int_X |f|^p \mu(dx) \right)^{1/p} < \infty. \tag{2}$$

The expression $\|f\|_p$ is called the *norm* of f. For $p = 2$ it is also useful to define the *scalar product* (f, g) of elements $f, g \in \mathcal{L}_2(\Pi, \mu)$:

$$(f, g) := \int_X fg\, \mu(dx). \tag{3}$$

This integral exists and is finite, since $|fg| \leq |f|^2 + |g|^2$. The expression $\|f - g\|_p$ defines a distance in $\mathcal{L}_p(\Pi, \mu)$ between elements $f, g \in \mathcal{L}_p(\Pi, \mu)$. It is "almost" a metric on $\mathcal{L}_p(\Pi, \mu)$, in the sense that, although the equality $\|f - g\|_p = 0$ implies that $f = g$ only almost everywhere with respect to μ, nevertheless $\|f - g\|_p = \|g - f\|_p$ and the *triangle inequality* holds:

$$\|f + g\|_p \leq \|f\|_p + \|g\|_p.$$

Indeed, by Hölder's inequality

$$\begin{aligned} \|f + g\|_p^p &\leq \int_X |f| \cdot |f + g|^{p-1} \mu(dx) + \int_X |g| \cdot |f + g|^{p-1} \mu(dx) \\ &\leq \|f\|_p \cdot \|f + g\|_p^{p-1} + \|g\|_p \cdot \|f + g\|_p^{p-1}. \end{aligned}$$

If $f_n, f \in \mathcal{L}_p(\Pi, \mu)$ and $\|f_n - f\|_p \to 0$ as $n \to \infty$, we will naturally say that f_n *converges to* f *in* $\mathcal{L}_p(\Pi, \mu)$. If $\|f_n - f_m\|_p \to 0$ as $n, m \to \infty$, we will call f_n a *Cauchy sequence in* $\mathcal{L}_p(\Pi, \mu)$. The following result is useful.

1. LEMMA. *If f_n is a Cauchy sequence in $\mathcal{L}_p(\Pi, \mu)$, then there exists a subsequence $f_{n(k)}$ such that $f_{n(k)}$ has a limit μ-a.e. as $k \to \infty$.*

PROOF. Take a sequence $n(k)$, $k = 1, 2, \ldots$, such that $n(k) \to \infty$ and $\|f_{n(k+1)} - f_{n(k)}\|_p \leq 2^{-k}$. Then, by Chebyshev's inequality,

$$\mu\Big(\big|f_{n(k+1)} - f_{n(k)}\big| > k^{-2}\Big) = \mu\Big(\big|f_{n(k+1)} - f_{n(k)}\big|^p > k^{-2p}\Big) \leq k^{2p}2^{-kp}.$$

Since the series $\sum k^{2p}2^{-kp}$ converges, it follows by the Borel-Cantelli Lemma that for μ–almost any x we have $|f_{n(k+1)}(x) - f_{n(k)}(x)| \leq k^{-2}$ for sufficiently large k. It remains only to observe that the series $\sum k^{-2}$ is also convergent and to apply Cauchy's criterion. □

The next lemma states that $\mathcal{L}_p(\Pi, \mu)$ is a pre-Banach space.

2. LEMMA. (a) *$\mathcal{L}_p(\Pi, \mu)$ is a linear space, that is, if $a, b \in (-\infty, \infty)$, $f, g \in \mathcal{L}_p(\Pi, \mu)$, then $af + bg \in \mathcal{L}_p(\Pi, \mu)$.*

(b) *$\mathcal{L}_p(\Pi, \mu)$ is a complete space, that is, for any Cauchy sequence $f_n \in \mathcal{L}_p(\Pi, \mu)$, there exists an $\mathfrak{A}$-measurable function f for which* (1) *is true; in addition, any $\mathfrak{A}^\mu$-measurable function f that satisfies* (1) *for some sequence $f_n \in \mathcal{L}_p(\Pi, \mu)$ is an element of $\mathcal{L}_p(\Pi, \mu)$.*

PROOF. Part (a) is almost obvious, since by the triangle inequality

$$\big\|af + bg - (af_n + bg_n)\big\|_p \leq |a| \cdot \big\|f - f_n\big\|_p + |b| \cdot \big\|g - g_n\big\|_p.$$

As to (b), without loss of generality we may assume that the f_n's are $\mathfrak{A}$-measurable. Take the sequence $f_{n(k)}$ of Lemma 1 and define $f(x)$ to be the limit of $f_{n(k)}(x)$ for those x for which the limit exists, and zero otherwise. Then (see Sec. 1.1) f is $\mathfrak{A}$-measurable and by Fatou's Lemma

$$0 = \lim_{r\to\infty} \sup_{m,n\geq r} \big\|f_n - f_m\big\|_p \geq \lim_{r\to\infty} \sup_{n\geq r} \varliminf_{k\to\infty} \big\|f_n - f_{n(k)}\big\|_p \geq \varlimsup_{n\to\infty} \big\|f_n - f\big\|_p.$$

It remains to deduce from (1), written for $f_n \in \mathcal{L}_p(\Pi, \mu)$, that $f \in \mathcal{L}_p(\Pi, \mu)$. To that end, it will obviously suffice to observe that $\|f - f_{nk}\|_p \leq \|f - f_n\|_p + \|f_n - f_{nk}\|_p$ and to select suitable f_{nk}. □

3. EXERCISE. Prove that if Π is a *σ-algebra*, then $\mathcal{L}_p(\Pi, \mu)$ is simply the set of all Π^μ-measurable functions f that satisfy (2), where Π^μ is the completion of Π with respect to $\mathfrak{A}^\mu, \mu$.

We now proceed to the main contents of this section. Let $(\Omega, \mathcal{F}, P)$ be a probability space and suppose that to every $\Delta \in \Pi_0$ there is assigned a random variable $\zeta(\Delta) = \zeta(\omega, \Delta)$.

4. DEFINITION. We will say that ζ is a *random orthogonal measure with structural measure* μ if (a) $\mathbf{E}|\zeta(\Delta)|^2 < \infty$ for any $\Delta \in \Pi_0$, (b) $\mathbf{E}\zeta(\Delta_1)\zeta(\Delta_2) = \mu(\Delta_1 \cap \Delta_2)$ for all $\Delta_1, \Delta_2 \in \Pi_0$.

We will always assume that ζ satisfies the assumptions of Definition 4, postponing the discussion of the definition itself until Remark 8. Note that by Exercise 3 we have $\zeta(\Delta) \in \mathcal{L}_2(\mathcal{F}, P)$ for any $\Delta \in \Pi_0$.

5. THEOREM. (a) *Let $f \in \mathcal{L}_2(\Pi, \mu)$. Then there exists a random variable α such that*

$$\|\alpha\|_2 = \|f\|_2, \qquad (\alpha, \zeta(\Delta)) = (f, I_\Delta)$$

for any $\Delta \in \Pi_0$, where the norm and the scalar product on the left are naturally taken in $\mathcal{L}_2(\mathcal{F}, P)$, and on the right in $\mathcal{L}_2(\Pi, \mu)$.

(b) *Let $f \in \mathcal{L}_2(\Pi, \mu)$, $\beta \in \mathcal{L}_2(\mathcal{F}, P)$ and*

$$(\beta, \zeta(\Delta)) = (f, I_\Delta)$$

for any $\Delta \in \Pi_0$. Then $\|\beta\|_2 \geq \|f\|_2$.

(c) *Under the assumptions of* (b) *let $g \in \mathcal{L}_2(\Pi, \mu)$, $\gamma \in \mathcal{L}_2(\mathcal{F}, P)$ and*

$$\|\beta\|_2 = \|f\|_2, \qquad (\gamma, \zeta(\Delta)) = (g, I_\Delta)$$

for any $\Delta \in \Pi_0$. Then $(\beta, \gamma) = (f, g)$. If also $\|\gamma\|_2 = \|g\|_2$, then

$$\|\beta - \gamma\|_2^2 = \|\beta\|_2^2 - 2(\beta, \gamma) + \|\gamma\|_2^2 = \|f - g\|_2^2, \tag{4}$$

and, in particular, if $f = g$ (μ-a.e.), then $\beta = \gamma$ (a.s.).

PROOF. (a) Let f_n be a defining sequence for f. For every n

$$f_n(x) = \sum_{i=1}^{m(n)} c_{in} I_{\Delta(i,n)}(x),$$

where $m(n) < \infty$, $c_{in} \in (-\infty, \infty)$, $\Delta(i, n) \in \Pi_0$. Define

$$\alpha_n = \sum_{i=1}^{m(n)} c_{in} \zeta(\Delta(i, n)).$$

It follows immediately from Definition 4 that $(\alpha_n, \zeta(\Delta)) = (f_n, I_\Delta)$ for any $\Delta \in \Pi_0$. Substituting here $\Delta = \Delta(i, n)$, multiplying by c_{in} and summing over i, we obtain $\|\alpha_n\|_2 = \|f_n\|_2$. In addition, the difference $f_n - f_k$ has the same form as f_n, and $\alpha_n - \alpha_k$ has the same form as α_n. Hence $\|\alpha_n - \alpha_k\|_2 = \|f_n - f_k\|_2$. Further, $\|f_n - f_k\|_2 \leq \|f_n - f\|_2 + \|f - f_k\|_2 \to 0$ as $n, k \to \infty$, and therefore α_n is a Cauchy sequence in $\mathcal{L}_2(\mathcal{F}, P)$. By Lemma 2 (b) there exists a random variable $\alpha \in \mathcal{L}_2(\mathcal{F}, P)$ such that $\|\alpha_n - \alpha\|_2 \to 0$. By the Cauchy-Bunyakovskiĭ inequality and the triangle inequality we obtain for $\Delta \in \Pi_0$

$$\begin{aligned} |(\alpha, \zeta(\Delta)) - (f, I_\Delta)| &\leq |(\alpha, \zeta(\Delta)) - (\alpha_n, \zeta(\Delta))| + |(f_n, I_\Delta) - (f, I_\Delta)| \\ &\leq \|\alpha - \alpha_n\|_2 \|\zeta(\Delta)\|_2 + \|f_n - f\|_2 \|I_\Delta\|_2, \end{aligned}$$

$$\big|\|\alpha\|_2 - \|f\|_2\big| \leq \big|\|\alpha\|_2 - \|\alpha_n\|_2\big| + \big|\|f_n\|_2 - \|f\|_2\big| \leq \|\alpha - \alpha_n\|_2 + \|f_n - f\|_2.$$

Since the right-hand sides of these inequalities vanish as $n \to \infty$, we have found the desired random variable α.

(b) It follows from the assumptions and from the structure of α_n that $(\alpha_n, \beta) = (f_n, f)$, where α_n, f_n are as in (a). Consequently (cf. (4)),

$$0 \le \|\beta - \alpha_n\|_2^2 = \|\beta\|_2^2 - 2(f_n, f) + \|f_n\|_2^2, \tag{5}$$

and since $(f_n, f) \to \|f\|_2^2$, $\|f_n\|_2^2 \to \|f\|_2^2$, we see from (5), letting $n \to \infty$, that $\|\beta\|_2 \ge \|f\|_2$.

(c) It follows from (b) that for any $\lambda \in (-\infty, \infty)$

$$\begin{aligned}\|\beta\|_2^2 + 2\lambda(f,g) + \lambda^2\|g\|_2^2 = \|f + \lambda g\|_2^2 &\le \|\beta + \lambda\gamma\|_2^2 \\ &= \|\beta\|_2^2 + 2\lambda(\beta,\gamma) + \lambda^2\|\gamma\|_2^2.\end{aligned}$$

Combining like terms in the last inequality, dividing by λ and letting $\lambda = 0$, we immediately obtain $(f, g) = (\beta, \gamma)$. Since all other assertions of (c) are obvious, the theorem is proved. □

In view of this theorem, the following definition is natural.

6. DEFINITION. Let $f \in \mathcal{L}_2(\Pi, \mu)$, let η be a random variable on $(\Omega, \mathcal{F})$, $\eta \in \mathcal{L}_2(\mathcal{F}, P)$, $\|\eta\|_2 = \|f\|_2$, and $(\eta, \zeta(\Delta)) = (f, I_\Delta)$ for any $\Delta \in \Pi_0$. Then we will call η a *stochastic integral of f with respect to $\zeta(dx)$* and write

$$\int_X f(x)\,\zeta(dx) = \eta. \tag{6}$$

The notation (6) is not entirely legitimate. Indeed, any function equal to η almost surely is obviously also a stochastic integral of f with respect to $\zeta(dx)$. Therefore, the left-hand side of (6), which involves only f, ζ, is not uniquely determined by f, ζ; as a function of ω it can be equated by (6) to many different functions of ω. We must therefore stipulate that, whenever the left-hand side of (6) is being considered, we will treat it as *a random variable* that satisfies the assumptions of Definition 6. This random variable may change from one appearance of $\int f\,\zeta(dx)$ to another. Such conventions are not uncommon in mathematics. For example, two functions denoted by the same symbol $o(t)$ may be absolutely different, but nevertheless this notation is extremely efficient.

We stress that, by Theorem 5, if $f \in \mathcal{L}_2(\Pi, \mu)$, there exists a stochastic integral of f with respect to $\zeta(dx)$, and two different stochastic integrals of f with respect to $\zeta(dx)$ can differ only on a set of probability zero.

Furthermore, it turns out to be convenient to regard the stochastic integral of f with respect to $\zeta(dx)$ as a result of application of a certain operator J to f. This operator should map $\mathcal{L}_2(\Pi, \mu)$ to $\mathcal{L}_2(\mathcal{F}, P)$ and map any function $f \in \mathcal{L}_2(\Pi, \mu)$ to one of its stochastic integrals with respect to $\zeta(dx)$, that is, to a random variable η satisfying the assumptions of Definition 6 (which exists by Theorem 5). We can then write (6) in the more compact form $Jf = \eta$ and, moreover, use the terminology commonly applied to operators. We will call J the *stochastic integration operator with respect to ζ*. Parts (c)–(e) of the following theorem shows, in particular, why the word "integral" in Definition 6 is natural.

7. THEOREM. *There exists an operator J on $\mathcal{L}_2(\Pi, \mu)$ that assigns to every function $f \in \mathcal{L}_2(\Pi, \mu)$ a unique random variable $Jf \in \mathcal{L}_2(\mathcal{F}, P)$ and possesses the following properties*:

(a) $\|Jf\|_2 = \|f\|_2$ (*that is, J is isometric*),

(b) $(Jf, \zeta(\Delta)) = (f, I_\Delta)$ *for all* $\Delta \in \Pi_0$,

(c) $JI_\Delta = \zeta(\Delta)$ (*a.s.*) *for any* $\Delta \in \Pi_0$,

(d) $J(af + bg) = aJf + bJg$ (*a.s.*) *for any* $a, b \in (-\infty, \infty)$, $f, g \in \mathcal{L}_2(\Pi, \mu)$ (*that is, J is linear*),

(e) $Jf = \int_X f(x)\zeta(dx)$ (*a.s.*) *for any* $f \in \mathcal{L}_2(\Pi, \mu)$.

In addition, if J_1 and J_2 are two operators on $\mathcal{L}_2(\Pi, \mu)$ with values in the set of random variables belonging to $\mathcal{L}_2(\mathcal{F}, P)$ and if J_1 and J_2 possess properties (a) *and* (b), *then $J_1 f = J_2 f$ (a.s.) for any $f \in \mathcal{L}_2(\Pi, \mu)$.*

PROOF. For any $f \in \mathcal{L}_2(\mathcal{F}, P)$, let Jf denote a fixed random variable α, which exists by Theorem 5 (a). Then parts (a), (b) are obvious, and (c) follows immediately from Definition 4 and the uniqueness part of Theorem 5 (c).

Under the assumptions of (d), obviously, $(aJf + bJg, \zeta(\Delta)) = (af + bg, I_\Delta)$ if $\Delta \in \Pi_0$, and $(Jf, Jg) = (f, g)$ by Theorem 5 (c). Hence (cf. (4)), $\|aJf + bJg\|_2 = \|af + bg\|_2^2$. Thus, (d) follows from the last assertion of Theorem 5. The latter obviously implies the last part of our theorem, and it remains only to note that part (e) is valid by Definition 6. □

8. REMARK. Returning to Definition 4, we note that if $\Delta_1, \Delta_2 \in \Pi_0$, $\Delta_1 \cap \Delta_2 = \emptyset$, then $(\zeta(\Delta_1), \zeta(\Delta_2)) = \mu(\Delta_1 \cap \Delta_2) = 0$. This is the reason for the word "orthogonal." The word "measure" is justified by the following: if $\Delta, \Delta_n \in \Pi_0$, $n = 1, 2, \dots$, $\Delta_n \cap \Delta_m = \emptyset$ for $n \neq m$, and $\Delta = \cup_n \Delta_n$, then

$$\Big(I_\Delta - \sum_{i=1}^n I_{\Delta_i}\Big)^2 = I_\Delta - \sum_{i=1}^n I_{\Delta_i} = I_{\Delta \setminus \cup_{i \le n} \Delta_i},$$

$$\mathbf{E}\Big|\zeta(\Delta) - \sum_{i=1}^n \zeta(\Delta_i)\Big|^2 = \mathbf{E}\Big|J\Big(I_\Delta - \sum_{i=1}^n I_{\Delta_i}\Big)\Big|^2 = \mu\Big(\Delta \setminus \bigcup_{i=1}^n \Delta_i\Big) \longrightarrow 0,$$

which means, by definition, that the series $\sum_i \zeta(\Delta_i)$ converges to $\zeta(\Delta)$ *in the sense* of $\mathcal{L}_2(\mathcal{F}, P)$.

The simplest but rather useless example of an orthogonal measure is obtained by taking $X = \Omega$, $\Pi = \mathfrak{A} = \mathcal{F}$, $\mu = P$, $\zeta(\Delta) = I_\Delta$. In that situation $\zeta(\omega, \Delta)$ is indeed a measure for any ω. The stochastic integral of f with respect to this random orthogonal measure is equal to f (a.s.). In the general case a random orthogonal measure is not a measure depending on the parameter ω.

9. PROBLEM. Let w_t be a one-dimensional Wiener process on $(\Omega, \mathcal{F}, P)$ for $t \in [0, 1] = X$, $a_n = 1/n$, $\Pi = \{(a_{n+1}, a_n]\colon n = 1, 2, \dots\}$, $\mathfrak{A} = \sigma(\Pi)$, $\zeta((a, b]) = w_b - w_a$ for $(a, b] \in \Pi$. Show that ζ is a random orthogonal measure whose structural measure is equal to Lebesgue measure. Using independence of the increments of w_t, prove also that

$$\mathbf{E}\exp\Big(-\sum_n |\zeta((a_{n+1}, a_n])|\Big) = 0 \tag{7}$$

and deduce that if $A \subset \Omega$ and we can define a function $\nu(\omega, \Gamma)$ for $\omega \in A$, $\Gamma \in \sigma(\Pi)$, which is countably-additive with respect to Γ and coincides with $\zeta(\omega, (a, b])$ if $\Gamma = (a, b] \in \Pi$, then A is a null set.

The last property of the integral with respect to ζ, which we will need further on, is contained in the following remark.

10. REMARK. If $\mathbf{E}\zeta(\Delta) = 0$ for any $\Delta \in \Pi_0$, then for any $f \in \mathcal{L}_2(\Pi, \mu)$

$$\mathbf{E} \int_X f \zeta(dx) = 0. \tag{8}$$

Indeed, equality (8) for $f \in S(\Pi)$ can be verified directly; for arbitrary $f \in \mathcal{L}_2(\Pi, \mu)$ it follows from the inequality

$$\left| \mathbf{E} \int_X f\zeta(dx) \right| = \left| \mathbf{E} \int_X (f - f_n)\zeta(dx) \right| \leq \left\| \int_X (f - f_n)\zeta(dx) \right\|_2 = \|f - f_n\|_2$$

which, in turn, follows from Theorem 7 (e), (d), (a) and the Cauchy-Bunyakovsky inequality for $f_n \in S(\Pi)$.

We now proceed to the question of when $\mathcal{L}_p(\Pi, \mu)$ and $\mathcal{L}_p(\mathfrak{A}, \mu)$ coincide, which is important in applications.

11. THEOREM. *Let* $\mathfrak{A}_1 = \sigma(\Pi)$. *Assume that* Π *is a* π*-system and that there exists a sequence* $\Delta(1), \Delta(2), \dots \in \Pi_0$ *such that* $\Delta(n) \subset \Delta(n+1)$, $X = \cup_n \Delta(n)$. *Then* $\mathcal{L}_p(\Pi, \mu) = \mathcal{L}_p(\mathfrak{A}_1, \mu)$.

PROOF. Let Σ denote the family of all subsets A of X such that

$$I_A I_{\Delta(n)} \in \mathcal{L}_p(\Pi, \mu)$$

for any n. By assumption, $\mu(\Delta(n)) < \infty$, and hence using the triangle inequality and Lemma 2, one readily sees that Σ is a λ-system. Since $\Sigma \supset \Pi$, because Π is a π-system, it follows that $\Sigma \supset \mathfrak{A}_1$. Consequently, it follows from the definition of $\mathcal{L}_p(\mathfrak{A}_1, \mu)$ and Lemma 2 that $I_{\Delta(n)} f \in \mathcal{L}_p(\Pi, \mu)$ for $f \in \mathcal{L}_p(\mathfrak{A}_1, \mu)$ for any n. Finally, a straightforward application of the Dominated Convergence Theorem shows that $\|I_{\Delta(n)} f - f\|_p \to 0$ as $n \to \infty$. Hence $f \in \mathcal{L}_p(\Pi, \mu)$ if $f \in \mathcal{L}_p(\mathfrak{A}_1, \mu)$ and $\mathcal{L}_p(\mathfrak{A}_1, \mu) \subset \mathcal{L}_p(\Pi, \mu)$. Since the reverse inclusion is obvious, the theorem is proved. □

The construction of a stochastic integral with respect to a random orthogonal measure is not specific to the probability theory. We have considered the case in which $\zeta(\Delta) \in \mathcal{L}_2(\mathcal{F}, P)$ is a *real-valued* function, where P is a *probability* measure. Our arguments could be repeated almost literally in the case of an arbitrary measure and a complex-valued function ζ. It would then turn out that the Fourier integral of $\mathcal{L}_2$ functions is a particular case of integrals with respect to random orthogonal measures. In this connection we offer the reader the following problem.

12. PROBLEM. Let Π be the set of all intervals $(a, b]$, where $a, b \in (-\infty, \infty)$, $a < b$. For $\Delta = (a, b] \in \Pi$, define a function $\zeta(\Delta) = \zeta(\omega, \Delta)$ on $(-\infty, \infty)$ by

$$\zeta(\Delta) = \frac{1}{i\omega}\left(e^{i\omega b} - e^{i\omega a}\right) = \int_\Delta e^{i\omega x}\, dx.$$

Define $\mathcal{L}_p = \mathcal{L}_p(\Pi, \ell) = \mathcal{L}_p(\mathfrak{B}(-\infty, \infty), \ell)$ as at the beginning of this section, but now considering complex-valued f, f_n. It is natural to define the scalar product (f, g) in $\mathcal{L}_2$ by a formula like (3), with $\bar{g}$ substituted for g. Using a change of variable, prove that the number $(\zeta(\Delta_1), \zeta(\Delta_2))$ equals its complex conjugate, that is, it is real, and that $\|\zeta(\Delta)\|_2^2 = c\,\ell(\Delta)$ for $\Delta_1, \Delta_2, \Delta \in \Pi$, where c is a constant independent of Δ. Observing that $\zeta(\Delta_1 \cup \Delta_2) = \zeta(\Delta_1) + \zeta(\Delta_2)$ if $\Delta_1, \Delta_2, \Delta_1 \cup \Delta_2 \in \Pi$, $\Delta_1 \cap \Delta_2 = \emptyset$, deduce that in this case $(\zeta(\Delta_1), \zeta(\Delta_2)) = 0$. Using the fact that $\Delta_1 = (\Delta_1 \setminus \Delta_2) \cup (\Delta_1 \cap \Delta_2)$ and adding an interval between Δ_1, Δ_2 if they do not intersect, prove that $(\zeta(\Delta_1), \zeta(\Delta_2)) = c\,\ell(\Delta_1 \cap \Delta_2)$ for any $\Delta_1, \Delta_2 \in \Pi$ and, consequently, that we can construct an integral with respect to ζ such that the *Parseval equality* holds for any $f \in \mathcal{L}_2$:

$$c\|f\|_2^2 = \left\| \int f\zeta(dx) \right\|_2^2.$$

Keeping in mind that for $f \in S(\Pi)$, we obviously have

$$\int f\zeta(dx) = \int_{-\infty}^{\infty} f(x) e^{i\omega x}\, dx \quad \text{(a.e.)},$$

use Lemma 1 to generalize this equality to all $f \in \mathcal{L}_2 \cap \mathcal{L}_1$. Putting $f = \exp(-x^2)$ and using Lemma I.4.10, prove that $c = 2\pi$. Finally, use Fubini's Theorem to prove that for $f \in \mathcal{L}_1$, $-\infty < a < b < \infty$,

$$\int_a^b \left(\int_{-\infty}^{\infty} \bar{f}(\omega) e^{i\omega x}\, d\omega \right) dx = \int_{-\infty}^{\infty} \frac{1}{i\omega} \left(e^{i\omega b} - e^{i\omega a}\right) \bar{f}(\omega)\, d\omega.$$

In other words, if $f \in \mathcal{L}_1 \cap \mathcal{L}_2$, then $(\zeta(\Delta), f) = c(I_\Delta, g)$, where

$$\bar{g}(x) = c^{-1} \int \bar{f}(\omega) \zeta(x, d\omega)$$

and (cf. Definition 6) this leads to the *inversion formula for the Fourier transformation*:

$$f(\omega) = \int g(x) \zeta(\omega, dx).$$

Generalize this formula from the case $f \in \mathcal{L}_1 \cap \mathcal{L}_2$ to all $f \in \mathcal{L}_2$.

2. Itô's stochastic integral with respect to a Wiener process

Let $(w_t, \mathcal{F}_t)$ be a one-dimensional Wiener process for $t \geq 0$ on a probability space $(\Omega, \mathcal{F}_t, P)$. Put $X = \Omega \times (0, \infty)$ and define a measure $\mu = P \times \ell$ on the σ-algebra $\mathfrak{A} := \mathcal{F} \otimes \mathfrak{B}((0, \infty))$, where ℓ is one-dimensional Lebesgue measure on $(0, \infty)$. As in Sec. 1 we can also define Π, ζ, and so on.

1. Definition. Let $\mathfrak{M}$ denote the set of all bounded Markov times with respect to $\{\mathcal{F}_t\}$. For $\tau \in \mathfrak{M}$ define

$$(1)\qquad \begin{aligned} &(0, \tau]\!] := \left\{(\omega, t) \in X : 0 < t \leq \tau(\omega)\right\}, \qquad \zeta((0, \tau]\!]) := w_\tau, \\ &\Pi := \left\{(0, \tau]\!] : \tau \in \mathfrak{M}\right\}, \qquad \mathcal{P} := \sigma(\Pi). \end{aligned}$$

Functions defined on X, that is, for $t > 0$, $\omega \in \Omega$, and measurable with respect to the σ-algebra $\mathcal{P}$ will be called *predictable functions*, and the elements of $\mathcal{P}$ will be called *predictable sets*.

The first notation in (1) will also be used for arbitrary Markov times τ. Note that $(0,\tau]\!] = \cup_n (0, \tau \wedge n]\!] \in \mathcal{P}$ for an arbitrary Markov time τ. It is convenient to treat $(0,\tau]\!]$ as a subgraph of the function $\tau(\omega)$. The indicator of this set, $I(\omega, t)$, is the simplest example of a predictable function. It is easy to see that it is left-continuous in t (it is in this sense that it is predictable), and, moreover, is a random variable for any t. Therefore, $I(\omega,t)$ is $\mathfrak{A}$-measurable (cf. Lemma I.5.7) and the set $(0,\tau]\!] = \{(\omega,t): I(\omega,t) = 1\}$ is an element of $\mathfrak{A}$. In particular, $\Pi \subset \mathfrak{A}$, and since $\mathfrak{A}$ is a σ-algebra, it follows that $\mathcal{P} \subset \mathfrak{A}$.

The following assertion yields a great many examples of predictable functions.

2. LEMMA. (a) *Let τ be a Markov time and g an $\mathcal{F}_\tau$-measurable random variable. Then $gI_{\tau<t}$ is a predictable function.*

(b) *Let $f(t) = f(\omega,t)$ be a real-valued function defined for $\omega \in \Omega$, $t \in [0,\infty)$, left-continuous in t on $(0,\infty)$ and $\mathcal{F}_t$-adapted for $t \in [0,\infty)$. Then it is predictable.*

PROOF. (a) Take a Borel set $B \subset E_1$ and define $\gamma(\omega) = \tau(\omega)$ on the set $A := \{\omega : g(\omega) \in B\} \in \mathcal{F}_\tau$. Outside A let $\gamma(\omega) = \infty$. We have already seen (in the proof of Lemma II.8.5) that γ is a Markov time. Hence $(0,\tau]\!], (0,\gamma]\!] \in \mathcal{P}$ and, as is readily seen,

$$\Big\{(\omega,t) \in X;\ g(\omega) I_{\tau(\omega)<t} \in B\Big\}$$
$$= \big(X \setminus (0,\gamma]\!]\big) \bigcup \Big((0,\tau]\!] \bigcap \big\{(\omega,t) \in X : 0 \in B\big\}\Big).$$

This proves part (a). Part (b) is a direct corollary of (a) and the fact that $f(\varkappa_n(t-)) \to f(t)$ on X, where $\varkappa_n(t) = 2^{-n}[2^n t]$ and for $t > 0$

$$f(\varkappa_n(t-)) = \sum_{k=0}^{\infty} f(k2^{-n})\big(I_{k2^{-n}<t} - I_{(k+1)2^{-n}<t}\big). \quad \square$$

This implies, for example, that $w_t(\omega)$ is predictable.

3. LEMMA. (a) $\Pi, \mathcal{P} \subset \mathfrak{A}$,

(b) ζ *is a random orthogonal measure with structural measure* μ,

(c) $\mathcal{L}_p(\Pi,\mu) = \mathcal{L}_p(\mathcal{P},\mu)$.

PROOF. Part (a) was proved before Lemma 2. To prove (b) it suffices to observe that, by Theorem II.6.9 and the definition of μ (cf. Fubini's Theorem I.5.5), for $\tau, \sigma \in \mathfrak{M}$ we have

$$\mathbf{E}\,\zeta\big((0,\tau]\!]\big)\zeta\big((0,\sigma]\!]\big) = \mathbf{E}\, w_\tau w_\sigma = \mathbf{E}\,\tau \wedge \sigma = \mu\big((0,\tau\wedge\sigma]\!]\big) = \mu\big((0,\tau]\!] \cap (0,\sigma]\!]\big).$$

Finally, part (c) follows by Theorem 1.11 (and Exercise II.4.8) from the fact that $X = \cup_{n=1}^{\infty} (0,n]\!]$. $\square$

By the results of the previous section, any function $f(\omega,t)$ in $\mathcal{L}_p(\mathcal{P},\mu)$ has a well-defined stochastic integral with respect to ζ. In the present situation, this integral is denoted by

(2)
$$\int_0^\infty f(t)\,dw_t$$

and called the *Itô stochastic integral of the function* f with respect to dw. This notation is quite logical, since, by Theorem 1.7, we have

(3)
$$\int_0^\infty I_{t\le\tau}\,dw_t = \zeta\left((0,\tau]\!]\right) = w_\tau \quad \text{(a.s.)}$$

for $\tau \in \mathfrak{M}$.

Note that by Remark 1.10 and Theorem II.6.9

(4)
$$\mathbf{E}\int_0^\infty f(t)\,dw_t = 0$$

for any $f \in \mathcal{L}_2(\mathcal{P},\mu)$.

One more property which shows that the notation (2) is natural is established in the following lemma, which allows us to factor "constants" out of stochastic integrals.

4. LEMMA. *Let $f \in \mathcal{L}_2(\mathcal{P},\mu)$; let τ be a Markov time and g a bounded $\mathcal{F}_\tau$-measurable variable. Then*

(5)
$$g\int_0^\infty I_{\tau<t} f(t)\,dw_t = \int_0^\infty g I_{\tau<t} f(t)\,dw_t \quad (a.s.).$$

PROOF. The existence of the stochastic integrals in (5) follows from Lemma 2 and the properties of measurable functions. Furthermore, if $g = I_A$, $f = I_{(0,\sigma]\!]}$, where $\sigma \in \mathfrak{M}$, and we put $\gamma(\omega) = \tau(\omega)$ if $\omega \in A$, $\gamma(\omega) = \infty$ if $\omega \notin A$, then

$$g I_{\tau<t} f(t) = I_{t\le\sigma} - I_{t\le\sigma\wedge\gamma},$$

$$\int_0^\infty g I_{\tau<t} f(t)\,dw_t = w_\sigma - w_{\sigma\wedge\gamma} = I_A(w_\sigma - w_{\sigma\wedge\tau}) \quad \text{(a.s.)},$$

$$g\int_0^\infty I_{\tau<t} f(t)\,dw_t = g\int_0^\infty \left(I_{t\le\sigma} - I_{t\le\sigma\wedge\tau}\right)dw_t = g(w_\sigma - w_{\sigma\wedge\tau}) \quad \text{(a.s.)}.$$

This gives (5) in this special case. If, as before, $g = I_A$, but $f \in S(\Pi)$, then (5) follows from the linearity of the stochastic integration operator. If $g = I_A$, $f \in \mathcal{L}_2(\mathcal{P},\mu)$, equality (5) follows from the fact that the stochastic integration operator is isometric, since, for example,

(6)
$$\int_X \left|g I_{\tau<t} f(t) - g I_{\tau<t} f_n(t)\right|^2 \mu(d\omega dt) \le \int_X \left|f(t) - f_n(t)\right|^2 \mu(d\omega dt).$$

Now fix $f \in \mathcal{L}_2(\mathcal{P}, \mu)$. Since (5) is true for the indicators of sets in $\mathcal{F}_\tau$, it is also true for $g \in S(\mathcal{F}_\tau)$. But any bounded $\mathcal{F}_\tau$-measurable g can be approximated uniformly by such functions, and it remains only to apply estimates similar to (6). □

5. COROLLARY. *Let $\tau_1, \dots, \tau_n \in \mathfrak{M}$, $0 = \tau_0 \le \tau_1 \le \dots \le \tau_n$, and let g_i be $\mathcal{F}_{\tau_i}$-measurable bounded variables, $i = 0, \dots, n-1$. Let $f(t) = g_i$ for $t \in (\tau_i, \tau_{i+1}]$, $i = 0, \dots, n-1$, so that $g_i = f(\tau_i+)$, and let $f(t) = 0$ for $t > \tau_n$. Then $f \in \mathcal{L}_2(\mathcal{P}, \mu)$ and*

$$\int_0^\infty f(t)\, dw_t = \sum_{i=0}^{n-1} g_i(w_{\tau_{i+1}} - w_{\tau_i}) = \sum_{i=0}^{n-1} f(\tau_i+)(w_{\tau_{i+1}} - w_{\tau_i}) \quad (a.s.).$$

This follows immediately from the equalities

$$f(t) = \sum_{i=0}^{n-1} g_i I_{\tau_i < t \le \tau_{i+1}},$$

$$g_i I_{\tau_i < t \le \tau_{i+1}} = g_i I_{\tau_i < t} I_{t \le \tau_{i+1}}, \qquad I_{\tau_i < t \le \tau_{i+1}} = I_{t \le \tau_{i+1}} - I_{t \le \tau_i},$$

$$\int_0^\infty g_i I_{\tau_i < t \le \tau_{i+1}}\, dw_t = g_i \int_0^\infty I_{\tau_i < t \le \tau_{i+1}}\, dw_t \quad \text{(a.s.)}.$$

6. PROBLEM. Deduce from Corollary 5, the isometric property of the stochastic integration operator and the Quadratic Variation Theorem II.3.6 that, for any $T \subset (0, \infty)$,

$$w_T^2 = T + 2 \int_0^\infty I_{t \le T} w_t\, dw_t. \tag{7}$$

Formula (7) is a very special case of Itô's formula, which we will discuss in Sec. 8. Further on in this section we will consider some other special cases of Itô's formula and use them to establish Clark's Theorem. This material (except for Remark 7) will not be used later.

7. REMARK. If $f \in \mathcal{L}_2(\mathcal{P}, \mu)$ then, by definition, the stochastic integral (2) is a random variable ξ such that, for any $\tau \in \mathfrak{M}$ we have

$$\mathbf{E}\,\xi^2 = \int_X f^2 \mu(d\omega dt), \qquad \mathbf{E}\, w_\tau \xi = \int_X I_{(0,\tau]\!]} f\, \mu(d\omega dt). \tag{8}$$

We will use this observation to prove the following lemma.

8. LEMMA. *Let $g(x)$ be a real-valued, continuous, continuously differentiable function defined on $(-\infty, \infty)$; $T, K, \varepsilon \in (0, \infty)$ constants,*

$$h(t,x) := K(t+\varepsilon)^{-\frac{1}{2}} e^{\frac{1}{2(t+\varepsilon)} x^2}, \qquad |g(x)|^2, |g'(x)|^2 \le h(T, x)$$

for all x. Set $u(t,x) = \mathbf{E}\, g(x + w_{T-t})$ for $t \in [0, T]$, $x \in (-\infty, \infty)$. Then for any $s \in [0, T]$

$$g(w_T) = u(s, w_s) + \int_0^\infty I_{s < t \le T} u_x(t, w_t)\, dw_t \quad (a.s.). \tag{9}$$

In particular ($s = 0$),

(10) $$g(w_T) = \mathbf{E}g(w_T) + \int_0^\infty I_{t\le T} u_x(t, w_t)\,dw_t \quad (a.s.).$$

PROOF. By Theorem II.8.1 u is continuously differentiable with respect to x in $[0, T] \times E_1$ and $u_x(t, x) = \mathbf{E}g'(x + w_{T-t})$. Hence, by Exercise II.6.8 and Fubini's Theorem

$$|u_x(t,x)|^2 \le \mathbf{E}\left(g'(x + w_{T-t})\right)^2 \le \mathbf{E}h(T, x + w_{T-t}) = h(t, x),$$

$$\int_X I_{t\le T}|u_x(t, w_t)|^2\,\mu(d\omega dt) = \int_0^T \mathbf{E}\,|u_x(t, w_t)|^2\,dt \le h(0,0)T.$$

Thus the stochastic integrals in (9) and (10) exist. Applying Remark 7, Theorem II.9.6 and Fubini's Theorem, we get $\mathbf{E}g(w_T) = u(0,0)$, $g(w_T) = u(T, w_T)$ and

$$\begin{aligned}\mathbf{E}\left(g(w_T) - u(0,0)\right)^2 &= \mathbf{E}u^2(T, w_T) - 2u(0,0)\mathbf{E}g(w_T) + u^2(0,0)\\ &= \mathbf{E}u^2(T, w_T) - u^2(0,0) = \int_X I_{t\le T}|u_x(t, w_t)|^2\,\mu(d\omega dt).\end{aligned}$$

In addition, by Theorem II.9.6 and Lemma II.8.8, for any $\tau \in \mathfrak{M}$,

$$\begin{aligned}\mathbf{E}\,w_\tau(g(w_T) - u(0,0)) &= \mathbf{E}\,w_\tau u(T, w_T) = \mathbf{E}\,w_{\tau\wedge T}u(\tau\wedge T, w_{\tau\wedge T})\\ &= \mathbf{E}\int_0^{\tau\wedge T} u_x(t, w_t)\,dt = \int_X I_{t\le\tau}I_{t\le T}u_x(t, w_t)\,\mu(d\omega dt).\end{aligned}$$

This proves (10). Replacing $T, g(x)$ in this equality by $s, u(s, x)$, respectively, and using the fact that, by the strong Markov property (or by Lemma II.6.1), $\mathbf{E}u(s, x + w_{s-t}) = u(t, x)$ for $t < s$, we obtain an expression for $u(s, w_s)$ as the sum of $\mathbf{E}u(s, w_s) = u(0,0) = \mathbf{E}g(w_T)$ and a stochastic integral. It remains only to subtract this representation from (10). □

9. EXERCISE. Deduce from Lemma 8 that

(11) $$e^{aw_T - \frac{1}{2}a^2T} = 1 + a\int_0^\infty I_{t\le T}e^{aw_t - \frac{1}{2}a^2t}\,dw_t \quad \text{(a.s.)},$$

where a is a constant. Using Lemma 8, prove formula (7) as well.

10. EXERCISE. Prove that if $f(t)$ is *independent* of ω one-time continuously differentiable on $[0, \infty)$, and $T \in [0, \infty)$ is a constant, then $fI_{(0,T]\!]} \in \mathcal{L}_2(\mathcal{P}, \mu)$ and the formula of integration by parts is valid:

(12) $$\int_0^\infty f(t)I_{t\le T}\,dw_t = w_Tf(T) - \int_0^T w_tf'(t)\,dt \quad \text{(a.s.)}.$$

11. EXERCISE. Define the stochastic integral of a complex-valued function in the natural way; using Remark 7 and formula (11), which is convenient for the calculation of $\mathbf{E}\, w_\tau \exp iaw_T$, prove that for any $a \in (-\infty, \infty)$

$$e^{iaw_T} - 1 + \frac{1}{2}a^2 \int_0^T e^{iaw_t}\, dt = ia \int_0^\infty I_{t \le T} e^{iaw_t}\, dw_t \quad \text{(a.s.)}. \tag{13}$$

Substituting $a = 2\pi n/h$, where $n = 0, \pm 1, \dots$, $h \in (-\infty, \infty)$, multiplying by suitable coefficients c_n, summing over n, and using the linearity and the isometric property of the stochastic integral, prove that if $u(x)$ is a sufficiently smooth periodic function with period h, then the following special case of *Itô's formula* holds:

$$u(w_T) = u(0) + \int_0^\infty I_{t \le T} u_x(w_t)\, dw_t + \frac{1}{2}\int_0^T u_{xx}(w_t)\, dt \quad \text{(a.s.)}. \tag{14}$$

To end this section, we will prove a remarkable result which, however, will not be used again. It describes the set of values of the stochastic integration operator in one particular case.

12. THEOREM (Clark [6]). *Let $\mathcal{F} = \sigma(w_\cdot)$. Then for any $\xi \in \mathcal{L}_2(\mathcal{F}, P)$ there exists a function $f \in \mathcal{L}_2(\mathcal{P}, \mu)$ such that*

$$\xi = \mathbf{E}\xi + \int_0^\infty f(t)\, dw_t \quad (a.s.). \tag{15}$$

Moreover, any two functions of class $\mathcal{L}_2(\mathcal{P}, \mu)$ that satisfy (15) *coincide μ-a.e.*

PROOF. The last assertion follows obviously from the isometric property of the stochastic integration operator. To prove the first one, note that if $g_1(x), \dots, g_n(x)$ are continuously differentiable functions, bounded together with their first derivatives, $0 \le s_1 \le \cdots \le s_n < \infty$, and if $v(t,x) = \mathbf{E}\, g_n(x + w_{s_n - t})$ for $t \le s_n$, then by Lemmas 8, 4

$$\begin{aligned} \xi := g_1(w_{s_1}) \cdot \ldots \cdot g_n(w_{s_n}) &= g_1(w_{s_1}) \cdot \ldots \cdot g_{n-1}(w_{s_{n-1}}) v(s_{n-1}, w_{s_{n-1}}) \\ &+ \int_0^\infty g_1(w_{s_1}) \cdot \ldots \cdot g_{n-1}(w_{s_{n-1}}) I_{s_{n-1} < t \le s_n} v_x(t, w_t)\, dw_t \quad \text{(a.s.)}. \end{aligned} \tag{16}$$

We have thus chopped off the last factor. Since $v(t,x)$ is continuously differentiable with respect to x and bounded together with its derivative (cf. Theorem II.8.1), we can repeat this operation, considering, of course, only the first term on of the right of (16). Then, after a finite number of steps, ξ will be expressed as a sum of a constant and a stochastic integral. The constant is easily found by using (4). This proves the theorem for the particular case of ξ. Note that we can now drop the smoothness assumption on the functions g_i. All we need is their continuity and boundedness. Indeed, let us define ξ^ε in the same way as ξ, replacing g_i by $g_i^\varepsilon(x) := \mathbf{E}\, g_i(x + w_\varepsilon)$, $\varepsilon > 0$. It follows easily from Theorem II.8.1 that the derivatives $(g_i^\varepsilon)'$ are continuous and bounded functions of x for any $\varepsilon > 0$, $g_i^\varepsilon \to g_i$ as $\varepsilon \downarrow 0$, and by the Dominated Convergence Theorem

$$\left(\mathbf{E}\,|\xi^\varepsilon - \xi|\right)^2 \le \mathbf{E}\,|\xi^\varepsilon - \xi|^2 \longrightarrow 0, \qquad \mathbf{E}\left|(\xi^\varepsilon - \mathbf{E}\,\xi^\varepsilon) - (\xi - \mathbf{E}\,\xi)\right|^2 \longrightarrow 0 \tag{17}$$

as $\varepsilon \downarrow 0$. Next we observe that by the isometric property of the stochastic integration operator J, the sequence f_n is a Cauchy sequence if Jf_n is a Cauchy sequence. This implies that the range of J is closed, and since $\xi^\varepsilon - \mathbf{E}\,\xi^\varepsilon$ lies in the range of the stochastic integration operator, $\xi - \mathbf{E}\,\xi$ can also be written as a stochastic integral.

Furthermore, as in the proof of Theorem I.5.4, we pass from bounded continuous g_i to indicators of closed sets, showing that the first assertion of the theorem holds if $\xi(\omega) = I_A(\omega.)$, where $A = \{x. \in C\colon x_{s_1} \in \Gamma_1, \dots x_{s_n} \in \Gamma_n\}$ and Γ_i are closed sets. Let Π_1 denote the family of all sets A of this type. It is obvious that Π_1 is a π-system. Let Λ be the family of all elements $B \in \mathfrak{A}(C)$ such that the first assertion of the theorem holds for $\xi(\omega) = I_B(w.)$. Linearity and the isometric property of the stochastic integration operator immediately imply that Λ is a λ-system. Since by what has already been proved it contains Π_1, it follows that $\Lambda \supset \sigma(\Pi_1)$, where the last σ-algebra coincides with $\mathfrak{A}(C)$ (cf. Exercise II.5.2). In other words (cf. Exercise II.5.3), the theorem is proved for $\xi = I_D$ for any $D \in \mathcal{F}$.

It is clear that the theorem is also valid for $\xi = S(\mathcal{F})$. It remains only to use the definition of $\mathcal{L}_2(\mathcal{F}, P)$ and repeat the arguments related to (17). □

13. Remark. If $f, g \in \mathcal{L}_2(\mathcal{P}, \mu)$ and for all $\tau \in \mathfrak{M}$

$$\int_X I_{(0,\tau]\!]} f\, \mu(d\omega dt) = \int_X I_{(0,\tau]\!]} g\, \mu(d\omega dt),$$

then it easily follows from the lemma on π- and λ-systems that the indicator can be replaced by any $h \in \mathcal{L}_2(\mathcal{P}, \mu)$, and hence $f = g$ (μ-a.e.). In addition, as we know, $\mathbf{E}\,w_\tau = 0$ for $\tau \in \mathfrak{M}$, and by Clark's Theorem $\xi - \mathbf{E}\,\xi$, where $\xi \in \mathcal{L}_2(\sigma(w.), P)$, can be written as a stochastic integral, whose integrand, as follows from above, is *uniquely* determined by $\mathbf{E}\,w_\tau \xi$, $\tau \in \mathfrak{M}$. In particular, if $\mathcal{F} = \sigma(w.)$, $\xi \in \mathcal{L}_2(\mathcal{F}, P)$, $f \in \mathcal{L}_2(\mathcal{P}, \mu)$, then (15) holds if and only if, for all $\tau \in \mathfrak{M}$, only the second equality in (8) is true. This result is very helpful, for example, in proving (13).

14. Problem. Assume $\mathcal{F} = \sigma(w.)$ and denote by J the operator of stochastic integration with respect to dw_t. Prove that $J^{-1} = J^*$.

3. Itô's stochastic integral with variable upper limit

In the previous section, the definitions and notation of which we will continue to use here, we introduced Itô's stochastic integral over the half-line $(0, \infty)$. However, we are most interested in stochastic integrals over finite time intervals. If $s \geq 0$ and $f \in \mathcal{L}_2(\mathcal{P}, \mu)$, then, by analogy with the usual integral, it is natural to define a stochastic integral ξ_s of f with respect to dw over the interval $(0, s]$ as

$$\xi_s = \int_0^\infty f(t) I_{t \leq s}\, dw_t. \tag{1}$$

It is clear that $f I_{(0,s]\!]} \in \mathcal{L}_2(\mathcal{P}, \mu)$, and the right-hand side of (1) is meaningful. We would like to interpret the process ξ_s as a model of Brownian motion in a liquid with variable and random temperature (see Sec. II.2). The principal obstacle to this interpretation is that, generally speaking, the functions $\xi_s(\omega)$, as functions of s for fixed ω, may be discontinuous, since an arbitrary stochastic integration operator applied to $f I_{(0,s]\!]}$ may produce a function that is not continuous in s. Nevertheless, it turns out that by taking different operators for different s, we may be able to make the sample paths of ξ continuous.

1. DEFINITION. Let $f \in \mathcal{L}_2(\mathcal{P}, \mu)$ and let ξ_s be a stochastic process which is continuous in s, defined for $\omega \in \Omega$, $s \in [0, \infty)$ and such that (1) is true (a.s.) for any $s \in [0, \infty)$ and $\xi_0(\omega) \equiv 0$. Then we define for all $s \in [0, \infty), \omega$,

$$\int_0^s f(t)\, dw_t = \xi_s \tag{2}$$

and call the process ξ_s the *Itô stochastic integral of f (with variable upper limit)*.

The notation (2), as in Definition 1.6, needs explanation, since there may (and in practice always do) exist many different processes ξ such that equality (1) holds (a.s.) for any $s \geq 0$. From now on, therefore, we will always assume that the left-hand side of (2) is a *continuous* stochastic process, which may be different in different contexts. In addition, we will never trouble ourselves with the question of whether these processes may be equal at a fixed ω, since the answer, which is of no interest, depends on a variety of unimportant details related to a way in which a given particular version of the left-hand side of (2) is chosen.

2. REMARK. A justification for the somewhat imprecise notation (2) is given by the fact that if two continuous processes ξ_s, η_s equal the left-hand side of (1) for every fixed $s \in [0, \infty)$ (a.s.), they are *indistinguishable*, that is, $\xi_s = \eta_s$ a.s. for all $s \in [0, \infty)$ at once:

$$P\Big\{ \sup_{s \geq 0} |\xi_s - \eta_s| > 0 \Big\} = 0. \tag{3}$$

Indeed, by the continuity of ξ_s, η_s

$$\Big\{\omega\colon \sup_{s \geq 0} |\xi_s - \eta_s| > 0\Big\} = \bigcup_{r \in \rho} \Big\{\omega\colon |\xi_r - \eta_r| > 0\Big\}, \tag{4}$$

where ρ is the set of all rationals in $[0, \infty)$. Since $\xi_r = \eta_r$ (a.s.), the set on the right of (4) is a countable union of events of probability zero, and this gives (3) (cf. Corollary I.2.9).

We now proceed to the existence of Itô's stochastic integral with variable upper limit.

3. LEMMA. *Let $f \in S(\Pi)$. Then there exists a process ξ_s satisfying the requirements of Definition* 1. *In addition,*

(a) $\lim_{s \to \infty} \xi_s$ *exists (a.s.),*

$$\lim_{s \to \infty} \xi_s = \overline{\lim_{s \to \infty}}\, \xi_s =: \xi_\infty, \qquad \xi_\infty = \int_0^\infty f(t)\, dw_t \quad (a.s.);$$

(b) *for any Markov time τ*

$$\int_0^\tau f(t)\, dw_t := \xi_\tau = \int_0^\infty f(t) I_{t \leq \tau}\, dw_t \quad (a.s.); \tag{5}$$

(c) $(\xi_s, \mathcal{F}_s)$ *is a martingale for $s \geq 0$.*

PROOF. By the linearity of the stochastic integration operator, it suffices to consider the case $f = I_{t\le\gamma}$, $\gamma \in \mathfrak{M}$. In that case, take $\xi_s = w_{s\wedge\gamma}$. Then $f(t)I_{t\le s} = I_{t\le\gamma\wedge s}$. Hence (cf. (2.3)) ξ satisfies the conditions of Definition 1. Part (a) for ξ is evident, and $\xi_\infty = w_\gamma$. As to part (b), we note that the definition in (5) is consistent (for $\tau = \infty$) by virtue of (a), and that the equality in (5) follows immediately from (2.3). Finally, the function $w_{s\wedge\gamma}$ is $\mathcal{F}_{s\wedge\gamma}$-measurable (Exercise II.4.17) and, since $s\wedge\gamma \le s$, it is $\mathcal{F}_s$-measurable (Exercise II.4.16). Thus ξ_s is a $\mathcal{F}_s$-adapted process, and by Theorem II.6.9 it is a martingale with respect to $\{\mathcal{F}_s\}$. □

We will now prove two properties of submartingales, which will be used below.

4. THEOREM. *Let $(\xi_s, \mathcal{F}_s)$ be a submartingale for $s \ge 0$. Let $p \in (1,\infty)$, $T, c \in (0,\infty)$. Then*

(a) (*the Doob–Kolmogorov inequalities*)

$$P\Big\{\max_{[0,T]}\xi_s \ge c\Big\} \le \frac{1}{c}\mathbf{E}\,\xi_T I_{\max_{[0,T]}\xi_s\ge c} \le \frac{1}{c}\mathbf{E}\,(\xi_T)_+, \tag{6}$$

$$P\Big\{\sup_{[0,\infty)}\xi_s > c\Big\} \le \frac{1}{c}\varliminf_{T\to\infty}\mathbf{E}\,\xi_T I_{\max_{[0,T]}\xi_s\ge c} \le \frac{1}{c}\sup_{T\ge 0}\mathbf{E}\,(\xi_T)_+; \tag{7}$$

(b) (*Doob's moment inequality*) *if $\xi_s \ge 0$ for all ω, s, then*

$$\mathbf{E}\sup_{s\ge 0}\xi_s^p \le \Big(\frac{p}{p-1}\Big)^p \lim_{s\to\infty}\mathbf{E}\,\xi_s^p, \qquad \mathbf{E}\sup_{s\ge 0}\xi_s^2 \le 4\lim_{s\to\infty}\mathbf{E}\,\xi_s^2. \tag{8}$$

PROOF. (a) Since the events under the P sign in (6) become larger with increasing T and their union over T contains the event on the left of (7), we conclude that (7) follows from (6) as $T \to \infty$. The second inequality in (6) is obvious. Hence, only the first one needs a proof. Define

$$\tau = \inf\{s \ge 0\colon \xi_s \ge c\}, \qquad A = \{\omega\colon \max_{[0,T]}\xi_s \ge c\}.$$

Then τ is a Markov time (cf. Example II.4.6), and by Chebyshev's inequality and Lemma II.8.5 (see also Exercise II.4.16),

$$\begin{aligned} P(A) = P\{\xi_\tau I_{\tau\le T} \ge c\} &\le \frac{1}{c}\mathbf{E}\,\xi_\tau I_{\tau\le T} \\ &= \frac{1}{c}\mathbf{E}\,\xi_{\tau\wedge T} I_{\tau\le\tau\wedge T} \le \frac{1}{c}\mathbf{E}\,\xi_T I_{\tau\le\tau\wedge T} = \frac{1}{c}\mathbf{E}\,\xi_T I_A. \end{aligned}$$

(b) Denoting $\eta = \sup\{\xi_s\colon s \ge 0\}$ in (7) and noting that $P(\zeta^p > c) = P(\zeta > c^{1/p})$, we integrate (7) with respect to c, apply Fatou's Lemma, Fubini's Theorem (see also Lemma I.5.7), and Hölder's inequality. We obtain for any $n \in (0,\infty)$

$$\begin{aligned} \mathbf{E}\,(\eta\wedge n)^p &= \int_0^{n^p} P\{\eta > c^{1/p}\}\,dc \le \varliminf_{T\to\infty}\mathbf{E}\,\xi_T\int_0^{n^p} c^{-1/p} I_{\eta\ge c^{1/p}}\,dc \\ &= \frac{p}{p-1}\varliminf_{T\to\infty}\mathbf{E}\,\xi_T(\eta\wedge n)^{p-1} \le \frac{p}{p-1}\big(\mathbf{E}\,(\eta\wedge n)^p\big)^{\frac{p-1}{p}}\Big(\lim_{T\to\infty}\mathbf{E}\,\xi_T^p\Big)^{1/p}, \end{aligned}$$

where the last limit exists by Lemma II.8.5.

The inequality between the outermost terms and the fact that $\mathbf{E}(\eta\wedge n)^p$ ($\le n^p$) is finite obviously imply that

$$\mathbf{E}(\eta\wedge n)^p \le \Big(\frac{p}{p-1}\Big)^p \lim_{T\to\infty} \mathbf{E}\,\xi_T^p.$$

Letting $n\to\infty$ and using the Monotone Convergence Theorem we obtain the first inequality in (8). Since the second one is a particular case of the first one with $p=2$, the theorem is proved. □

5. THEOREM. *Let $f\in\mathcal{L}_2(\mathcal{P},\mu)$. Then there exists a process ξ_s that satisfies the conditions of Definition* 1. *Moreover, any such process also possesses properties* (a), (b) *of Lemma* 3. *If, additionally, $(\Omega,\mathcal{F},P)$ is a complete probability space and the σ-algebras $\mathcal{F}_t$ are complete with respect to $\mathcal{F}, P$, this process possesses also property* (c) *of Lemma* 3.

PROOF. Take $f^n\in S(\Pi)$ such that $f^n\to f$ in $\mathcal{L}_2(\mathcal{P},\mu)$. Let ξ^n be the processes of Lemma 3 corresponding to f^n. By Lemma 3, the processes $\xi^n-\xi^m$ are martingales, by Lemma II.8.5 the processes $|\xi^n-\xi^m|$ are submartingales. Hence, by Theorem 4,

$$\begin{aligned}
&\mathbf{E}\sup_{s\ge0}\left|\xi_s^n-\xi_s^m\right|^2 \le 4\sup_{s\ge0}\mathbf{E}\left|\xi_s^n-\xi_s^m\right|^2\\
(9)\qquad &=4\sup_{s\ge0}\mathbf{E}\Big(\int_0^\infty [f^n(t)-f^m(t)]I_{t\le s}\,dw_t\Big)^2\\
&=4\sup_{s\ge0}\int_X \left|f^n(t)-f^m(t)\right|^2 I_{t\le s}\mu(d\omega dt)=4\int_X\left|f^n(t)-f^m(t)\right|^2\mu(d\omega dt).
\end{aligned}$$

Hence, by the choice of f^n, there is a subsequence $n(k)$ (cf. the proof of Lemma 1.1) such that with probability one,

$$(10)\qquad \sum_{k=0}^\infty \sup_{s\ge0}\left|\xi_s^{n(k+1)}-\xi_s^{n(k)}\right|(\omega)<\infty.$$

Let Ω' be the set of all ω that satisfy (10). Clearly, $\Omega'\in\mathcal{F}$, $P(\Omega')=1$, and for any $\omega\in\Omega'$ the sequence $\xi_s^{n(k)}$ converges as $k\to\infty$ uniformly in $s\in[0,\infty)$. If $\omega\in\Omega'$, let $\xi_s(\omega)$ denote the limit of this sequence. If $\omega\notin\Omega'$, let $\xi_s(\omega):=0$ for all $s\in[0,\infty)$. Obviously, ξ_s is a continuous stochastic process, $\xi_0=0$. In addition, for every s

$$(11)\qquad \mathbf{E}\left|\xi_s^{n(k)}-\int_0^\infty f(t)I_{t\le s}\,dw_s\right|^2=\int_X\left|f^{n(k)}(t)-f(t)\right|^2 I_{t\le s}\,\mu(d\omega dt);$$

and the last expressions tends to zero as $n\to\infty$. Together with Fatou's Lemma, this proves the first part of the theorem.

The first part of the Lemma 3 (a) is obvious for the process ξ, since $\xi^{n(k)}$ converges to ξ uniformly in the half line $[0,\infty)$ (a.s.). By Fatou's Lemma this, together with (11) for $s=\infty$, implies the second part of Lemma 3 (a). Finally, the second equality in (5) will follow from (11) if we substitute τ for s, which is possible since the function $I_{t\le\tau}$ is predictable and $fI_{(0,\tau]\!]}\in\mathcal{L}_2(\mathcal{P},\mu)$.

In addition, $\mathbf{E}\,\xi_\tau=0$ for any Markov time. This follows immediately from (2.4) and the last equality in (5). If $\mathcal{F}_s$ are complete, the function ξ_s, as a limit of the convergent (a.s.) sequence of $\mathcal{F}_s$-measurable functions $\xi_s^{n(k)}$, turns out to be $\mathcal{F}_s$-measurable and the process ξ_s turns out to be a martingale with respect to $\{\mathcal{F}_s\}$.

The properties of the above process ξ that have already been proved, are valid *for any* process that satisfies the conditions of Definition 1, since by Remark 2 any such process is indistinguishable from ξ. □

6. REMARK. The martingale property of the stochastic integral is very important. To make it hold in the sense of Definition II.8.4, in Theorem 5 we *have* to require $\mathcal{F}_s$ to be complete. This is because ξ_s must be $\mathcal{F}_s$-measurable, and the set Ω', which is defined by the values of $\xi_s^{n(k)}$ for all $s \geq 0$, needs not to be an element of, say, $\mathcal{F}_0$. In this connection one should note that if $(w_t, \mathcal{F}_t)$ is a Wiener process on $(\Omega, \mathcal{F}, P)$, then $(w_t, \mathcal{F}_t^P)$ is a Wiener process on $(\Omega, \mathcal{F}^P, P)$, where $\mathcal{F}_t^P$ is the completion of $\mathcal{F}_t$ with respect to $\mathcal{F}^P, P$. Moreover, σ-algebras $\mathcal{F}_t^P$, $\mathcal{F}^P$ are complete with respect to $\mathcal{F}^P, P$ and $\{\mathcal{F}_t^P\}$ is a filtration of σ-algebras. These are quite elementary consequences of the definitions and results of Sec. I.3, true not only for one-dimensional but also for multidimensional Wiener processes.

We can now construct Itô stochastic integrals with respect to $(w_t, \mathcal{F}_t^P)$. We will then have new $\mathfrak{M}$, Π, $\mathcal{P}$, but since $\mathcal{F}_t \subset \mathcal{F}_t^P$, the new $\mathfrak{M}$, Π, $\mathcal{P}$ will contain the old ones, and by Theorem 1.7 the new stochastic integral will coincide (a.s.) with the old one for functions that are stochastically integrable in the old sense.

By the way, it follows from this argument and from the last assertion of Theorem 5 that the process ξ_s of Definition 1 is always a martingale with respect to $\{\mathcal{F}_s^P\}$ for $s \geq 0$ on $(\Omega, \mathcal{F}^P, P)$.

Moreover, we have shown that there *exist* a complete probability space $(\Omega, \mathcal{F}, P)$ and a Wiener process $(w_t, \mathcal{F}_t)$ on it, $t \geq 0$, for which the σ-algebras $\mathcal{F}_t$ are complete with respect to $\mathcal{F}, P$.

To end this rather tedious discussion of the $\mathcal{F}_s$-measurability of ξ_s, we note that although equality (5) is quite natural, certain efforts were necessary to prove it. The reader will appreciate these efforts while solving the following problem.

7. PROBLEM. Use the predictability of w_t to deduce that $f(t) := I_{(0,\infty)}(w_t) I_{(0,1)}(t)$ is a predictable function and is an element of $\mathcal{L}_2(\mathcal{P}, \mu)$, whose stochastic integral, say ξ, has the mean value zero. Find an error (or errors) in the following arguments. Trying to calculate ξ, let us represent the open set $\{t > 0 \colon w_t > 0\}$ as a countable union of disjoint intervals (α_n, β_n). Clearly,

$$I_{(0,\infty)}(w_t) I_{(0,1)}(t) = \sum_n I_{(1\wedge\alpha_n, 1\wedge\beta_n)}(t). \tag{12}$$

In addition, for almost all (ω, t),

$$I_{(1\wedge\alpha_n, 1\wedge\beta_n)}(t) = I_{t \leq 1\wedge\beta_n} - I_{t \leq 1\wedge\alpha_n}. \tag{13}$$

Integrating with respect to dw_t we see that the integral on the left-hand side of (13) is $w_{1\wedge\beta_n} - w_{1\wedge\alpha_n}$, and this expression is different from zero only if the interval (α_n, β_n) contains the point 1, in which case it equals w_1. Hence by (12), the variable ξ, which is the sum of the series whose terms are the integrals on the left-hand sides of (13), equals w_1 if $1 \in (\alpha_n, \beta_n)$ for some n, and $\xi = 0$ if there is no such n. In other words, $\xi = (w_1)_+$; but this is impossible, since $\mathbf{E}\,\xi = 0 < \mathbf{E}\,(w_1)_+$.

4. Extending the set of Itô-integrable functions. The notion of a local martingale

One corollary of formula (3.5) is that for a Markov time τ and a number $s \geq 0$,

$$\xi_s = \xi_{s\wedge\tau} = \int_0^\infty f(t) I_{t \leq s\wedge\tau}\, dw_t \quad \text{(a.s.)} \tag{1}$$

on the set $\{\omega \colon s \leq \tau(\omega)\}$. In order to define the right-hand side, we need values of f only on $(0, \tau]\!]$; it follows that in order to define a stochastic integral

$$\int_0^s f(t)\, dw_t \tag{2}$$

on a subgraph of a Markov time τ, we again need f on the subgraph only, and the right-hand side of (1) is meaningful not only if $f \in \mathcal{L}_2(\mathcal{P}, \mu)$ but also for $f I_{(0,\tau]\!]} \in \mathcal{L}_2(\mathcal{P}, \mu)$. This makes it possible to define the stochastic integral *locally* and to enlarge the set of Itô-integrable functions.

In this section we will retain the definitions and notation of Secs. 2 and 3.

1. DEFINITION. We write $f \in \mathcal{L}_{p,\text{loc}}(\mathcal{P}, \mu)$ if there exists a sequence of Markov times τ_n, $n = 1, 2, \dots$, such that $\tau_n \to \infty$ as $n \to \infty$ (a.s.), $\tau_1 \leq \tau_2 \leq \dots$, and $f^n \in \mathcal{L}_p(\mathcal{P}, \mu)$ for all n, where $f^n(t) := f(t) I_{t \leq \tau_n}$. Any such sequence $\{\tau_n\}$ will be called a *localizing* sequence for f.

2. DEFINITION. Let $f \in \mathcal{L}_{2,\text{loc}}(\mathcal{P}, \mu)$, and let ξ_s be a stochastic process, continuous in s and defined for $\omega \in \Omega$, $s \in [0, \infty)$; assume $\xi_0 = 0$. Suppose that for any *finite* Markov time τ such that $f I_{(0,\tau]\!]} \in \mathcal{L}_2(\mathcal{P}, \mu)$

$$\xi_\tau = \int_0^\infty f(t) I_{t \leq \tau}\, dw_t \quad \text{(a.s.)}. \tag{3}$$

Then for all $s \in [0, \infty)$, ω we set

$$\int_0^s f(t)\, dw_t = \xi_s \tag{4}$$

and call ξ_s the *Itô stochastic integral of f (with variable upper limit)*.

Obviously, $\mathcal{L}_2(\mathcal{P}, \mu) \subset \mathcal{L}_{2,\text{loc}}(\mathcal{P}, \mu)$, and by Theorem 3.5, if $f \in \mathcal{L}_2(\mathcal{P}, \mu)$, then its stochastic integral in the sense of Definition 3.1 is also a stochastic integral in the sense of Definition 2. We have thus enlarged the scope of the old notion, rather than introduced a new concept. As to the legitimacy of (4), one can repeat the comments about (3.2) in the previous section. In particular, any two different versions of the left-hand side of (4) are indistinguishable, as follows in this case from relations of the type

$$\Big\{\omega\colon \sup_{s\geq 0}|\xi_s-\eta_s|>0, \tau_n\uparrow\infty\Big\}\subset\bigcup_n\Big\{\omega\colon \sup_{s<\infty}|\xi_{s\wedge\tau_n}-\eta_{s\wedge\tau_n}|>0\Big\}$$
$$=\bigcup_n\bigcup_{r\in\rho}\Big\{\omega\colon |\xi_{r\wedge\tau_n}-\eta_{r\wedge\tau_n}|>0\Big\}.$$

3. THEOREM. *Let $f\in\mathcal{L}_{2,\mathrm{loc}}(\mathcal{P},\mu)$. Then there exists a process ξ that satisfies the conditions of Definition 2.*

PROOF. Take a localizing sequence $\{\tau_n\}$ for f and denote

$$\xi_s^n=\int_0^s f(t)I_{t\leq\tau_n}\,dw_t,\qquad \Omega'=\bigcap_{n\geq 1}\bigcap_{m\geq n}\bigcap_{r\in\rho}\Big\{\omega\colon \xi_r^n=\xi_{r\wedge\tau_n}^m\Big\}\bigcap\Big\{\omega\colon \tau_k\uparrow\infty\Big\},$$

where ρ is the set of all rational points in $[0,\infty)$. For $\omega\in\Omega'$, $s\in[0,\infty)$, take any n such that $\tau_n(\omega)\geq s$, and set $\xi_s(\omega)=\xi_s^n(\omega)$. This definition is consistent, since if $\omega\in\Omega'$, $\tau_n(\omega)\geq s$, $\tau_m(\omega)\geq s$, and, say, $m\geq n$, then $\xi_r^n(\omega)=\xi_r^m(\omega)$ for all $r\in\big[0,\tau_n(\omega)\big]\cap\rho$, and due to the continuity of ξ^n, ξ^m, for all $r\in\big[0,\tau_n(\omega)\big]$. In particular, $\xi_s^n(\omega)=\xi_s^m(\omega)$. If $\omega\notin\Omega'$, we let $\xi_s(\omega):=0$ for all $s\in[0,\infty)$. Obviously, $\Omega'\in\mathcal{F}$, ξ_s is a continuous stochastic process, $\xi_0=0$. By Theorem 3.5, for $m\geq n$ (a.s.)

$$\xi_{r\wedge\tau_n}^m=\int_0^{r\wedge\tau_n} f(t)I_{t\leq\tau_m}\,dw_t=\int_0^\infty f(t)I_{t\leq\tau_m\wedge\tau_n}I_{t\leq r}\,dw_t=\xi_r^n.$$

Hence $P(\Omega')=1$,

$$\xi_{s\wedge\tau_n}=\xi_{s\wedge\tau_n}^n \tag{5}$$

for all $s\in[0,\infty)$ and all $n=1,2,\dots$ at once a.s. (for example, on Ω'). Moreover, for any Markov time τ such that $fI_{(0,\tau]\!]}\in\mathcal{L}_2(\mathcal{P},\mu)$, we have

$$\begin{aligned}&\lim_{n\to\infty}\mathbf{E}\Big(\int_0^\infty f(t)I_{t\leq\tau\wedge\tau_n}\,dw_t-\int_0^\infty f(t)I_{t\leq\tau}\,dw_t\Big)^2\\&\qquad=\lim_{n\to\infty}\int_X f^2(t)I_{\tau\wedge\tau_n<t\leq\tau}\,\mu(d\omega dt).\end{aligned}\tag{6}$$

The last expression vanishes by the Dominated Convergence Theorem. If, in addition, τ is finite, then (a.s.)

$$\xi_\tau=\lim_{n\to\infty}\xi_{\tau\wedge\tau_n}=\lim_{n\to\infty}\int_0^{\tau\wedge\tau_n} f(t)I_{t\leq\tau_n}\,dw_t$$
$$=\lim_{n\to\infty}\int_0^\infty f(t)I_{t\leq\tau_n\wedge\tau}\,dw_t=\int_0^\infty f(t)I_{t\leq\tau}\,dw_t,$$

where the first equality (a.s.) follows from the continuity of ξ_s, the finiteness of τ and the fact that $\tau_n\uparrow\infty$ (a.s.) and the second equality follows from (5). The existence (a.s.) of the other limits follows from equalities (a.s.) of the expressions involved, whereas the last equality (a.s.) follows from Fatou's Lemma and (6). □

Theorem 3 establishes the existence of stochastic integral (2) for $f \in \mathcal{L}_{2,\mathrm{loc}}(\mathcal{P}, \mu)$. We will now derive some of its properties.

4. THEOREM. (a) *If* $f \in \mathcal{L}_{2,\mathrm{loc}}(\mathcal{P}, \mu)$ *and* τ *is a Markov time, then* $fI_{(0,\tau]\!]} \in \mathcal{L}_{2,\mathrm{loc}}(\mathcal{P}, \mu)$, *and for all* $s \in [0, \infty)$ *at once*

$$(7) \qquad \int_0^{s\wedge\tau} f(t)\, dw_t = \int_0^s f(t) I_{t\le\tau}\, dw_t \quad (a.s.).$$

(b) *If* $f, g \in \mathcal{L}_{2,\mathrm{loc}}(\mathcal{P}, \mu)$, $a, b \in (-\infty, \infty)$, *then* $af + bg \in \mathcal{L}_{2,\mathrm{loc}}(\mathcal{P}, \mu)$. *If also* $h = af + bg$ (μ-*a.e.*), *then* $h \in \mathcal{L}_{2,\mathrm{loc}}(\mathcal{P}, \mu)$, *and for all* $s \in [0, \infty)$ *at once*

$$(8) \qquad \int_0^s h(t)\, dw_t = a \int_0^s f(t)\, dw_t + b \int_0^s g(t)\, dw_t \quad (a.s.).$$

(c) *If* $f \in \mathcal{L}_{2,\mathrm{loc}}(\mathcal{P}, \mu)$ *and* τ *is a finite Markov time, then*

$$(9) \qquad \begin{aligned} \mathbf{E}\Big(\int_0^\tau f(t)\, dw_t\Big)^2 &\le \int_X f^2(t) I_{t\le\tau}\, \mu(d\omega dt) \\ &\le \mathbf{E} \sup_{s<\infty} \Big|\int_0^{s\wedge\tau} f(t)\, dw_t\Big|^2 \le 4 \int_X f^2(t) I_{t\le\tau}\, \mu(d\omega dt). \end{aligned}$$

PROOF. (a) That $fI_{(0,\tau]\!]}$ is $\mathcal{P}^\mu$-measurable is obvious. In addition, a localizing sequence $\{\tau_n\}$ for f is clearly a localizing sequence for $fI_{(0,\tau]\!]}$. This proves the first assertion in (a). To derive (7), note that by Definition 2 (a.s.)

$$\int_0^{s\wedge\tau\wedge\tau_n} f(t)\, dw_t = \int_0^\infty f(t) I_{t\le\tau} I_{t\le s\wedge\tau_n}\, dw_t = \int_0^{s\wedge\tau_n} f(t) I_{t\le\tau}\, dw_t.$$

As $n \to \infty$, this gives (7) for any $s \in [0, \infty)$ (a.s.); to obtain (7) for all s at once (a.s.) it remains only to use the arguments of Remark 3.2.

We now proceed to part (b). That $af + bg, h \in \mathcal{L}_{2,\mathrm{loc}}(\mathcal{P}, \mu)$ follows from the fact that if $\{\tau_n\}$, $\{\sigma_n\}$ are localizing sequences for f, g, then $\gamma_n := \tau_n \wedge \sigma_n$ is a localizing sequence for $f, g, af + bg, h$. To prove (8) for arbitrary fixed s (a.s.), simply take $s \wedge \gamma_n$ instead of s, use Definition 2, the linearity and isometric property of the stochastic integration operator on $(0, \infty)$, and then again apply Definition 2 and let $n \to \infty$. The equality for all s at once is proved as in Remark 3.2.

To prove (c) assume first that $f \in \mathcal{L}_2(\mathcal{P}, \mu)$. Then Theorem 3.5 and the isometric property of the stochastic integration operator yield an equality instead of the first inequality in (9) and this obviously implies the second inequality. In addition, by Remark 3.6, both processes in (7) are martingales with respect to $\mathcal{F}_s^P$. Therefore, the last inequality in (9) follows from this by Doob's inequality (cf. (3.9)).

In the general case, take a localizing sequence $\{\tau_n\}$ for f. By the previous arguments, inequalities (9) remain valid after substituting $fI_{t\le\tau_n}$ for f. Now let $n\to\infty$ and note that, by (a),

$$\int_0^{\tau} f(t)I_{t\le\tau_n}\,dw_t=\int_0^{\tau\wedge\tau_n} f(t)\,dw_t,$$

$$\sup_{s<\infty}\left|\int_0^{s\wedge\tau} f(t)I_{t\le\tau_n}\,dw_t\right|=\sup_{s<\infty,s\le\tau_n}\left|\int_0^{s\wedge\tau} f(t)\,dw_t\right| \tag{10}$$

$$\uparrow\sup_{s<\infty}\left|\int_0^{s\wedge\tau} f(t)\,dw_t\right|\quad\text{(a.s.)}.$$

To prove inequalities (9) in the general case, it remains now to apply Fatou's Lemma. □

5. REMARK. The simplest example of a function belonging to $\mathcal{L}_{2,\mathrm{loc}}(\mathcal{P},\mu)$ is $f\equiv 1$ (with $\tau_n:=n$). In this case there exists a process ξ that satisfies the conditions of Definition 2 (by Theorem 3). It follows from (3) for $\tau=s$ and from (2.3) for $\tau=s$ that $\xi_s=w_s$ for any $s\in[0,\infty)$ (a.s.). This equality obviously holds also for all $s\in[0,\infty)$ at once (a.s.). In particular, by Theorem 4 (a), for any Markov time τ and all s at once (a.s.)

$$\int_0^{s\wedge\tau} 1\,dw_t=\int_0^{s} I_{t\le\tau}\,dw_t=w_{s\wedge\tau}. \tag{11}$$

6. PROBLEM. If $s<1$, one can define the process $\eta_s=\int_0^s(1-t)^{-1}\,dw_t$. Let τ be the first moment when η_s hits the point (-1). Prove that $P\{\tau<1\}=1$, $f(t):=(1-t)^{-1}I_{t<\tau}\in\mathcal{L}_{2,\mathrm{loc}}(\mathcal{P},\mu)$, and for $s\ge 1$ a.s.

$$\int_0^{s} f(t)\,dw_t=-1.$$

(Hint: $\eta_{\psi(s)}$ is a Wiener process, where the functions $\psi(s)$ is the inverse of the function $\int_0^s(1-t)^{-2}\,dt$.)

This problem demonstrates that a stochastic integral is not always a martingale, even on a complete probability space with complete σ-algebras $\mathcal{F}_t$.

7. DEFINITION. Let α_s be a real-valued continuous $\mathcal{F}_s$-adapted process given for $t\ge 0$. We say that α_s *is a local martingale* (with respect to $\{\mathcal{F}_s\}$), or that $(\alpha_s,\mathcal{F}_s)$ *is a local martingale*, if there exist Markov times $\tau_n\to\infty$ (a.s.) such that $\tau_1\le\tau_2\le\ldots$ and the processes $\alpha_{s\wedge\tau_n}-\alpha_0$ are martingales for all n. Such Markov times τ_n are called *localizing times* for α. Similarly one can define *local sub-* and *supermartingales*.

8. THEOREM. *Let $(\Omega, \mathcal{F}, P)$ be a complete probability space, let the σ-algebras $\mathcal{F}_s$ be complete with respect to $\mathcal{F}, P$, and let $f \in \mathcal{L}_{2,\text{loc}}(\mathcal{P}, \mu)$. Then the process ξ_s, which satisfies the conditions of Definition 2 is a local martingale.*

Indeed, it suffices to take a localizing sequence τ_n for f, substitute $\tau_n = \tau$ in (7) and use Theorem 3.5.

A few simple results are yet worth mentioning.

9. THEOREM. (a) *A martingale is a local martingale.*

(b) *If α_s is a local martingale and β a finite $\mathcal{F}_0$-measurable variable, then $\alpha_s + \beta$ is a local martingale.*

(c) *If α_s is a local martingale and there exists a random variable β such that $\mathbf{E}\,\beta < \infty$, and $|\alpha_s| \leq \beta$ for all ω, s, then α_s is a martingale.*

(d) *If α_s is a local martingale and τ a Markov time, then the* stopped *process $\alpha_{s\wedge\tau}$ is also a local martingale.*

(e) *A continuous $\mathcal{F}_s$-adapted process α_s is a local martingale if and only if, for $\tau(n) = \inf\{s \geq 0\colon |\alpha_s - \alpha_0| \geq n\}$, the bounded process $\alpha_s^n = \alpha_{s\wedge\tau(n)} - \alpha_0$ is a martingale for any $n \geq 1$.*

(f) *A continuous $\mathcal{F}_s$-adapted process α_s is a local martingale if there exists a sequence of Markov times $\gamma(n) \to \infty$ (a.s.) such that $\alpha_{s\wedge\gamma(n)}$ is a local martingale for any n.*

(g) *If α_s, β_s are local martingales, then $\alpha_s + \beta_s$ is also a local martingale.*

(h) *Assertions* (a)–(g) *remain valid upon substitution of "submartingale" or "supermartingale" for the word "martingale."*

(i) *A continuous $\mathcal{F}_s$-adapted nondecreasing process is a local submartingale.*

(j) *If α_s is a local (sub- or) supermartingale, τ a Markov time, and f a finite $\mathcal{F}_\tau$-measurable function and $f \geq 0$, then $\beta_s = (\alpha_s - \alpha_{s\wedge\tau})f$ is a local (sub- or) supermartingale. If α_s also is a local martingale, then β_s is a local martingale for any finite $\mathcal{F}_\tau$-measurable f.*

(k) *Let α_s^n be local martingales, $n = 1, 2, \ldots$, and α_s a continuous $\mathcal{F}_s$-adapted process such that, for any $T \in [0, \infty)$, $\varepsilon > 0$,*

$$\lim_{n\to\infty} P\{\sup_{s\leq T} |\alpha_s^n - \alpha_s| \geq \varepsilon\} = 0. \tag{12}$$

Then α_s is a local martingale. A similar assertion holds for sub- and supermartingales.

PROOF. For (a), one can take $\tau_n \equiv \infty$; (b) is obvious; (c) is a simple corollary of Definition 7 and the Dominated Convergence Theorem applied to equations like $\mathbf{E}\,\alpha_{\tau\wedge\tau_n} = \mathbf{E}\,\alpha_{\sigma\wedge\tau_n}$. Assertion (d) follows from Definition 7 (and Exercises II.4.16, II.4.17). The reader can easily prove assertion (e) by combining (a)–(d) and noting that $\tau(n) \uparrow \infty$ as $n \to \infty$. Under the assumptions of (f) we see that, if α_s^n is the process from (e), then $|\alpha^n_{s\wedge\gamma(k)}| \leq n$ and $\alpha^n_{s\wedge\gamma(k)}$ is a local martingale by (b)–(d). Letting $k \to \infty$ and using the Dominated Convergence Theorem, we see that α_s^n is a martingale. This, together with (e), gives (f). To prove (g), it suffices to observe that if τ_n, σ_n localize α_s, β_s respectively, then $\tau_n \wedge \sigma_n$ will localize $\alpha_s, \beta_s, \alpha_s + \beta_s$. The proof of (h) is obtained by duplicating the previous arguments. Assertion (i) follows, for example, from (e), (h).

To prove (j) for the case of a local supermartingale α_s, by considering α_s^n from (e) instead of α_s, we obtain $\beta_{s\wedge\tau(n)}$ instead of β_s. Using (f), we can now reduce everything to the case in which α_s is a bounded supermartingale. In addition, putting $\gamma(n) = \tau$

for $f > n$, $\gamma(n) = \infty$ for $f \leq n$, and noting that

$$\beta_{s\wedge\gamma(n)} = \big(\alpha_s - \alpha_{s\wedge\tau}\big) f I_{f\leq n},$$

we may assume that f is a bounded function. Applying Lemma II.8.5 (a) and using the formula

$$\beta_r = \beta_t + \big(\alpha_{r\vee\tau} - \alpha_{t\vee\tau}\big) f,$$

where $r \geq t$, $r, t \in \mathfrak{M}$, we conclude that $\mathbf{E}\beta_r \leq \mathbf{E}\beta_t$, that is, β_s is a supermartingale. Similar arguments work when α_s is a local submartingale. If α_s is a local martingale, we make use also of the relation $f = f_+ - f_-$.

Assertion (k) needs be proved only for local submartingales. By (12), there is a sequence $n(k)$, $k = 1, 2, \ldots$, such that, if $\beta_s^k := \alpha_s^{n(k)}$,

$$P\{2 \sup_{s\leq k} |\beta_s^k - \alpha_s| \geq k^{-1}\} \leq k^{-2}. \tag{13}$$

Hence, putting $\tilde{\beta}_s^k := \beta_s^k - \beta_0^k$, $\tilde{\alpha}_s := \alpha_s - \alpha_0$,

$$\sigma(k) = \inf\{s \geq 0 \colon |\tilde{\beta}_s^k - \tilde{\alpha}_s| \geq k^{-1}\},$$

and noting that the event $\{\sigma(k) \leq k\}$ is contained in the event from (13), we obtain $P\{\sigma(k) \leq k\} \leq k^{-2}$. Since the series $\sum k^{-2}$ converges, it follows that for almost any ω, we have $\sigma(k) > k$ for sufficiently large k, and therefore $\sigma(k) \to \infty$ (a.s.).

Now take $\tau(n)$ as in (e) and put

$$\tilde{\beta}_s^{kn} = \tilde{\beta}^k_{s\wedge\sigma(k)\wedge\tau(n)}.$$

Since $|\tilde{\beta}_s^k - \tilde{\alpha}_s| \leq k^{-1}$ for $s \leq \sigma(k)$ and $|\tilde{\alpha}_s| \leq n$ for $s \leq \tau(n)$, it follows that $|\tilde{\beta}_s^{kn}| \leq n + 1$ and $\tilde{\beta}_\tau^{kn} \to \tilde{\alpha}_{\tau\wedge\tau(n)}$ (a.s.) as $k \to \infty$ for any $\tau \in \mathfrak{M}$. It remains to note that $\tilde{\beta}_s^{kn}$ is a submartingale and conclude, using the Dominated Convergence Theorem, that $\tilde{\alpha}_{s\wedge\tau(n)}$ is also a submartingale. □

We end this section with a brief discussion of the properties of stochastic integrals with *variable lower limit*.

10. Definition. Let τ be a Markov time and $f(t)I_{\tau<t} \in \mathcal{L}_{2,\mathrm{loc}}(\mathcal{P},\mu)$. Then we define

$$\int_\tau^s f(t)\,dw_t = \int_0^s f(t) I_{\tau<t}\,dw_t.$$

We do not assume in this definition that $f \in \mathcal{L}_{2,\mathrm{loc}}(\mathcal{P},\mu)$. In addition, when $\tau = 0$ this definition is consistent with the old one ($f(t)I_{0<t} = f(t)$ on X). Finally, it also follows from Theorem 4 that, if $f \in \mathcal{L}_{2,\mathrm{loc}}(\mathcal{P},\mu)$, then (a.s.)

$$\int_\tau^s f(t)\,dw_t = \int_0^s f(t)I_{\tau<t\leq s}\,dw_t = \int_0^s f(t)I_{\tau\wedge s<t\leq s}\,dw_t$$

$$= \int_0^s f(t)I_{t\leq s}\,dw_t - \int_0^s f(t)I_{t\leq\tau\wedge s}\,dw_t = \int_0^s f(t)\,dw_t - \int_0^{\tau\wedge s} f(t)\,dw_t.$$

5. Quadratic variation of stochastic integral and pseudopredictable functions

Let $(\Omega, \mathcal{F}, P)$ be a *complete* probability space and $(w_t, \mathcal{F}_t)$ a one-dimensional Wiener process on $(\Omega, \mathcal{F}, P)$ defined for all $t \in [0, \infty)$, such that the σ-algebras $\mathcal{F}_t$ are *complete* with respect to $\mathcal{F}, P$. We again use the notations and definitions of Secs. 2–4. Recall, in particular, that $\mathfrak{A} = \mathcal{F} \otimes \mathfrak{B}((0, \infty))$, $\mu = P \times \ell$.

Let $f \in \mathcal{L}_{2,\text{loc}}(\mathcal{P}, \mu)$ and let ξ be the Itô stochastic integral of f with variable upper limit, that is, ξ is the process of Definition 4.2. It turns out that many properties of ξ are closely related to the properties of the process

$$\langle \xi \rangle_s := \int_0^s f^2(t)\, dt, \tag{1}$$

which is called the *quadratic variation of* ξ. The meaning of these words will become clear only later in Sec. IV.1; for the moment we will discuss only the existence of $\langle \xi \rangle_s$ and its properties.

Note that in Definition 1 and Lemma 2 (below) we do not use the completeness of $\mathcal{F}$, $\mathcal{F}_t$ and existence of a Wiener process w_t.

1. DEFINITION. Let $\mathcal{P}_\mu$ denote the set of all $A \in \mathcal{P}^\mu$ for each of which

$$\{t \colon (\omega, t) \in A\} \in \mathfrak{B}^\ell((0, \infty)) \tag{2}$$

for all $\omega \in \Omega$. It is obvious that $\mathcal{P}_\mu$ is a σ-algebra. We will call its elements *pseudopredictable sets*, and $\mathcal{P}_\mu$-measurable functions will be called *pseudopredictable functions*.

If f is a pseudopredictable function, the right-hand side of (1) is defined as the Lebesgue integral *for any* ω. This shows the convenience of working with pseudopredictable functions. The following lemma states, in addition, that the restriction of the class of $\mathcal{P}^\mu$-measurable functions to the class of $\mathcal{P}_\mu$-measurable functions is in fact inessential.

2. LEMMA. (a) $\mathcal{P} \subset \mathcal{P}_\mu \subset \mathcal{P}^\mu$.

(b) *If g is an arbitrary $\mathcal{P}^\mu$-measurable function, then $g(\omega, t)$ is $\mathfrak{B}^\ell((0, \infty))$-measurable as a function of t (a.s.).*

(c) *For any $\mathcal{P}^\mu$-measurable function g, there exists a predictable (hence, pseudopredictable) function h such that $g = h$ (μ-a.e.). If, in addition, g is a pseudopredictable function, then a.s.*

$$\int_0^\infty |g(t) - h(t)|\, dt = 0. \tag{3}$$

PROOF. By Lemma 2.3, $\mathcal{P} \subset \mathfrak{A}$. Hence, by Fubini's Theorem, $I_A(\omega, t)$ is a Borel function of t for any ω, provided $A \in \mathcal{P}$. But in that case (2) is true, since the left-hand side of (2) is just $\{t \colon I_A(\omega, t) = 1\}$. This proves (a).

Now let g be a $\mathcal{P}^\mu$-measurable function. By Theorem I.3.3, there exists a $\mathcal{P}$-measurable function h such that $g = h$ (μ-a.e.). By Definition I.2.1, this means that there exists a set $B \in \mathfrak{A}$ such that $\{(\omega, t) \colon g \neq h\} \subset B$ and $\mu(B) = 0$. It follows from the last equality by Fubini's Theorem that $\ell(\{t \colon (\omega, t) \in B\}) = 0$ (a.s.). Thus, for almost any ω, $g(\omega, t)$ as a function of t differs from the Borel function $h(\omega, t)$ only on a set of Lebesgue measure zero. This obviously implies (b) and (c). □

We will now prove several useful properties of pseudopredictable functions (now the completeness of $\mathcal{F}$ and $\mathcal{F}_t$ becomes essential).

3. LEMMA. *Let g be a pseudopredictable function, $p \in [1, \infty)$.*
(a) *If $g \geq 0$, then the integral*

$$\int_0^s g(t)\,dt \quad \left(:= \int_0^\infty I_{t \leq s} g(t)\,dt \right) \tag{4}$$

is $\mathcal{F}_s$-measurable for $s \in [0, \infty)$, $\mathcal{F}$-measurable for $s = \infty$, and

$$\mathbf{E} \int_0^\infty g(t)\,dt = \int_X g(\omega, t)\,\mu(d\omega dt). \tag{5}$$

If the integral (4) is finite for all $\omega \in \Omega$, $s \in [0, \infty)$, then it is continuous in s and for any $c > 0$

$$\tau_c(g) := \inf \left\{ s \geq 0 \colon \int_0^s g(t)\,dt \geq c \right\} \tag{6}$$

is a Markov time.
(b) *$g \in \mathcal{L}_p(\mathcal{P}, \mu)$ if and only if*

$$\mathbf{E} \int_0^\infty |g(t)|^p\,dt < \infty.$$

(c) *$g \in \mathcal{L}_{p,\mathrm{loc}}(\mathcal{P}, \mu)$ if for all $\omega \in \Omega$, $s \in [0, \infty)$*

$$\int_0^s |g(t)|^p\,dt < \infty. \tag{7}$$

Moreover, in that case we can take $\tau_n = \tau_n(|g|^p)$ as a localizing sequence for g and

$$\int_0^{\tau_n} |g(t)|^p\,dt \leq n, \qquad \mathbf{E} \int_0^{\tau_n} |g(t)|^p\,dt \leq n. \tag{8}$$

(d) *If $g \in \mathcal{L}_{p,\mathrm{loc}}(\mathcal{P}, \mu)$, then there exists a sequence $g_n \in S(\Pi)$ such that, for any $s \in [0, \infty)$ and $\varepsilon > 0$ we have*

$$\lim_{n \to \infty} P\left\{ \int_0^s |g(t) - g_n(t)|^p\,dt > \varepsilon \right\} = 0. \tag{9}$$

PROOF. (a) Let Σ denote the family of all predictable sets A such that the expression (4) with $g = I_A$ is $\mathcal{F}_s$-measurable for $s \in [0, \infty)$ and $\mathcal{F}$-measurable for $s = \infty$. If $\sigma \in \mathfrak{M}$ and $A = (0, \sigma]\!]$, then (4) is equal to $s \wedge \sigma$, which has the desired measurability properties. Therefore $\Pi \subset \Sigma$, and since Σ is obviously a λ-system, we obtain $\mathcal{P} = \sigma(\Pi) \subset \Sigma$. Thus (4) has the desired measurability properties when $g = I_A$ for any $A \in \mathcal{P}$. This is also true for all simple predictable functions and, using the standard approximation of measurable functions by simple functions and the Monotone Convergence Theorem, one can easily extend the result to all nonnegative predictable functions. To complete the proof, take the function h of Lemma 2 (c), $h \geq 0$, and note that, by (3),

$$\int_0^s g(t)\,dt = \int_0^s h(t)\,dt \quad \text{(a.s.)}, \tag{10}$$

and that by the completeness of $\mathcal{F}_s, \mathcal{F}$, functions that are equal a.s. to $\mathcal{F}_s$- or $\mathcal{F}$-measurable functions are also measurable in the appropriate sense. In addition, by Fubini's Theorem, equality (5) remains valid upon substitution of h for g. Since $h = g$ (μ-a.e.) and (10) is true (a.s.), we can return to g in this version of (5).

If the integral (4) is finite, it is continuous in s by the Dominated Convergence Theorem, and in that case $\tau_c(g)$ is a Markov time (see Example II.4.6).

Part (b) follows immediately from (a) (and Exercise 1.3). Moreover, by (a) and the fact that $(0, \tau_n]\!] \in \mathcal{P}$ under the assumptions of (c), we obtain

$$\mathbf{E} \int_0^{\tau_n} |g(t)|^p\,dt = \int_X |g|^p I_{(0,\tau_n]\!]}\,\mu(d\omega dt). \tag{11}$$

Since the inequalities in (8) are obvious and clearly $\tau_n \to \infty$, this proves (c).

To prove (d) take a localizing sequence σ_n for g. Then $gI_{(0,\sigma_n]\!]} \in \mathcal{L}_p(\mathcal{P}, \mu)$ and, by Lemma 2.3 (c), for every n there exists $g_n \in S(\Pi)$ such that (see (5))

$$\mathbf{E} \int_0^\infty |gI_{(0,\sigma_n]\!]} - g_n|^p\,dt \leq \frac{1}{n}.$$

Hence, by Chebyshev's inequality,

$$\begin{aligned} P\Big\{\int_0^s |g - g_n|^p\,dt \geq \varepsilon\Big\} &\leq P\{\sigma_n < s\} \\ &+ P\Big\{\sigma_n \geq s, \int_0^s |gI_{(0,\sigma_n]\!]} - g_n|^p\,dt \geq \varepsilon\Big\} \leq P\{\sigma_n < s\} + \frac{1}{n\varepsilon}. \end{aligned} \tag{12}$$

Finally, since the events $\{\sigma_n < s\}$ decrease when n increases and their intersection is contained in $\{\sigma_n \not\to \infty\}$, it follows that $P\{\sigma_n < s\} \to 0$ and (9) is a corollary of (12). □

4. EXERCISE. Prove that if $g \in \mathcal{L}_{p,\mathrm{loc}}(\mathcal{P}, \mu)$, then for any $s \in [0, \infty)$ inequality (7) is valid (a.s.).

5. DEFINITION. By H_p we denote the set of all pseudopredictable functions g with values in $[-\infty, \infty]$ satisfying inequality (7) for all $\omega \in \Omega$ and $s \in [0, \infty)$.

Part (c) of Lemma (3) means, in particular, that $H_p \subset \mathcal{L}_{p,\text{loc}}(\mathcal{P}, \mu)$. By Theorem 4.3, if $f \in H_2$, there exists a stochastic integral ξ_s of function f with variable upper limit, and by Theorem 4.8 $(\xi_s, \mathcal{F}_s)$ is a local martingale.

6. THEOREM. *Let $f \in H_2$, let ξ_s be the process of Definition 4.2, $\langle\xi\rangle_s$ the process defined by (1), τ a Markov time,*

$$A := \left\{\omega\colon \sup_{s<\infty} |\xi_s| < \infty\right\}, \qquad B := \left\{\omega\colon \langle\xi\rangle_\infty < \infty\right\},$$
$$C := \left\{\omega\colon \lim_{s\to\infty} \xi_s \text{ exists and is finite}\right\},$$

Then

(a)

$$\mathbf{E}\,\langle\xi\rangle_\tau \le \mathbf{E} \sup_{s<\infty} \left|\xi_{s\wedge\tau}\right|^2 \le 4\mathbf{E}\,\langle\xi\rangle_\tau. \tag{13}$$

(b) *If $\mathbf{E}\,\langle\xi\rangle_\tau < \infty$, then $(\xi_{s\wedge\tau}, \mathcal{F}_s)$ is a martingale for $s \ge 0$,*

$$\xi_\tau := \overline{\lim_{s\to\infty}}\, \xi_{s\wedge\tau} = \underline{\lim_{s\to\infty}}\, \xi_s$$

(a.s.) on the set $\{\omega\colon \tau(\omega) = \infty\}$ and the Wald identities *hold:*

$$\mathbf{E}\,\xi_\tau = 0, \qquad \mathbf{E}\,\xi_\tau^2 = \mathbf{E}\,\langle\xi\rangle_\tau.$$

(c) *$I_A = I_B = I_C$ (a.s.).*

PROOF. (a) For finite τ, inequalities (13) are just some of the inequalities in (4.9) written in different notations (cf. (5)). For arbitrary τ, they are easily obtained by using Fatou's Lemma, after substituting $\tau \wedge n$ for τ and letting $n \to \infty$ (cf. the last expression in (4.10)).

In part (b) we have $\mathbf{E}\,\langle\xi\rangle_\tau < \infty$, so that $f I_{(0,\tau]]}$ belongs to $\mathcal{L}_2(\mathcal{P}, \mu)$ (cf. (5)) and by Theorem 4.4

$$\xi_{s\wedge\tau} = \int_0^s f(t) I_{t\le\tau}\, dw_t \quad \text{(a.s.)}. \tag{14}$$

Now the assertions of (b) follow at once from Theorem 3.5, formula (2.4) and the isometric property of the stochastic integration operator on $(0, \infty)$.

If $\tau_n = \tau_n(f^2)$, then obviously $\xi_s = \xi_{s\wedge\tau_n}$ on the set $\{\tau_n = \infty\}$, and it follows from (14) by Theorem 3.5 that on the same set ξ_s has a finite limit as $s \to \infty$ (a.s.). This limit exists and is finite on $B = \cup_n\{\tau_n = \infty\}$ (a.s.) as well, or, equivalently, $I_B \le I_C$ (a.s.). The inequality $I_C \le I_A$ is obvious.

Now let $\sigma(n) = \inf\{s \ge 0\colon |\xi_s|^2 \ge n\}$, $n = 1, 2, \dots$. The first inequality in (13) implies $\mathbf{E}\,\langle\xi\rangle_{\sigma(n)} \le n$, $\langle\xi\rangle_{\sigma(n)} < \infty$ (a.s.), and $\langle\xi\rangle_\infty < \infty$ (a.s.) on the set $\{\sigma(n) = \infty\}$. Since $A = \cup_n\{\sigma(n) = \infty\}$, it follows that $\langle\xi\rangle_\infty < \infty$ (a.s.) on A, $I_A \le I_B$ (a.s.). Finally, $I_B \le I_C \le I_A \le I_B$ (a.s.). □

7. REMARK. In Theorem 6.11 we will strengthen the last assertion of this theorem.

6. Passage to a limit within the Itô stochastic integral. Itô's inequalities. Convergence in probability

As in the theory of usual integral, convergence theorems play an important role in the theory of stochastic integral. Let $(\Omega, \mathcal{F}, P)$ be a complete probability space, $(w_t, \mathcal{F}_t)$ a one-dimensional Wiener process on $(\Omega, \mathcal{F}, P)$, defined for $t \geq 0$, where the σ-algebras $\mathcal{F}_t$ are complete with respect to $\mathcal{F}, P$ (see Remark 3.6). The basic tools for proving limit theorems for stochastic integrals are certain inequalities. We call the first of them *Itô's inequality*, since it is essentially contained in his fundamental paper [21] (1951).

1. LEMMA. *Let* $f \in H_2$ *(see Sec. 5),*

$$(1) \qquad \xi_s := \int_0^s f(t)\, dw_t, \qquad \langle \xi \rangle_s := \int_0^s f^2(t)\, dt,$$

let $\delta, \varepsilon > 0$ *be numbers,* γ *a Markov time. Then*

$$(2) \qquad P\Big\{ \sup_{s<\infty} |\xi_{s\wedge\gamma}|^2 \geq \varepsilon \Big\} \leq P\{\langle \xi \rangle_\gamma \geq \delta\} + \frac{\delta}{\varepsilon}.$$

PROOF. Define $\tau = \tau_\delta(f^2)$, where $\tau_\delta(f^2)$ is taken from (5.6). The left-hand side of (2) is equal to

$$(3) \qquad \begin{aligned} &P\Big\{ \sup_{s\leq\gamma, s<\infty} |\xi_{s\wedge\tau}|^2 \geq \varepsilon,\ \gamma \leq \tau \Big\} + P\Big\{ \sup_{s<\infty} |\xi_{s\wedge\gamma}|^2 \geq \varepsilon,\ \gamma > \tau \Big\} \\ &\qquad \leq P\Big\{ \sup_{s<\infty} \Big| \int_0^s f(t) I_{t\leq\tau}\, dw_t \Big|^2 > \varkappa \Big\} + P\{\gamma > \tau\}, \end{aligned}$$

where $\varkappa \in (0, \varepsilon)$, and in derivation of the inequality we have used Theorem 4.4. The last term in (3) does not exceed the first term on the right of (2), since the events in question stand in the appropriate relation. To estimate the first term on the right of (3), note that $f I_{(0,\tau]\!]} \in \mathcal{L}_2(\mathcal{P}, \mu)$ (Lemma 5.3 (c)), the stochastic integral of a function of class $\mathcal{L}_2(\mathcal{P}, \mu)$ is a martingale (Theorem 3.5), and the square of a martingale is a submartingale (Lemma II.8.5). Then, by the Doob–Kolmogorov inequality (and (5.11)) we see that the term in question is at most the supremum over $T \geq 0$ of the expression

$$\frac{1}{\varkappa} \mathbf{E} \Big(\int_0^\infty f(t) I_{t\leq\tau\wedge T}\, dw_t \Big)^2 = \frac{1}{\varkappa} \mathbf{E} \int_0^{\tau\wedge T} f^2(t)\, dt \leq \frac{\delta}{\varkappa}$$

for any $\varkappa \in (0, \varepsilon)$. □

2. REMARK. The above proof follows Itô's reasoning of [21] (1951) and is presented here only for methodological reasons. To estimate the first term on the left of (3), instead of using the Doob–Kolmogorov inequality directly, we could have established a particular case of it for ξ_s^2. It is evident from the proof of the Doob-Kolmogorov

inequality that in that case we would have needed only the inequality

$$\mathbf{E}\Big(\int_0^\sigma f(t)\,dw_t\Big)^2 \le \mathbf{E}\int_0^\sigma f^2(t)\,dt \tag{4}$$

for all $\sigma \in \mathfrak{M}$, $f \in H_2$. Incidentally, if the right-hand side of (4) is finite, then (4) follows from Theorem 5.6 (b). If it is infinite, then (4) is obvious.

According to this remark, the proof of the next lemma differs from that of Lemma 1 in inessential details only. It is, therefore, natural to call the inequalities established in Lemma 3 the *generalized Itô inequalities*.

3. LEMMA. *Let η_t, ζ_t be continuous $\mathcal{F}_t$-adapted processes for $t \ge 0$. Assume that $\eta_t \ge 0$ and for all $\tau \in \mathfrak{M}$*

$$\mathbf{E}\eta_\tau \le \mathbf{E}\zeta_\tau. \tag{5}$$

Assume also that $\zeta_0 \le \delta_0$ for some constant $\delta_0 \in [0,\infty)$. Then for all numbers $\varepsilon > 0$, $\delta \ge \delta_0$ and Markov times γ

$$\begin{aligned} P\{\sup_t \eta_{t\wedge\gamma} \ge \varepsilon\} &\le P\{\sup_t \zeta_{t\wedge\gamma} \ge \delta\} + \frac{1}{\varepsilon}\mathbf{E}\,[\delta \wedge \sup_t \zeta_{t\wedge\gamma}] \\ &\le P\{\sup_t \zeta_{t\wedge\gamma} \ge \delta\} + \frac{\delta}{\varepsilon}. \end{aligned} \tag{6}$$

PROOF. By considering $\sup\{\zeta_s : s \le t\}$ rather than ζ_t, we can reduce the general case to that of increasing ζ_t. We will indeed limit ourselves to this case, denoting $\zeta_\infty := \lim_{t\to\infty}\zeta_t$. In addition, by replacing τ in (5) by $\tau \wedge n$ and letting $n \to \infty$, we conclude via Fatou's Lemma and the relation $\zeta_{\tau\wedge n} \le \zeta_\tau$ that inequality (5) holds for all Markov times under the stipulation that $\eta_\infty := 0$.

Now take $\varkappa \in (0,\varepsilon)$ and set

$$\tau = \inf\{t \ge 0 : \eta_t \ge \varkappa\}, \qquad \sigma = \inf\{t \ge 0 : \zeta_t \ge \delta\}.$$

Here τ, σ are Markov times and $\zeta_{\gamma\wedge\sigma} = \zeta_\gamma \wedge \zeta_\sigma \le \zeta_\gamma \wedge \delta \le \delta$ since $\zeta_0 \le \delta$ and $\zeta_\sigma \le \delta$. Hence, by the assumption and Chebyshev's inequality

$$\begin{aligned} P\{\sup_t \eta_{t\wedge\gamma} \ge \varepsilon\} &\le P\{0 < \tau < \gamma\} + P\{\eta_0 \ge \varkappa\} \\ &= P\{0 < \tau < \gamma, \sigma < \gamma\} + P\{0 < \tau < \gamma \le \sigma\} + P\{\eta_0 \ge \varkappa\} \\ &\le P\{\sigma < \gamma\} + P\{\eta_{\tau\wedge\gamma\wedge\sigma} \ge \varkappa\} \le P\{\zeta_\gamma \ge \delta\} + \frac{1}{\varkappa}\mathbf{E}\eta_{\tau\wedge\gamma\wedge\sigma} \\ &\le P\{\zeta_\gamma \ge \delta\} + \frac{1}{\varkappa}\mathbf{E}\zeta_{\tau\wedge\gamma\wedge\sigma} \le P\{\zeta_\gamma \ge \delta\} + \frac{1}{\varkappa}\mathbf{E}(\delta \wedge \zeta_\gamma). \end{aligned}$$

Letting $\varkappa \uparrow \varepsilon$ in the inequality between the outermost terms, we obtain the first inequality of (6) for our increasing process ζ. Since the second inequality is obvious, the lemma is proved. □

Clearly, Itô's inequality (2) is obtained from (6) by taking $\eta_t = |\xi_t|^2$, $\zeta_t = \langle\xi\rangle_t$, $\delta_0 = 0$, which is possible by virtue of (4).

The following definition will allow us to state the theorem on passage to a limit within stochastic integral in the most natural way.

4. DEFINITION. Let $\zeta, \zeta_1, \zeta_2, \dots$ be random d-dimensional vectors on $(\Omega, \mathcal{F}, P)$. We will say that the sequence ζ_n *converges in probability* to ζ as $n \to \infty$, writing

$$\zeta_n \xrightarrow{P} \zeta, \quad \text{or } P\text{-}\lim_{n\to\infty} \zeta_n = \zeta \quad \text{(a.s.)}, \tag{7}$$

if $P\{|\zeta_n - \zeta| \geq \varepsilon\} \to 0$ as $n \to \infty$ for any $\varepsilon > 0$.

5. REMARK. The second formula of (7) is meaningful, in the sense that if $\zeta_n \xrightarrow{P} \zeta$, $\zeta_n \xrightarrow{P} \eta$, then $\zeta = \eta$ (a.s.). Indeed, in that case, for any $\varepsilon > 0$,

$$\{\omega : |\zeta - \eta| \geq 2\varepsilon\} \subset \{\omega : |\zeta - \zeta_n| \geq \varepsilon\} \cup \{\omega : |\zeta_n - \eta| \geq \varepsilon\},$$

$$P\{|\zeta - \eta| \geq 2\varepsilon\} \leq P\{|\zeta - \zeta_n| \geq \varepsilon\} + P\{|\eta - \zeta_n| \geq \varepsilon\} \longrightarrow 0,$$

that is, the left-hand side of the last expression vanishes and $\zeta = \eta$ (a.s.).

6. THEOREM. *Let $f, f_1, f_2, \dots \in H_2$, and let γ be a Markov time. Then*

$$P\text{-}\lim_{n\to\infty} \int_0^{\gamma} |f(t) - f_n(t)|^2 \, dt = 0 \quad (a.s.) \tag{8}$$

if and only if

$$P\text{-}\lim_{n\to\infty} \sup_s \left| \int_0^{s\wedge\gamma} f(t)\, dw_t - \int_0^{s\wedge\gamma} f_n(t)\, dw_t \right| = 0 \quad (a.s.). \tag{9}$$

PROOF. To derive (9) from (8) replace f in Itô's inequality (2) by $f - f_n$ and let $n \to \infty$ and then $\delta \downarrow 0$. To derive (8) from (9), do the same for the inequality

$$P\left\{ \int_0^{\gamma} f^2(t)\, dt \geq \varepsilon \right\} \leq P\left\{ \sup_s \left| \int_0^{s\wedge\gamma} f(t)\, dw_t \right|^2 \geq \delta \right\} + \frac{\delta}{\varepsilon},$$

which follows from Theorem 5.6 (a) and Itô's inequalities (6) with

$$\eta_t = \int_0^t f^2(s)\, ds, \qquad \zeta_t = \sup_{s\leq t} \left| \int_0^s f(u)\, dw_u \right|^2, \qquad \delta_0 = 0. \quad \square \tag{10}$$

7. REMARK. It is clear that if (9) is true, then, for any $s \in [0, \infty)$,

$$\int_0^{s\wedge\gamma} f_n(t)\, dw_t \xrightarrow{P} \int_0^{s\wedge\gamma} f(t)\, dw_t.$$

The proof of Theorem 6 was based on generalized Itô's inequalities, which can also be used for other purposes. For example, integrating the Doob-Kolmogorov inequality in Sec. 3 we established Doob's moment inequality. Observing that for a random variable $\zeta \geq 0$ and a number $\alpha \in (0,1)$, by Fubini's Theorem,

$$\int_0^\infty P\{\zeta \geq \varepsilon^{1/\alpha}\} d\varepsilon = \int_0^\infty P\{\zeta^\alpha \geq \varepsilon\} d\varepsilon = \mathbf{E}\zeta^\alpha,$$

$$\int_0^\infty \varepsilon^{-1/\alpha} \mathbf{E}\,(\varepsilon^{1/\alpha} \wedge \zeta) d\varepsilon = \mathbf{E}\left(\int_{\zeta^\alpha}^\infty \zeta \varepsilon^{-1/\alpha}\, d\varepsilon + \int_0^{\zeta^\alpha} d\varepsilon \right) = \frac{1}{1-\alpha} \mathbf{E}\zeta^\alpha,$$

and substituting $\varepsilon^{1/\alpha}$ for ε and δ in (6), we arrive at the following result.

8. THEOREM. *Under the assumptions of Lemma* 3, *let* $\delta_0 = 0$ *and* $\zeta_t \geq 0$. *Then for any* $\alpha \in (0,1)$

$$\mathbf{E} \sup_t \eta^\alpha_{t\wedge\gamma} \leq \frac{2-\alpha}{1-\alpha} \mathbf{E} \sup_t \zeta^\alpha_{t\wedge\gamma}.$$

A particular case of this theorem, which is used most frequently, is obtained by applying it with $\alpha = 1/2$ to the pair η, ζ from (10) and the pair $\eta = |\xi|^2, \zeta = \langle\xi\rangle$ (see Theorem 5.6 and (4)).

9. COROLLARY. *If* $f \in H_2$, *then* (*in the notations from* (1)) *for any Markov time* γ, *we have the* Davis inequality:

$$\frac{1}{3} \mathbf{E} \sup_t |\xi_{t\wedge\gamma}| \leq \mathbf{E}\left(\langle\xi\rangle_\gamma^{\frac{1}{2}}\right) \leq 3\mathbf{E} \sup_t |\xi_{t\wedge\gamma}|.$$

10. EXERCISE. Prove that if the middle expression in the last inequality is finite, then $\mathbf{E}\,\xi_\gamma = 0$ (cf. Theorems 5.6, II.7.6).

We have so far applied Lemma 3 only with $\delta_0 = 0$. In the following theorem we will see that Lemma 3 is also useful when $\delta_0 > 0$.

11. THEOREM. (a) *Let* η_s *be a local supermartingale,* $\eta_s \geq 0$, *and* $\mathbf{E}\,\eta_0 < \infty$. *Then for any* $c > 0$

$$P\Big\{ \sup_s \eta_s \geq c \Big\} \leq \frac{1}{c} \mathbf{E}\,\eta_0. \tag{11}$$

In particular, the left-hand side of (11) *vanishes as* $c \to \infty$ *and, consequently, almost all sample paths of* η *are bounded on* $[0, \infty)$.

(b) *Let* $f \in H_2$. *Using the notation* (1), *let* A, B, C *be as in Theorem* 5.6 *and*

$$D = \Big\{\omega\colon\ \sup_s \xi_s < \infty\Big\}.$$

Then $I_A = I_B = I_C = I_D$ (*a.s.*).

PROOF. (a) There is a sequence of Markov times $\tau(n) \uparrow \infty$ (a.s.) such that $\mathbf{E}\,\eta_0 \geq \mathbf{E}\,\eta_{\tau\wedge\tau(n)}$ for any $\tau \in \mathfrak{M}$. Hence, by Fatou's Lemma, $\mathbf{E}\,\eta_0 \geq \mathbf{E}\,\eta_\tau$, and (11) follows from Lemma 3 with $\zeta_s \equiv \delta_0 = \mathbf{E}\,\eta_0$, $\delta \downarrow \mathbf{E}\,\eta_0$, $\gamma = \infty$.

(b) It will suffice to prove that $I_D = I_A$ (a.s.). Denote $\tau(n) = \inf\{s\colon \xi_s \geq n\}$. Then the process $n - \xi_{s\wedge\tau(n)}$ is a nonnegative local martingale. It is bounded (a.s.) by (a). In particular, ξ_s is bounded as a function of s a.s. on $\{\omega\colon \tau(n) = \infty\}$ and on

$D = \cup_n\{\tau(n) = \infty\}$. This means that $I_D \le I_A$ (a.s.). Since the inequality $I_A \le I_D$ is obvious, the theorem is proved. □

We will now use the generalized Itô inequality to derive some useful results concerning what one might call unilaterally bounded local martingales.

12. THEOREM. (a) *Let m_t^n, $n = 1, 2, \dots$ be local martingales with $m_0^n = 0$ such that*

$$\sup_{t \le T}(m_t^n)_+ \xrightarrow{P} 0 \tag{12}$$

for any $T \in [0, \infty)$. Then for any $T \in [0, \infty)$

$$\sup_{t \le T} |m_t^n| \xrightarrow{P} 0. \tag{13}$$

(b) *Assume that the processes η_t^n do not decrease as functions of t, $\eta_t^n \ge 0$, $\eta_t^n \xrightarrow{P} 0$ for every t, and let m_t^n be local martingales, $m_0^n = 0$. Suppose that $m_t^n \le \eta_t^n$ for all ω, t or $m_t^n + \eta_t^n \ge 0$ for all ω, t. Then* (13) *is true for all $T \in [0, \infty)$.*

(c) *Let the processes η_t^n be continuous in t, $\eta_t^n \ge 0$, and let m_t^n be local martingales, $m_0^n = 0$, and $m_t^n \le \eta_t^n$ for all ω, t or $m_t^n + \eta_t^n \ge 0$ for all ω, t. Then for all $T, a, b \in (0, \infty)$*

$$P\Big\{\sup_{t \le T} |m_t^n| \ge a\Big\} \le P\Big\{2\sup_{t \le T} \eta_t^n \ge b\Big\} + \frac{b}{a} \tag{14}$$

and the processes m_t^n are bounded in probability *on $[0, T]$ uniformly in n, that is,*

$$\lim_{c \to \infty} \sup_n P\Big\{\sup_{[0,T]} |m_t^n| \ge c\Big\} = 0, \tag{15}$$

if the processes η_t^n are bounded in probability on $[0, T]$ uniformly in n.

PROOF. Part (b) follows immediately from (a), since if $m_t^n \le \eta_t^n$, then $(m_t^n)_+ \le \eta_t^n$, and we obtain (12). Similarly, if $\eta_t^n + m_t^n \ge 0$, then $(-m_t^n)_+ \le \eta_t^n$.

To prove (c), we note that for suitable Markov times τ we have $\mathbf{E}\, m_\tau^n = 0$, and

$$\mathbf{E}\,|m_\tau^n| = 2\mathbf{E}\,(m_\tau^n)_+ \le 2\mathbf{E}\, \sup_{t \le \tau}(m_t^n)_+. \tag{16}$$

By Fatou's Lemma, the inequality between the outer terms can be extended to all $\tau \in \mathfrak{M}$. Together with Lemma 3, this gives (14) with η_t^n replaced by $(m_t^n)_+$. If $m_t^n \le \eta_t^n$, inequality (14) is now obvious. Taking suprema over n in (14), letting $a \to \infty$ and then $b \to \infty$, we obtain (15). The case $m_t^n + \eta_t^n \ge 0$ is dealt with by substituting $(-m^n)$ for m^n.

Finally, to prove (a) we need only remember that η_t^n can be replaced in (14) by $(m_t^n)_+$, then let $n \to \infty$ and $b \downarrow 0$, in that order. □

We will often need some properties of convergence in probability. The most important ones are collected in the following lemma.

13. LEMMA. (a) *$\zeta_n \xrightarrow{P} \zeta$ if and only if we can extract from any subsequence $\{\zeta_{n(k)}\}$ of the sequence $\{\zeta_n\}$ a subsequence $\{\zeta_{m(k)}\}$ such that $\zeta_{m(k)} \to \zeta$ (a.s.) as $k \to \infty$.*

(b) *If $\zeta_n \to \zeta$ (a.s.), then $\zeta_n \xrightarrow{P} \zeta$.*

(c) *If $\zeta_n \xrightarrow{P} \zeta$ and $f(x)$ is a continuous function on E_d, then $f(\zeta_n) \xrightarrow{P} f(\zeta)$.*

(d) *If* ξ_n, η_n *are random variables and* $\xi_n \xrightarrow{P} \xi$ *and* $\eta_n \xrightarrow{P} \eta$, *then* $\xi_n+\eta_n \xrightarrow{P} \xi+\eta$, $\xi_n\eta_n \xrightarrow{P} \xi\eta$.

(e) *If* $\mathbf{E}\left(1 \wedge |\zeta_n - \zeta|^\alpha\right) \to 0$, *where* $\alpha > 0$, *then* $\zeta_n \xrightarrow{P} \zeta$.

(f) *If* $\zeta_n \xrightarrow{P} \zeta$, $|\zeta_n| \le \eta$, $\mathbf{E}\eta < \infty$, *then* $\mathbf{E}\,|\zeta_n - \zeta| \to 0$.

(g) *If* $\zeta_n \xrightarrow{P} \zeta$, $p > m > 0$, *and* $\sup_n \mathbf{E}\,|\zeta_n|^p < \infty$, *then* $\mathbf{E}\,|\zeta_n - \zeta|^m \to 0$.

PROOF. (a) Considering $\zeta_n - \zeta$ instead of ζ_n, we reduce matters to the case $\zeta = 0$. Furthermore, if $\zeta_n \xrightarrow{P} 0$, we can find a sequence $n(k)$ such that $P\{|\zeta_{n(k)}| \ge k^{-1}\} \le k^{-2}$. Since the series $\sum k^{-2}$ converges, it follows by the Borel-Cantelli Lemma that for almost every ω there exists $k_0(\omega) < \infty$ such that $|\zeta_{n(k)}| \le k^{-1}$ for all $k \ge k_0(\omega)$. In particular, $\zeta_{n(k)} \to 0$ (a.s.). Any subsequence of $\{\zeta_n\}$ can be considered in the same way, so this proves necessity. To prove sufficiency, fix $\varepsilon > 0$ and select a sequence $\zeta_{n(k)}$ such that

$$\lim_{k\to\infty} P\{|\zeta_{n(k)}| \ge \varepsilon\} = \varlimsup_{n\to\infty} P\{|\zeta_n| \ge \varepsilon\}. \tag{17}$$

Extracting a subsequence $\zeta_{m(k)} \to 0$ (a.s.) from $\zeta_{n(k)}$, we note that the right-hand side of (17) remains unchanged upon substitution of $m(k)$ for $n(k)$. By Chebyshev's inequality and the Dominated Convergence Theorem,

$$P\{|\zeta_{m(k)}| > \varepsilon\} = P\{(2\varepsilon) \wedge |\zeta_{m(k)}| > \varepsilon\} \le \frac{1}{\varepsilon}\mathbf{E}\left[(2\varepsilon) \wedge |\zeta_{m(k)}|\right] \longrightarrow 0.$$

Thus the left-hand side of (17) vanishes, proving (a).

Parts (b)–(d) are obvious corollaries of (a). Part (e) follows from Chebyshev's inequality. To prove (f), it suffices to use (a) and the Dominated Convergence Theorem and extract a subsequence whose limit is the upper limit of $\mathbf{E}\,|\zeta_n - \zeta|$. Under the assumptions of (g), $\mathbf{E}\,|\zeta|^p < \infty$ by Fatou's Lemma and by (a). Hence the only case to be studied in details is $\zeta = 0$. But then (g) follows easily from (c), (f), and the inequalities

$$|\zeta_n|^m \le \left(c \wedge |\zeta_n|\right)^m + I_{|\zeta_n|>c}|\zeta_n|^m \le \left(c \wedge |\zeta_n|\right)^m + c^{m-p}|\zeta_n|^p,$$

$$\mathbf{E}\,|\zeta_n|^m \le \mathbf{E}\left(c \wedge |\zeta_n|\right)^m + c^{m-p}\sup_k \mathbf{E}\,|\zeta_k|^p,$$

which hold for any $c > 0$. □

Theorem 6 and Lemma 13 are helpful in proving many properties of stochastic integrals, when one starts from the properties of integrals of functions belonging to $\mathcal{S}(\Pi)$ or $\mathcal{L}_2(\mathcal{P},\mu)$, and also applies Lemma 5.3(d). Here is an example. We will generalize Lemma 2.4, which states when a "constant" may be taken through the stochastic integral sign.

14. THEOREM. *Let* $f \in H_2$; *let* τ *be a Markov time and* g *a finite* $\mathcal{F}_\tau$*-measurable random variable. Then* $gI_{\tau<t}f(t) \in H_2$, *and for all* $s \in [0,\infty)$ *at once (a.s.),*

$$\int_\tau^s gf(t)\,dw_t = \int_0^s gI_{\tau<t}f(t)\,dw_t = g\int_0^s I_{\tau<t}f(t)\,dw_t = g\int_\tau^s f(t)\,dw_t. \tag{18}$$

PROOF. The first assertion of the theorem follows directly from the definition of H_2 and Lemma 2.2. By the continuity in s of all the expressions in (18), it will suffice to prove (18) for any fixed s. If g is bounded and $f \in \mathcal{L}_2(\mathcal{P}, \mu)$, the middle equality is simply Lemma 2.4. It is now easy to pass to the limit, first from bounded g to finite one, using Theorem 6, Remark 7, and Lemma 13 (d); we then go from $f \in \mathcal{L}_2(\mathcal{P}, \mu)$ to $f \in H_2$, using Lemma 5.3 (d) as well. The remaining equalities in (18) follow from Definition 4.10. □

Lemmas 5.3 (d) and 13 (d) also yield the following generalization of Lemma II.8.5 (c).

15. THEOREM. *Let η_s be a local submartingale, $g \in H_1$, $g \geq 0$. Then*

$$\varkappa_s := \eta_s \int_0^s g(t)\,dt - \int_0^s \eta_t g(t)\,dt = \int_0^s (\eta_s - \eta_t) g(t)\,dt \tag{19}$$

is also a local submartingale. The word "submartingale" can be replaced throughout by "supermartingale," and if η_s is a local martingale, then $\varkappa_s$ is a local martingale for any $g \in H_1$.

PROOF. It follows easily from Lemmas 5.3(d), 13(d) and Theorem 4.9(k) that the only case needing consideration is $g \in S(\Pi)$. In that case, for any $T \in [0, \infty)$ the Riemann sums for the last integral in (19), constructed for the partitions $\{0, 1/n, 2/n, \dots\} \cap [0, s]$, converge to the integral as $n \to \infty$ uniformly in $s \in [0, T]$ for any ω. It remains to note that the terms in the Riemann sums can be assumed to have the form $(\eta_s - \eta_{s\wedge(k/n)})g(k/n)(1/n)$; and then apply Theorem 4.9 (j), (k) and Lemma 13 (b). □

We have defined convergence in probability only for sequences. The concept can be defined similarly for random vectors that depend on a continuously varying parameter, or, in general, a parameter ranging over some directed set. For example, if ζ_{nk} are random d-dimensional vectors, $n, k = 1, 2, \dots$, then the notation $\zeta_{nk} \xrightarrow{P} 0$ as $n, k \to \infty$ means that $P\{|\zeta_{nk}| \geq \varepsilon\} \to 0$ as $n, k \to \infty$ for all $\varepsilon > 0$. In this connection we offer the reader an exercise, which will be used later in Sec. 10 and Chapter V.

16. EXERCISE. State and prove an analog of part (b) of Theorem 12 for the case when the martingales and increasing processes depend on n, k, where $n, k = 1, 2, \dots$ and $n, k \to \infty$.

17. EXERCISE. We say that a sequence of random d-dimensional vectors ζ_n, $n = 1, 2, \dots$ is *bounded in probability* if

$$\lim_{c \to \infty} \sup_n P\{|\zeta_n| \geq c\} = 0.$$

Noting that $P\{|\zeta_n| \geq c\}$ decreases with increasing c, prove that if $\zeta_n \xrightarrow{P} 0$, then the ζ_n's are bounded in probability. (It is convenient here to argue by contradiction.) Using the inclusion $\{\omega\colon |\zeta_n + \eta_n| \geq 2c\} \subset \{\omega\colon |\zeta_n| \geq c\} \cup \{\omega\colon |\eta_n| \geq c\}$, prove that if ζ_n, η_n are bounded in probability, then their sum is also bounded in probability.

We present two more results about convergence in probability. The first of them concerns sequences that are, so to speak, *Cauchy in probability*, and will be used in Sec. 10 and later. The second one will be needed in Chapter V.

18. LEMMA. *Let α_s^n be continuous $\mathcal{F}_s$-adapted d-dimensional stochastic processes, defined for $s \in [0,\infty)$, $n = 1,2,\dots$, such that for any $T \in [0,\infty)$*

$$\sup_{s \le T} |\alpha_s^n - \alpha_s^k| \xrightarrow{P} 0 \tag{20}$$

as $n, k \to \infty$. Then there exists a continuous $\mathcal{F}_s$-adapted process α_s such that for any $T \in [0,\infty)$

$$\sup_{s \le T} |\alpha_s^n - \alpha_s| \xrightarrow{P} 0. \tag{21}$$

PROOF. Select a sequence $n(r)$, $r = 1,2,\dots$, such that $n(r) \to \infty$ as $r \to \infty$ and

$$P\Big\{ \sup_{s \le r} |\alpha_s^{n(r+1)} - \alpha_s^{n(r)}| > r^{-2} \Big\} \le r^{-2}.$$

Then by the Borel–Cantelli Lemma (cf. the proof of Lemma 13) a.s.

$$\sum_{r=1}^{\infty} \sup_{s \le r} |\alpha_s^{n(r+1)} - \alpha_s^{n(r)}| < \infty, \tag{22}$$

and (by Cauchy's test, for example), if Ω' is the set on which (22) holds, then the sequence $\alpha_s^{n(r)}$ converges on Ω', uniformly on any finite interval of $[0,\infty)$, to a function α_s, which is continuous in s. Define $\alpha_s \equiv 0$ for $\omega \notin \Omega'$. Since $P(\Omega') = 1$, it follows that $\Omega' \in \mathcal{F}_s$ for any s, and α_s is a continuous $\mathcal{F}_s$-adapted process. Finally, (21) follows from the inequality $|\alpha_s^r - \alpha_s| \le |\alpha_s^r - \alpha_s^{n(r)}| + |\alpha_s^{n(r)} - \alpha_s|$, the relations $P(\Omega') = 1$ and (20), and Lemma 13. □

19. LEMMA. *Let $R, T \in [0,\infty)$ be constants, $\tau^n \le T$ nonnegative random variables and $q_t^n = q_t^n(\omega)$ d-dimensional $\mathfrak{A}^\mu$-measurable functions such that $|q_t^n| I_{t \le \tau^n} \le R$ and $q_t^n I_{t \le \tau^n}$ converge to zero* in measure μ, *that is, for any $\varepsilon > 0$, and $n \to \infty$*

$$\mu\Big((\omega, t) \colon |q_t^n| I_{t \le \tau^n} \ge \varepsilon\Big) \longrightarrow 0 \tag{23}$$

as $n \to \infty$.

Let $f(t,x) = f(\omega,t,x)$ be a real-valued function defined on $\Omega \times (0,\infty) \times E_d$, continuous in x at zero and such that $f(t,0) = 0$ and suppose that for all ω the integral

$$\int_0^T \sup_{|x| \le R} |f(t,x)|\, dt$$

is defined and finite.

Finally, assume that for any ω the expression

$$\zeta^n := \int_0^{\tau^n} f(t, q_t^n)\, dt := \int_0^{\infty} I_{t \le \tau^n} f(t, q_t^n)\, dt$$

is meaningful and ζ^n are random variables. Then $\zeta^n \xrightarrow{P} 0$.

PROOF. Since convergence in measure possesses properties similar to those of convergence in probability, there exists a subsequence $n(k)$ such that $q_t^{n(k)} I_{t \le \tau^{n(k)}} \to 0$ (μ-a.e.). Covering the $\mathfrak{A}^\mu$-measurable set of all (ω, t) for which this convergence does not hold by a $\mathfrak{A}$-measurable set of measure zero and applying Fubini's Theorem, we conclude, as in the proof of Lemma 5.2, that for almost all ω the above convergence holds for almost all t. Hence, by the Dominated Convergence Theorem, $\zeta^{n(k)} \to 0$ for almost all ω. Since any subsequence of sequence ζ^n may be treated in this way, our lemma is proved owing to Lemma 13 (a). □

20. EXERCISE. Prove that convergence $\zeta_n \xrightarrow{P} \zeta$ is equivalent to convergence in the sense of the *metric*: $\rho(\zeta_n, \zeta) := \mathbf{E}\left(1 \wedge |\zeta_n - \zeta|\right) \to 0$, that is convergence in probability is metrizable.

21. PROBLEM. Give an example of a sequence that converges in probability but not in the sense of almost sure convergence. By using Lemma 13 (a) prove that if there were a real-valued function $f(\xi, \eta)$ defined on the set of all pairs of random variables (ξ, η), such that $f(\zeta_n, \eta) \to 0$ if and only if $\zeta_n \to \zeta$ almost surely, then almost sure convergence would be equivalent to convergence in probability. Hence conclude that almost sure convergence is not necessarily metrizable.

7. Itô's integral with respect to a multidimensional Wiener process

Let $(\Omega, \mathcal{F}, P)$ be a complete probability space, d_1 an integer, $d_1 \ge 1$, and $(w_t, \mathcal{F}_t)$ a d_1-dimensional Wiener process on $(\Omega, \mathcal{F}, P)$, defined for $t \ge 0$, where the σ-algebras $\mathcal{F}_t$ are complete with respect to $\mathcal{F}, P$. Fix an integer $d \ge 1$.

1. DEFINITION. Let $\sigma(t) = \left(\sigma^{ij}(t) \colon i = 1, \dots, d,\ j = 1, \dots, d_1\right)$ be a function on $\Omega \times (0, \infty)$ with values in the set of $(d \times d_1)$-matrices. We write $\sigma \in H_2$ if $\sigma^{ij} \in H_2$ for all i, j, and let

$$\int_0^s \sigma(t)\, dw_t \tag{1}$$

denote the stochastic process with values in E_d whose ith coordinate is

$$\int_0^s \sigma^i(t)\, dw_t := \sum_{j=1}^{d_1} \int_0^s \sigma^{ij}(t)\, dw_t^j. \tag{2}$$

Since H_2 is defined independently of any Wiener process, Definition 1 is legitimate. Moreover, we have defined the stochastic integral (1) with respect to the multidimensional Wiener process w_t in terms of stochastic integrals with respect to the one-dimensional processes w_t^j, and the properties of these integrals automatically yield much information about the properties of the integral (1). For brevity we will not formulate all these properties, and will discuss only few of them (cf. Theorem 5.6).

2. Theorem. *Let $\sigma \in H_2$ and let τ be a Markov time. Denote the process* (1) *by ξ_s and let*

$$
\langle\xi\rangle_s := \int_0^s \operatorname{tr}\sigma(t)\sigma^*(t)\,dt \left(= \sum_{i,j} \int_0^s \left(\sigma^{ij}(t)\right)^2 dt \right),
$$

(3)
$$
A := \left\{\omega\colon \sup_{s<\infty} |\xi_s| < \infty\right\}, \qquad B := \left\{\omega\colon \langle\xi\rangle_\infty < \infty\right\},
$$

$$
C := \left\{\omega\colon \lim_{s\to\infty} \xi_s \text{ exists}\right\}.
$$

Then

(a) *We have*

(4)
$$
\mathbf{E}\,\langle\xi\rangle_\tau \le \mathbf{E} \sup_{s<\infty} |\xi_{s\wedge\tau}|^2 \le 4\mathbf{E}\,\langle\xi\rangle_\tau;
$$

(b) *if $\mathbf{E}\,\langle\xi\rangle_\tau < \infty$, then $\left(\xi^i_{s\wedge\tau}, \mathcal{F}_s\right)$ are martingales for $s \ge 0$, $i = 1, \dots, d$, furthermore*

(5)
$$
\xi^i_\infty := \overline{\lim_{s\to\infty}}\, \xi^i_s = \underline{\lim_{s\to\infty}}\, \xi^i_s, \qquad i = 1, \dots, d \quad (a.s.)
$$

on the set $\{w\colon \tau = \infty\}$, and the Wald identities *hold*:

(6)
$$
\mathbf{E}\,\xi_\tau = 0, \qquad \mathbf{E}\,|\xi_\tau|^2 = \mathbf{E}\,\langle\xi\rangle_\tau;
$$

(c) $I_A = I_B = I_C$ *(a.s.)*.

Proof. We begin with (b). The martingale property, the relations (5) and the first equality in (6) follow immediately from definition (2) and from Theorem 5.6. Writing $|\xi_\tau|^2$ as the sum of squares of ξ^i_s and squaring the sums in (2), we see that (6) will be proved if we can show that the terms in (2) with $s = \tau$ are not correlated. If $s = \tau$, it follows from Theorems 4.4 and 3.5 that the terms in question can be rewritten as stochastic integrals from 0 to ∞, and all we have to prove is that for $f, g \in \mathcal{L}_2(\mathcal{P}, \mu)$, $i \ne j$

(7)
$$
\mathbf{E}\left(\int_0^\infty f(t)\,dw^i_t \int_0^\infty g(t)\,dw^j_t \right) = 0.
$$

The bilinearity of the left-hand side of (7) in f and g, the Cauchy-Bunyakovskiĭ inequality and the isometric property of the stochastic integration operators now reduce the proof of (7) for fixed g to the case $f \in S(\Pi)$, then to the case $f = I_{(0,\tau]\!]}$, where $\tau \in \mathfrak{M}$, and finally to the case when also $g = I_{(0,\sigma]\!]}$, where $\sigma \in \mathfrak{M}$. In the latter case (7) follows from Theorem II.6.9, and (b) is proved.

Next, let τ^{ij}_n be localizing sequences for σ^{ij}. Then

$$
\tau_n := \min\{\tau^{ij}_n : i = 1, \dots, d,\ j = 1, \dots, d_1\}
$$

is clearly a localizing sequence for σ^{ij}, and for any Markov time τ the second equality in (6) holds upon substitution of $\tau \wedge \tau_n$ for τ. The first inequality of (4) follows from this and Fatou's Lemma. To prove the second inequality, we note that

$$
\left|\xi_{s\wedge\tau_n}\right| = \sup_i \left(\xi_{s\wedge\tau_n}, x_i\right),
$$

where $\{x_i\}$ is a countable dense subset of $\{x \in E_d : |x| = 1\}$. Hence by Lemma II.8.5, $|\xi_{s\wedge\tau_n}|$ is a submartingale. Therefore, the second inequality (4) with $\tau \wedge \tau_n$ substituted for τ follows from Doob's moment inequality and the second Wald identity. By Fatou's Lemma, we get (4) with no substitutions. Finally, the proof of (c) is a literal repetition of the corresponding part in the proof of Theorem 5.6. □

3. REMARK. In the previous section we derived (6.2) and the Davis inequality from Lemma 6.3. By Theorem 2, these proofs carry over literally to the multidimensional case. In particular, these inequalities hold with the same constants, independently of d, d_1.

The second Wald identity and Definition 4.7 imply the following corollary.

4. COROLLARY. *The process* $|\xi_s|^2 - \langle\xi\rangle_s$ *is a local martingale for which*

$$\tau_n = \inf\{s \geq 0 : \langle\xi\rangle_s \geq n\}$$

is a localizing sequence.

8. Itô's formula

In usual calculus, after introducing the concept of an integral, one considers rules of integration and uses them to compile a table of elementary integrals. Of all the integration rules, the most important are the formulas for change of variables and for integration by parts, both proved using the formula for the derivative of a composite function, in particular, the formula for the derivative of a product. The formula for the stochastic differential of a composite function is known as Itô's formula. We encountered particular cases of this formula in Sec. 2.

We return to the assumptions of the previous section.

1. DEFINITION. Let $b(t) = (b^i(t),\ i = 1, \ldots, d)$ be a E_d-valued function on $\Omega \times (0, \infty)$ belonging to H_1, that is, such that $b^i \in H_1$ for all $i = 1, \ldots, d$. Let $\sigma = (\sigma^{ij} : i = 1, \ldots, d,\ j = 1, \ldots, d_1) \in H_2$, and let ξ_s be a continuous $\mathcal{F}_s$-adapted process with values in E_d, such that for all $s \in [0, \infty)$ at once (a.s.)

$$\xi_s = \xi_0 + \int_0^s \sigma(t)\, dw_t + \int_0^s b(t)\, dt. \tag{1}$$

Then we will say that ξ has a *stochastic differential* (with respect to w) equal to $\sigma(s)\, dw_s + b(s)\, ds$ and write $d\xi_s = \sigma(s)\, dw_s + b(s)\, ds$.

It is important to note that the expression $\sigma(s)\, dw_s + b(s)\, ds$ must be treated only formally. We could just as well use notation $(\sigma(s), b(s))$, and it is only the convenience of operating with expressions like $\sigma(s)\, dw_s + b(s)\, ds$ that gives advantage to this representation as compared to all the others. Note also that the equality $d\xi_s = \sigma(s)\, dw_s + b(s)\, ds$ means that equality (1) is valid and nothing more.

We have introduced formal d-dimensional expressions $\sigma(s)\, dw_s + b(s)\, ds$. Now we will define rules of operations with them, paying no attention, whether these rules will give us stochastic differentials of some processes, or they will not. We will assume that when we multiply them by functions of (ω, s), add together, or calculate scalar products of such expressions, then the usual rules of opening brackets, combining like terms, and the following multiplication table

$$(2)\qquad dw_s^i\,dw_s^j=\delta^{ij}\,ds,\qquad ds\,dw_s^i=dw_s^i\,ds=(ds)^2=0,\qquad i,j=1,\dots,d_1$$

hold.

It turns out that stochastic differential is defined uniquely by a process ξ. To prove this statment (and also in some other cases) the following lemma is useful.

2. LEMMA. *Let η_s,ζ_s be continuous $\mathcal{F}_s$-adapted processes, defined on Ω for $s\in[0,\infty)$. Assume that for almost any ω functions η_s,ζ_s are nondecreasing, and $\eta_s-\zeta_s$ is a local martingale. Then $\eta_s-\zeta_s=\eta_0-\zeta_0$ for all $s\in[0,\infty)$ at once a.s.*

PROOF. Since we can always replace η_s,ζ_s by $\eta_s-\eta_0,\zeta_s-\zeta_0$, we will assume without limiting the generality that $\eta_0=\zeta_0=0$. Besides, let $\tau_n=\inf\{s\ge 0:\eta_s+\zeta_s\ge n\}$. Then $\eta_{s\wedge\tau(n)},\zeta_{s\wedge\tau(n)}$ are bounded and increasing and by Theorem 4.9 the process $\eta_{s\wedge\tau(n)}-\zeta_{s\wedge\tau(n)}$ is a local martingale. Since $\tau(n)\uparrow\infty$, it suffices to prove that for any n the process $\eta_{s\wedge\tau(n)}-\zeta_{s\wedge\tau(n)}$ is indistinguishable from zero. Therefore we assume that η,ζ are bounded. Note also, that in this case the process $\varkappa_s:=\eta_s-\zeta_s$ is a martingale.

Now fix $s\in[0,\infty)$, define $s_{kn}=sk2^{-n}$, $k=0,1,\dots,2^n$, $n=1,2,\dots$, and use the formula $a^2-b^2=2(a-b)b+(a-b)^2$. Then, setting

$$\varepsilon(n)=\max\big\{|\varkappa_u-\varkappa_v|:\,|u-v|\le s2^{-n},\ u,v\le s\big\},$$

noting that

$$\sum_k|\varkappa_{s_{k+1,n}}-\varkappa_{s_{k,n}}|\le\sum_k\Big(|\eta_{s_{k+1,n}}-\eta_{s_{k,n}}|+|\zeta_{s_{k+1,n}}-\zeta_{s_{k,n}}|\Big)=\eta_s+\zeta_s,$$

and using Lemma II.8.5 (a), we obtain

$$\begin{aligned}\mathbf{E}\,\varkappa_s^2&=\mathbf{E}\sum_k(\varkappa^2_{s_{k+1,n}}-\varkappa^2_{s_{k,n}})\\
(3)\qquad&=2\sum_k\mathbf{E}\,\varkappa_{s_{k,n}}(\varkappa_{s_{k+1,n}}-\varkappa_{s_{k,n}})+\mathbf{E}\sum_k(\varkappa_{s_{k+1,n}}-\varkappa_{s_{k,n}})^2\\
&\le\mathbf{E}\,\varepsilon(n)\sum_k|\varkappa_{s_{k+1,n}}-\varkappa_{s_{k,n}}|\le N\,\mathbf{E}\,\varepsilon(n),\end{aligned}$$

where N is a constant independent of n. Letting $n\to\infty$ we see by the Dominated Convergence Theorem and the continuity and boundedness of the sample paths of $\varkappa$ that $\mathbf{E}\,\varkappa_s^2=0$, $\varkappa_s=0$ (a.s.) for any s and for all s at once (a.s.). □

3. EXERCISE. Deduce from Lemma 2 and Theorem 7.2 (a) that if ξ has a stochastic differential and $d\xi_s=\sigma_i(s)\,dw_s+b_i(s)\,ds$, $i=1,2$, then $\sigma_1=\sigma_2$ and $b_1=b_2$ (μ-a.e.).

Later we will need a kind of integral with respect to processes having stochastic differentials.

4. LEMMA. *Let ξ be a one-dimensional process with stochastic differential: $d\xi_s = \sigma(s)\,dw_s + b(s)\,ds$, $f(t)$ an $\mathcal{F}_t$-adapted continuous stochastic process, $t_{kn} = k2^{-n}$, $n, k = 0, 1, 2, \ldots$. Then for any $T \in [0, \infty)$ the supremum over $s \in [0, T]$ of the absolute value of the expression*

$$\sum_k f(t_{kn})(\xi_{s\wedge t_{k+1,n}} - \xi_{s\wedge t_{k,n}}) - \int_0^s f(t)\sigma(t)\,dw_t - \int_0^s f(t)b(t)\,dt \tag{4}$$

tends to zero in probability as $n \to \infty$.

PROOF. By the properties of stochastic integrals, in particular, by Theorem 6.14, if $p > q > 0$, then

$$f(q)\Big(\int_0^{s\wedge p} \sigma(t)\,dw_t - \int_0^{s\wedge q} \sigma(t)\,dw_t\Big) = \int_0^s f(q) I_{q<t\leq p}\sigma(t)\,dw_t \quad \text{(a.s.)}.$$

A similar formula is valid for the usual integral. Hence for all s at once (a.s.) the sum with respect to k in (4) is equal to

$$\int_0^s f(\varkappa_n(t-))\sigma(t)\,dw_t + \int_0^s f(\varkappa_n(t-))b(t)\,dt,$$

where $\varkappa_n(t) = 2^{-n}[2^n t]$. Since f is continuous in t, our assertion follows from Theorem 6.6, Lemma 6.13 and the Dominated Convergence Theorem. □

The most important point in the derivation of Itô's formula is the following particular case of it, suggested by formula (2.7), according to which $d|w_s|^2 = 2w_s\,dw_s + (dw_s)^2$ for $d = 1$.

5. LEMMA. *Let ξ_s, η_s be scalar stochastic processes with stochastic differentials. Then the product $\xi_s\eta_s$ also has a stochastic differential and*

$$d(\xi_s\eta_s) = \eta_s\,d\xi_s + \xi_s\,d\eta_s + (d\xi_s)(d\eta_s). \tag{5}$$

In particular,

$$d(\xi_s^2) = 2\xi_s\,d\xi_s + (d\xi_s)^2. \tag{6}$$

PROOF. The relation (5) follows from (6) if we consider the coefficient of λ in the expression $d\left[(\xi_s + \lambda\eta_s)^2\right]$. To prove (6) we set $d\xi_s = \sigma(s)\,dw_s + b(s)\,ds$, where σ is an $(1 \times d_1)$-matrix, that is, a row-vector, and b takes values in $(-\infty, \infty)$. Let $\varepsilon(n, T)$ be the supremum over $s \in [0, T]$ of the absolute value of expression (4) for $f(t) \equiv \xi_t$. By Lemma 4 there exist integers $n(p)$ such that $n(p) \to \infty$ as $p \to \infty$ and for any $p = 1, 2, \ldots$

$$P\{\varepsilon(n(p), p) \geq p^{-1}\} \leq p^{-1}.$$

Consequently, the numbers $\varepsilon(n(p), p)$ tend to zero in probability, and some subsequence of them tends to zero a.s. Hence, the right-hand sides of the equality (cf. (3))

$$\xi_s^2 - \xi_0^2 - 2\sum_k \xi_{s\wedge t_{k,n}}\left(\xi_{s\wedge t_{k+1,n}} - \xi_{s\wedge t_{k,n}}\right) = \sum_k \left(\xi_{s\wedge t_{k+1,n}} - \xi_{s\wedge t_{k,n}}\right)^2 \tag{7}$$

converge uniformly over some subsequence on any finite interval of $[0,\infty)$ a.s. to

$$\eta_s := \xi_s^2 - \xi_0^2 - 2\int_0^s \xi_t\sigma(t)\,dw_t - 2\int_0^s \xi_t b(t)\,dt. \tag{8}$$

In addition, it is obvious that if $n \geq m$, $s_1 = i_1 2^{-m}$, $s_2 = i_2 2^{-m}$ and $s_1 > s_2$, then for $s = s_1$ the right-hand side of (7) contains more terms than for $s = s_2$. Hence $\eta_{s_2} \leq \eta_{s_1}$ (a.s.), and since this is true for any binary rationals $s_1 \geq s_2$, and η_s is a continuous process by definition, it follows that almost all sample paths of η_s are increasing.

By the rules for operating with stochastic differentials and the multiplication table (2), equality (6) is equivalent to

$$\eta_s = \int_0^s |\sigma(t)|^2\,dt, \tag{9}$$

which should hold for all $s \in [0,\infty)$ at once (a.s.). Both sides of (9) are $\mathcal{F}_s$-adapted, continuous in s and increasing in s (a.s.); hence, by Lemma 2, it remains only to prove that their difference is a local martingale.

A few simple manipulations, which use Theorem 6.14 for $\tau = 0$ and the fact that $I_{0<t} = 1$ on X, show that (a.s.)

$$\begin{aligned}\xi_s^2 - \xi_0^2 &- 2\int_0^s \xi_t\sigma(t)\,dw_t - 2\int_0^s \xi_t b(t)\,dt\\ &= (\xi_s - \xi_0)^2 - 2\int_0^s (\xi_t - \xi_0)\sigma(t)\,dw_t - 2\int_0^s (\xi_t - \xi_0)b(t)\,dt.\end{aligned}$$

In this sense expression (8) is independent of ξ_0 (a.s.), and considering $\xi_s - \xi_0$ instead of ξ_s we reduce the problem to the case $\xi_0 = 0$. In this case $\xi_s = \psi_s + \varphi_s$ (a.s.), where

$$\psi_s := \int_0^s \sigma(t)\,dw_t, \qquad \varphi_s := \int_0^s b(t)\,dt.$$

To make our final argument more compact, let us write $\alpha_s \sim \beta_s$, if $\alpha_s - \beta_s$ is a local martingale. By Theorems 4.8, 6.15, Corollary 7.4 and Fubini's Theorem we conclude

$$\eta_s \sim \psi_s^2 + 2\psi_s\varphi_s + \varphi_s^2 - 2\int_0^s \psi_t b(t)\,dt - 2\int_0^s \varphi_t b(t)\,dt,$$

$$\int_0^s \psi_t b(t)\,dt \sim \psi_s\varphi_s, \qquad \psi_s^2 \sim \int_0^s |\sigma(t)|^2\,dt,$$

$$2\int_0^s \varphi_t b(t)\,dt = 2\int_0^s \Big(\int_0^s I_{u\le t} b(u)\,du\Big) b(t)\,dt = 2\int_0^s\int_0^s I_{u\le t} b(u)b(t)\,dudt$$

$$(10) \qquad = \int_0^s\int_0^s I_{u\le t} b(u)b(t)\,dudt + \int_0^s\int_0^s I_{t\le u} b(u)b(t)\,dudt$$

$$= \int_0^s\int_0^s b(u)b(t)\,dudt = \varphi_s^2,$$

$$\eta_s \sim \int_0^s |\sigma(t)|^2\,dt. \quad \square$$

We end the preparations for the derivation of Itô's formula with a lemma, which will be proved at the end of the section.

6. LEMMA. *Let $u(x)$ be a real-valued function defined on E_d with continuous derivatives u_{x^i}, $u_{x^ix^j}$ on E_d for all $i, j = 1, \dots, d$. Then there exists a sequence of polynomials $u^n(x)$ in $x^1, \dots, x^d$ such that u^n, $u^n_{x^i}$, $u^n_{x^ix^j}$ converge to u, u_{x^i}, $u_{x^ix^j}$, respectively, for all $i, j = 1, \dots, d$, uniformly on any compact set.*

We now have all we need to prove *Itô's formula*.

7. THEOREM. *Let a function $u(x)$ satisfy the assumption of Lemma 6 and ξ_s a d-dimensional stochastic process that has a stochastic differential. Then the process $u(\xi_s)$ also has a stochastic differential and*

$$(11) \qquad du(\xi_s) = \sum_{i=1}^d u_{x^i}(\xi_s)\,d\xi_s^i + \frac{1}{2}\sum_{i,j=1}^d u_{x^ix^j}(\xi_s)\,d\xi_s^i d\xi_s^j.$$

PROOF. Observe that if $d\xi_s = \sigma(s)\,dw_s + b(s)\,ds$, then (11) means that

$$du(\xi_s) = \sum_{i=1}^d\sum_{k=1}^{d_1} u_{x^i}(\xi_s)(\sigma^{ik}(s)\,dw_s^k + b^i(s)\,ds)$$
$$+ \frac{1}{2}\sum_{i,j=1}^d\sum_{k=1}^{d_1} u_{x^ix^j}(\xi_s)\sigma^{ik}(s)\sigma^{jk}(s)\,ds.$$

Next, let $u^n(x)$ be the polynomials of Lemma 6. Note that, for all ω, s, i and j,

$$\sup_{t\le s}\big|u^n_{x^i}(\xi_t) - u_{x^i}(\xi_t)\big| + \sup_{t\le s}\big|u^n_{x^ix^j}(\xi_t) - u_{x^ix^j}(\xi_t)\big| \longrightarrow 0$$

as $n \to \infty$, since each sample path of ξ is continuous and therefore bounded on $[0, s]$. It follows that if formula (11) holds for u^n, it is true for u by Theorem 6.6 (on passage to the limit in a stochastic *integral*; see the remarks after Definition 1) and the Dominated Convergence Theorem.

Thus it suffices to prove formula (11) for polynomials; to do this, in turn, it obviously suffices to prove it for all linear functions and to show that if it holds for u and v, then it also holds for $u + v$ and uv.

Formula (11) is obvious for linear functions. It is also obvious that if (11) holds for u and v, it also holds for $u+v$. As to the product, Lemma 5 and the multiplication table (2) give

$$\begin{aligned} d(uv) &= u\,dv + v\,du + du dv \\ &= \sum_{i=1}^{d} \left(uv_{x^i}\, d\xi^i + vu_{x^i}\, d\xi^i \right) + \frac{1}{2} \sum_{i,j=1}^{d} \left[uv_{x^i x^j}\, d\xi^i d\xi^j + vu_{x^i x^j}\, d\xi^i d\xi^j \right] \\ &\quad + \sum_{i,j=1}^{d} u_{x^i} v_{x^j}\, d\xi^i d\xi^j = \sum_{i=1}^{d} (uv)_{x^i}\, d\xi^i + \frac{1}{2} \sum_{i,j=1}^{d} (uv)_{x^i x^j}\, d\xi^i d\xi^j, \end{aligned}$$

where the arguments ξ_s of all the functions and the subscript s of $d\xi_s^i$ are omitted for brevity. □

We will discuss this theorem and its applications in Chapter IV, and now we proceed to the proof of Lemma 6. At this point, courses of calculus usually prove the Weierstrass theorem on the uniform approximation of continuous functions by polynomials. Here, however, we have to approximate a function *together with its derivatives*, and that is the only reason to present the proof of Lemma 6 here. We will use the following result, which contains the Weierstrass theorem as a special case.

8. LEMMA. *If p, q are d-dimensional vectors, let $p \cdot q = (p^i q^i)$ denote the d-dimensional vector whose coordinates are the products of the coordinates of p and q; $p/q = (p^i/q^i)$, if none of the q^i is zero;*

$$p^{(q)} = (p^1)^{(q^1)} \cdot \ldots \cdot (p^d)^{(q^d)},$$

where $0^0 := 1$, if all the q^i are nonnegative integers. If also all p^i are nonnegative integers, we write $p \le q$ when $p^i \le q^i$ for all i. In that case we also define

$$C_q^p = \binom{q^1}{p^1} \cdot \ldots \cdot \binom{q^d}{p^d}.$$

Let $\mathbb{I}$ be the vector all of whose coordinates are 1, $Q = \{x \in E_d : 0 \le x^1, \ldots, x^d \le 1\}$. Finally, assume that $v^n(x)$ is a given sequence of continuous functions which converges to a function $v(x)$ uniformly on Q. Then for any sequences of integer-valued vectors $k = k(n)$ and vectors $m = m(n)$ such that $m \to \mathbb{I}$, $k^1 \wedge \cdots \wedge k^d \to \infty$, the sequence of Bernstein polynomials

$$P_{mk}(v^n, x) := \sum_{r \le k} v^n\left(\frac{m \cdot r}{k}\right) C_k^r x^{(r)} (\mathbb{I} - x)^{(k-r)}$$

also converges to v uniformly on Q.

PROOF. It is not difficult to see that we need only prove that $P_{mk}(v^n, x_n) \to v(x_0)$ for any $x_0 \in Q$ and sequence $x_n \in Q$ converging to x_0. Take a point $x_0 \in Q$ and a sequence $x_n \to x_0$, $x_n \in Q$; to simplify we will omit the subscript n of x_n. Let $\eta_j^1, \dots, \eta_j^d$, $j = 1, 2, \dots$, be independent random variables defined on some probability space, such that $P\{\eta_j^i = 1\} = x^i$, $P\{\eta_j^i = 0\} = 1 - x^i$ (cf. Exercise II.2.9). Define

$$S^i(k) = \sum_{j=1}^{k^i} \eta_j^i, \qquad S(k) = (S^1(k), \dots, S^d(k)).$$

Then it follows immediately from Theorem I.4.17 that $S^i(k)$, $i = 1, \dots, d$, are independent, and from Exercise II.2.9 we obtain

$$P\{S(k) = r\} = C_k^r x^{(r)} (\mathbb{I} - x)^{(k-r)}, \quad r \le k,$$
$$P_{mk}(v^n, x) = \mathbf{E} \sum_{r \le k} v^n\Big(\frac{m}{k} \cdot r\Big) I_{S(k)=r} = \mathbf{E}\, v^n\Big(\frac{m}{k} \cdot S(k)\Big).$$

In addition, it follows from Exercise II.2.9 and Lemma 6.13 (e) that $\big(S(k)/k\big) - x \xrightarrow{P} 0$, and since $x \to x_0$, $m \to \mathbb{I}$, we obtain (by Lemma 6.13 (d)) $S(k) \cdot m/k \xrightarrow{P} x_0$. Finally,

$$\Big|v^n\Big(\frac{m}{k} \cdot S(k)\Big) - v(x_0)\Big| \le \sup_Q |v^n(y) - v(y)| + \Big|v\Big(\frac{m}{k} \cdot S(k)\Big) - v(x_0)\Big|,$$

whence it follows, again by Lemma 6.13, that the first expression tends to zero in probability. It remains only to observe that

$$|P_{mk}(v^n, x) - v(x_0)| \le \mathbf{E} \Big|v^n\Big(\frac{m}{k} \cdot S(k)\Big) - v(x_0)\Big|$$

and to apply Lemma 6.13 once again. □

9. PROOF OF LEMMA 6. It will suffice to find, for each cube $\{-N \le x^1, \dots, x^d \le N\}$, a polynomial which, together with its first and second derivatives, differs from u and its derivatives in the cube by at most $1/N$. Each such cube can be considered separately, and since it can be mapped by a linear transformation onto Q, it will suffice to find a suitable polynomial in Q. We will show that the Bernstein polynomials $P_{mk}(u, x)$, where k is any function of n such that $k^1 \wedge \dots \wedge k^d \to \infty$ and m is any function of n such that $m \to \mathbb{I}$, can be used to that end.

By Lemma 8, $P_{mk}(u, x) \to u(x)$ uniformly on Q as $n \to \infty$. In addition, a few simple manipulations show that for any $i = 1, \dots, d$

$$\frac{\partial}{\partial x^i} P_{mk}(u, x) = P_{m_i k_i}(\Delta_{ikm} u, x), \tag{12}$$

where $k_i = k - e_i$, e_i is the ith basis vector, $m_i = m \cdot k_i/k$,

$$\Delta_{ikm} u(x) := k^i \Big[u\Big(x + \frac{m}{k} \cdot e_i\Big) - u(x)\Big] = m^i \int_0^1 u_{x^i}\Big(x + t\frac{m}{k} \cdot e_i\Big)\, dt.$$

Now, using the condition $m \to \mathbb{I}$ we easily see that $\Delta_{ikm}u \to u_{x^i}$ uniformly on Q as $n \to \infty$. Hence, Lemma 8 and the relations $k_i^1 \wedge \cdots \wedge k_i^d \to \infty$, $m \to \mathbb{I}$ imply that

$$\frac{\partial}{\partial x^i} P_{mk}(u, x) \longrightarrow u_{x^i}(x)$$

uniformly on Q.

The rest is obvious: differentiate (12) with respect to x^j and again use (12) to transform the derivative on the right-hand side. Again using the Newton-Leibniz formula and Lemma 8, we see that the derivatives of P_{mk} with respect to $x^i x^j$ converge uniformly to $u_{x^i x^j}$. □

10. REMARK. It is clear from the proof that if $u, u_{x^p}, u_{x^i x^j}$ exist and are continuous only for certain p, i, j, then there is a sequence of polynomials v^n such that v^n, $v^n_{x^p}$, $v^n_{x^i x^j}$ converge to $u, u_{x^p}, u_{x^i x^j}$ uniformly on every compact set for those values of p, i, j.

9. Martingale version of Itô's formula. Lévy's theorem

It sometimes becomes necessary in the theory of diffusion processes to express $u(\xi_s)$, where ξ_s is a given process and u a smooth function, as a sum of an usual integral with respect to ds and a local martingale, the exact form of the latter being unimportant. This representation, known as the martingale version of Itô's formula, will be established for some cases in the present section. The martingale version of Itô's formula is independent of the theory of stochastic integration, and we will prove it by methods other than those of Sec. 8. The helpfulness of this representation will be demonstrated in the proof of one of Lévy's Theorems.

Throughout this section $(\Omega, \mathcal{F}, P)$ will be a complete probability space and $\{\mathcal{F}_t\}$ a filtration of σ-algebras on it, defined for $t \geq 0$, such that σ-algebras $\mathcal{F}_t$ are complete with respect to $\mathcal{F}, P$.

1. DEFINITION. Let ξ_s be a d-dimensional stochastic process on $(\Omega, \mathcal{F}, P)$ for $s \in [0, \infty)$. We will say that ξ_s is a *local martingale* (with respect to $\{\mathcal{F}_s\}$) or that $(\xi_s, \mathcal{F}_s)$ is a local martingale for $s \geq 0$, if $(\xi_s^i, \mathcal{F}_s)$ are local martingales for $s \geq 0$ for all $i = 1, \ldots, d$. A local d-dimensional martingale $(\xi_s, \mathcal{F}_s)$ is said to be *admissible* if there exists a function $a = a(t)$ on $\Omega \times (0, \infty)$ with values in the set of symmetric nonnegative $(d \times d)$-matrices, such that $a \in H_1$ and for all $i, j = 1, \ldots, d$ the processes

$$\xi_s^i \xi_s^j - \int_0^s a^{ij}(t)\, dt, \qquad s \in [0, \infty) \tag{1}$$

are local martingales (with respect to $\{\mathcal{F}_s\}$). In that case we also write

$$\langle \xi^i, \xi^j \rangle_s := \int_0^s a^{ij}(t)\, dt, \qquad \langle \xi, \xi \rangle_s := \int_0^s a(t)\, dt, \qquad \langle \xi \rangle_s := \int_0^s \operatorname{tr} a(t)\, dt. \tag{2}$$

In particular, by Definition 8.1 it follows from (2) that in particular, $d\langle \xi, \xi \rangle_s = a(s)\, ds$, and so on. It is sometimes useful to keep in mind the obvious fact that $(\xi_s, \mathcal{F}_s)$ is a d-dimensional local martingale if and only if $((\lambda, \xi_s), \mathcal{F}_s)$ is a (one-dimensional) local martingale for any $\lambda \in E_d$. In addition, it follows easily from the formula

$$(A\lambda, \pi) = \frac{1}{4}\Big[\big(A(\lambda + \pi), \lambda + \pi\big) - \big(A(\lambda - \pi), \lambda - \pi\big)\Big],$$

where A is a symmetric matrix and π, λ vectors of an appropriate dimension, that $(\xi_s, \mathcal{F}_s)$ is an admissible local martingale if and only if there exists a matrix-valued function a which satisfies the conditions of Definition 1, such that for any $\lambda \in E_d$ the one-dimensional process

$$(\lambda, \xi_s)^2 - \int_0^s (a(t)\lambda, \lambda)\,dt$$

is a local martingale.

Throughout this section (and the next one), $(\xi_s, \mathcal{F}_s)$ will be an admissible local d-dimensional martingale for $s \geq 0$ and $d\langle \xi, \xi \rangle_s = a(s)\,ds$.

2. EXERCISE. Using Lemma 8.2, prove that if processes (1) are local martingales for $a = a_1$, $a = a_2$, then $a_1 = a_2$ (μ-a.e.) and the right-hand sides of (2), written for $a = a_1$ and $a = a_2$, coincide for all $s \in [0, \infty)$ at once a.s. (that is, they are indistinguishable).

This exercise shows to what extent the notations (2) are well defined. The uncertainty here is the same as in constructions of stochastic integrals; once pointed out, it cannot cause any misunderstanding.

3. REMARK. The basic example of an admissible local martingale is obtained in the following way. Let $\sigma(t)$ be a function on $\Omega \times (0, \infty)$ with values in the set of $(d \times d_1)$-matrices, $\sigma \in H_2$, and let $(w_t, \mathcal{F}_t)$ be a d_1-dimensional Wiener process on $(\Omega, \mathcal{F}, P)$ for $t \geq 0$. Then by Theorems 4.8, 4.9 (g), the process

$$\xi_s := \int_0^s \sigma(t)\,dw_t \tag{3}$$

is a local martingale with respect to $\{\mathcal{F}_s\}$. Moreover, by Corollary 7.4 the process

$$(\lambda, \xi_s)^2 - \int_0^s |\lambda^*\sigma(t)|^2 dt = (\lambda, \xi_s)^2 - \int_0^s (a(t)\lambda, \lambda)\,dt$$

where $a := \sigma\sigma^*$ and $\lambda \in E_d$ is any constant vector, is a local martingale. Consequently, ξ_s is an admissible local martingale, so that formulas (2) are applicable; the last of these formulas agrees with (7.3).

4. REMARK. Coming back to Definition 1, it turns out that not only the processes from (1) are local martingales, but also the processes

$$\xi^i_{s\wedge\tau}\xi^j_{s\wedge\sigma} - \int_0^{s\wedge\tau\wedge\sigma} a^{ij}(t)\,dt, \qquad (\xi^i_{s\wedge\sigma} - \xi^i_{s\wedge\gamma})(\xi^j_{s\wedge\tau} - \xi^j_{s\wedge\gamma}) - \int_{s\wedge\gamma}^{s\wedge\sigma\wedge\tau} a^{ij}(t)\,dt,$$

where τ, σ, γ are arbitrary Markov times such that $\tau \geq \gamma$, $\sigma \geq \gamma$.

Indeed, the processes of the second group are easily expressed in terms of those processes of the first group. For the first group we observe that

$$\xi^i_{s\wedge\tau}\xi^j_{s\wedge\sigma} = \xi^i_{s\wedge\tau\wedge\sigma}\xi^j_{s\wedge\tau\wedge\sigma} + \xi^i_{s\wedge\tau}(\xi^j_{s\wedge(\sigma\vee\tau)} - \xi^j_{s\wedge\tau}) + \xi^j_{s\wedge\sigma}(\xi^i_{s\wedge(\sigma\vee\tau)} - \xi^i_{s\wedge\sigma}),$$

and we immediately obtain our assertion from the conditions of Definition 1 by Theorem 4.9 (j), (d) applicability of which to the last two terms of this formula is based on the fact that they remain unchanged upon replacement of the first factors by $\xi^i_\tau I_{\tau<\infty}$, $\xi^j_\sigma I_{\sigma<\infty}$.

The following theorem deals with what is known as the *martingale version of Itô's formula.*

5. THEOREM. *Let $b(t)$ be a function on $\Omega \times (0,\infty)$ with values in E_d, belonging to H_1,*

$$\eta_s := \int_0^s b(t)\,dt, \qquad \zeta_s := \xi_s + \eta_s,$$

so that, by Definition 8.1, $d\eta_s = b(s)\,ds$. Let $u(x)$ be a real-valued function defined on E_d, having continuous derivatives u_{x^i}, $u_{x^ix^j}$ on E_d for all $i,j = 1,\dots,d$. Then the stochastic process

$$\alpha(s) := u(\zeta_s) - \int_0^s L_t^{a,b} u(\zeta_t)\,dt,$$

where

$$L_t^{a,b}u(x) := \frac{1}{2}\sum_{i,j=1}^d a^{ij}(t)u_{x^ix^j}(x) + \sum_{i=1}^d b^i(t)u_{x^i}(x),$$

is a local martingale.

PROOF. For $s \ge t$, by Taylor's formula

$$u(\zeta_s) - u(\zeta_t) = \big[u(\xi_s+\eta_t) - u(\xi_t+\eta_t)\big] + \big[u(\xi_s+\eta_s) - u(\xi_s+\eta_t)\big]$$

$$= m^1(s,t) + m^2(s,t) + \Delta^1(s,t) + \int_t^s L_r^{a,b}u(\zeta_r)\,dr + \Delta^2(s,t),$$

where

$$m^1(s,t) = \sum_{i=1}^d u_{x^i}(\zeta_t)(\xi_s^i - \xi_t^i),$$

$$m^2(s,t) = \frac{1}{2}\sum_{i,j=1}^d u_{x^ix^j}(\zeta_t)\Big[(\xi_s^i-\xi_t^i)(\xi_s^j-\xi_t^j) - \int_t^s a^{ij}(r)\,dr\Big],$$

$$\Delta^1(s,t) = \frac{1}{2}\sum_{i,j=1}^d \Big[u_{x^ix^j}(\theta^1) - u_{x^ix^j}(\zeta_t)\Big](\xi_s^i-\xi_t^i)(\xi_s^j-\xi_t^j),$$

$$\Delta^2(s,t) = \int_t^s \Big\{\frac{1}{2}\sum_{i,j=1}^d \Big[u_{x^ix^j}(\zeta_t) - u_{x^ix^j}(\zeta_r)\Big]a^{ij}(r)$$

$$+ \sum_{i=1}^d \Big[u_{x^i}(\theta^2) - u_{x^i}(\zeta_r)\Big]b^i(r)\Big\}dr,$$

θ^1, θ^2 are some points of E_d depending on ω, t, s, such that

$$|\theta^1 - \zeta_t| \le \big|(\xi_s+\eta_t) - (\xi_t+\eta_t)\big| = |\xi_s - \xi_t|,$$
$$|\theta^2 - \zeta_t| \le \big|(\xi_s+\eta_t) - (\xi_s+\eta_s)\big| = |\eta_s - \eta_t|.$$

It is evident that if we take $\varepsilon > 0$, ω, s, t such that the contents of the square brackets in the expressions for Δ^1, Δ^2 are less than ε in absolute value, then by the inequalities

$$\left|(\xi_s^i - \xi_t^i)(\xi_s^j - \xi_t^j)\right| \le (\xi_s^i - \xi_t^i)^2 + (\xi_s^j - \xi_t^j)^2, \qquad \left|a^{ij}(r)\right| \le a^{ii}(r) + a^{jj}(r),$$

we obtain

$$\left|\Delta^1(s,t) + \Delta^2(s,t)\right| \le N_0\varepsilon\left[\varkappa(s) - \varkappa(t)\right] + N_1\varepsilon m^3(s,t),$$

where N_0, N_1 are constants depending only on d, and

$$\varkappa(s) = \int_0^s \left(\operatorname{tr} a(t) + |b(t)|\right) dt, \qquad m^3(s,t) = |\xi_s - \xi_t|^2 - \int_t^s \operatorname{tr} a(r)\, dr.$$

For $x \in E_d$, $\lambda > 0$ let $w(x,\lambda)$ denote the maximum of the numbers $|v(y) - v(x)|$, where $v = u_{x^i}, u_{x^ix^j}$, $i,j = 1,\dots,d$, $|y - x| \le \lambda$. Note that $w(x,\lambda)$ is continuous in x, λ, put $\gamma_0 = 0$ and for $n = 0, 1, \dots$ denote

$$\gamma_{n+1} = \inf\left\{s \ge 0\colon 2w\left(\zeta_{s\wedge\gamma_n}, |\xi_s - \xi_{s\wedge\gamma_n}| + |\eta_s - \eta_{s\wedge\gamma_n}|\right) \ge \varepsilon\right\}.$$

The times γ_n, $n \ge 1$, are the first exit times of continuous $\mathcal{F}_t$-adapted processes from $(-\infty, \varepsilon)$. Therefore γ_n are Markov times. It is clear that $\xi_s = \xi_{s\wedge\gamma_n}$, $\eta_s = \eta_{s\wedge\gamma_n}$ for $s \le \gamma_n$. Hence $\gamma_{n+1} \ge \gamma_n$. Moreover, if $\gamma_n < \infty$ for some n, ω, then for $t = \gamma_n \le r \le s \le \gamma_{n+1}$ it is also true that

$$\zeta_{s\wedge\gamma_n} = \zeta_t$$
$$|\theta^1 - \zeta_t| \le |\xi_s - \xi_t| \le |\xi_s - \xi_{s\wedge\gamma_n}| + |\eta_s - \eta_{s\wedge\gamma_n}|,$$
$$\left|u_{x^ix^j}(\theta^1) - u_{x^ix^j}(\zeta_t)\right| \le w\left(\zeta_{s\wedge\gamma_n}, |\xi_s - \xi_{s\wedge\gamma_n}| + |\eta_s - \eta_{s\wedge\gamma_n}|\right) \le \frac{\varepsilon}{2} < \varepsilon,$$
$$|\zeta_r - \zeta_t| \le |\xi_r - \xi_{r\wedge\gamma_n}| + |\eta_r - \eta_{r\wedge\gamma_n}|,$$
$$\left|u_{x^ix^j}(\zeta_t) - u_{x^ix^j}(\zeta_r)\right| \le w\left(\zeta_{r\wedge\gamma_n}, |\xi_r - \xi_{r\wedge\gamma_n}| + |\eta_r - \eta_{r\wedge\gamma_n}|\right) \le \varepsilon,$$
$$\left|u_{x^i}(\theta^2) - u_{x^i}(\zeta_r)\right| \le \left|u_{x^i}(\theta^2) - u_{x^i}(\zeta_t)\right| + \left|u_{x^i}(\zeta_t) - u_{x^i}(\zeta_r)\right| \le \varepsilon.$$

By the previous arguments, for any $s \in [0,\infty)$, $n = 1, 2, \dots$,

$$\alpha(s\wedge\gamma_n) - \alpha(0) = \sum_{i=0}^{n-1}\left[\alpha(s\wedge\gamma_{i+1}) - \alpha(s\wedge\gamma_i)\right] \le N_0\varepsilon\varkappa(s) + \sum_{i=0}^{n-1} m_i(s), \tag{4}$$

where $m_i(s) = (m^1 + m^2 + N_1\varepsilon m^3)(s\wedge\gamma_{i+1}, s\wedge\gamma_i)$ are local martingales by Remark 4 and Theorem 4.9 (d), (j) (see the reasoning of Remark 4). For fixed n there exists a sequence of Markov times $\bar\gamma(k) = \bar\gamma(k,n)$, $k = 1, 2, \dots$, which localize $m_i(s)$ simultaneously for $i = 0, \dots, n-1$. Define

$$\tau(p) = \inf\left\{s \ge 0\colon \left|\alpha(s) - \alpha(0)\right| + \varkappa(s) \ge p\right\}, \qquad p = 1, 2, \dots.$$

Take an arbitrary moment $\tau \in \mathfrak{M}$, substitute $s = \bar\gamma(k)\wedge\tau(p)\wedge\tau$ in (4) and calculate the expectations of the outermost terms. We obtain

$$\mathbf{E}\left[\alpha\left(\bar\gamma(k)\wedge\tau(p)\wedge\tau\wedge\gamma_n\right) - \alpha(0)\right] \le N_0\varepsilon\mathbf{E}\,\varkappa\left(\bar\gamma(k)\wedge\tau(p)\wedge\tau\right) \le N_0\varepsilon p.$$

Letting $k \to \infty$ and then $n \to \infty$ and noting that obviously $\gamma_n \uparrow \infty$ while $\left|\alpha(s) - \alpha(0)\right| \le p$ for $s \le \tau(p)$, we obtain, by the Dominated Convergence Theorem, $\mathbf{E}\left[\alpha(\tau(p)\wedge\tau) - \alpha(0)\right] \le N_0\varepsilon p$. Since this inequality holds for any $\varepsilon > 0$, the left-hand side cannot be positive. Substitution of $(-u)$ for u shows that the same holds for the process α built up from $(-u)$, and therefore $\mathbf{E}\left[\alpha(\tau(p)\wedge\tau)-\alpha(0)\right]=0$. This proves the theorem. □

Using Theorem 5 and Remark 4 we now prove the following remarkable P. Lévy's theorem.

6. THEOREM. *If $a^{kj}=\delta^{kj}$ for all k, j (μ-a.e.) and $\xi_0=0$, then ξ_s is a d-dimensional Wiener process with respect to $\{\mathcal{F}_s\}$ for $s\in[0,\infty)$.*

PROOF. By Lemma 5.2 (c) we may assume without loss of generality that $a^{kj}=\delta^{kj}$ for all k, j, ω, t. Then we take constants $t_1,\dots,t_n\in[0,\infty)$, $\lambda_1,\dots,\lambda_n\in E_d$ and define

$$\eta_s=\sum_{k=1}^{n}(\lambda_k,\xi_{s\wedge t_k}),\qquad \zeta_s=\frac{1}{2}\sum_{k,r=1}^{n}(\lambda_k,\lambda_r)(t_k\wedge t_r\wedge s).$$

It is easily verified with the help of Remark 4 that η_s is an admissible (one-dimensional) local martingale and $\langle\eta\rangle_s=2\zeta_s$. By using only this fact, considering the two-dimensional process $(\eta_s,\zeta_s)=(\eta_s,0)+(0,\zeta_s)$, and applying Theorem 5 separately to the real and imaginary parts of the process $\varkappa_s:=\exp(i\eta_s+\zeta_s)$, we readily see that $\varkappa_s$ is a (complex) local martingale. Since it is bounded for $s\in[0,T]$ and $\omega\in\Omega$ for any $T\in[0,\infty)$, it follows that $\varkappa_s$ is a martingale. Hence, $\mathbf{E}\exp i\eta_s=\exp(-\zeta_s)$ and for $s\geq t_1\vee\dots\vee t_n$

$$\mathbf{E}\exp i\sum_{k=1}^{n}(\lambda_k,\xi_{t_k})=\exp\Big(-\frac{1}{2}\sum_{k,r=1}^{n}(\lambda_k,\lambda_r)(t_k\wedge t_r)\Big).$$

Recalling our definitions and Theorem I.4.12, we see from the last equality and the continuity of ξ_s that ξ_s is a d-dimensional Wiener process. In addition, since $\varkappa$ is a martingale, it follows for $n=2$, $t=t_1<t_2=r$, $\lambda_2=-\lambda_1=\lambda$, $A\in\mathcal{F}_t$, that $\mathbf{E}\,I_A\varkappa_r=\mathbf{E}\,I_A\varkappa_t$, that is,

$$\mathbf{E}\,I_A e^{i(\lambda,\xi_r-\xi_t)}=P(A)e^{-\frac{1}{2}\lambda^2(r-t)}=P(A)\,\mathbf{E}\,e^{i(\lambda,\xi_r-\xi_t)}.$$

From the equality of the outer terms for any $\lambda\in E_d$, as in the proofs of Theorems I.4.12, I.5.4, we see that $\exp i(\lambda,x)$ in this expression can be replaced by any continuous bounded function, and then that it can be replaced by the indicator of any Borel set $\Gamma\subset E_d$. In other words, $P(A,\xi_r-\xi_t\in\Gamma)=P(A)P(\xi_r-\xi_t\in\Gamma)$, and by Definition II.4.9 this means that $(\xi_s,\mathcal{F}_s)$ is a Wiener process. □

7. PROBLEM. Let $(w_t,\mathcal{F}_t)$ be a one-dimensional Wiener process. In the previous proof, take $d=1$, $\xi_s\equiv w_s$. Let $T\in[0,\infty)$ and let $\tau\in\mathfrak{M}$. Applying Theorem 5 to transform $\varkappa_s w_s$, prove that the process

$$\varkappa_s\Big[iw_s+\sum_{k=1}^{n}\lambda_k(t_k\wedge s)\Big]\tag{5}$$

is a local martingale. Using Theorem II.3.12, deduce from this that, the process (5) is a martingale. Find $\mathbf{E}\,|\varkappa_T|^2$ and conclude by Lemma II.8.8. and Remark 2.7 that

$$\varkappa_T=1+i\int_0^\infty I_{t\leq T}\sum_{k=1}^{n}\lambda_k I_{t\leq t_k}\varkappa_t\,dw_t\quad\text{(a.s.)}.$$

This formula is similar to formula (2.11) and, of course, can be also derived from Theorem 8.7.

8. PROBLEM. Under the assumptions of the previous problem, let α be a random variable such that $\alpha \in \mathcal{L}_2(\mathcal{F}, P)$ and $\mathbf{E}\,\alpha w_\tau = 0$ for all $\tau \in \mathfrak{M}$. Deduce from Theorem 1.5 and the previous arguments that $\mathbf{E}\,\alpha \varkappa_T = \mathbf{E}\,\alpha = \mathbf{E}\,\alpha\,\mathbf{E}\,\varkappa_T$. Repeating the final part of the proof of Theorem 6 and using the lemma on π- and λ-systems, conclude that the random variable $\alpha - \mathbf{E}\,\alpha$ is orthogonal in $\mathcal{L}_2(\mathcal{F}, P)$ to the set of all $\mathcal{F}^w_\infty$-measurable variables, so that if $\mathcal{F} = \mathcal{F}^w_\infty$ and $\mathbf{E}\,\alpha = 0$, then $\alpha = 0$ (a.s.). Finally, deduce from the linearity and isometric property of the operator J of stochastic integration over $(0, \infty)$ with respect to dw that if $\mathcal{F} = \mathcal{F}^w_\infty$, then the orthogonal complement in $\mathcal{L}_2(\mathcal{F}, P)$ of the set $J\mathcal{L}_2(\mathcal{P}, \mu)$ contains only constants (a.s.). Use this result to get another proof of Clark's theorem 2.12.

Note also that as is clear from the aforesaid, this section (except for Remark 3 and Problems 7, 8) could have been placed immediately after Sec. II.4, provided that necessary notions and results from the theory of martingales and pseudopredictable functions were available. To solve Problems 7, 8, we would then additionally need Theorem II.6.9, the theory of Sec. 1, and its application to objects (2.1).

10. Stochastic integral with respect to an admissible local martingale

In this section we retain the definitions and notation of the previous section, considering a somewhat different approach to constructing stochastic integrals and deriving Itô's formula, in comparison with that used in Secs. 1–8. We will need the following generalization of Lemma 5.3 (d).

1. LEMMA. *Let $H_2(a)$ denote the set of all pseudopredictable functions $f = f(t)$ on $\Omega \times (0, \infty)$ with values in E_d, for each of which $(af, f) \in H_1$. Then $H_2(a)$ is a linear space, and for any $f \in H_2(a)$ there exists a sequence $f_n \in H_2(a)$ such that $f_n \in S(\Pi)$, that is, f_n has the form*

$$f_n = \sum_{i=1}^{k(n)} c_{in} I_{(0,\tau(i,n)]\!]}, \tag{1}$$

where $k(n) < \infty$, $c_{in} \in E_d$, $\tau(i, n) \in \mathfrak{M}$, and for any $T \in [0, \infty)$

$$\int_0^T \Big(a(t)(f(t) - f_n(t)), f(t) - f_n(t)\Big)\,dt \xrightarrow{P} 0. \tag{2}$$

PROOF. The first assertion of the lemma follows directly from the inequality $(a(f+g), f+g) \le 2(af, f) + 2(ag, g)$. Using the same inequality, the equality

$$f(t) - f_n(t) = \big(f(t) - f(t)I_{|f(t)| \le n}\big) + \big(f(t)I_{|f(t)| \le n} - f_n(t)\big),$$

Lemma 6.13 and the Dominated Convergence Theorem, we see that we need only prove the second assertion for bounded functions f. In this case note that $(a\lambda, \lambda) \le |\lambda|^2 \operatorname{tr} a$ for any $\lambda \in E_d$, and the left-hand side of (2) is bounded from above by the (finite) expression

$$\int_0^T \big|f(t) - f_n(t)\big|^2 \operatorname{tr} a(t)\,dt = \sum_{j=1}^d \int_0^T \big|f^j(t) - f^j_n(t)\big|^2 \operatorname{tr} a(t)\,dt. \tag{3}$$

We now define a measure ν on $X = \Omega \times (0, \infty)$ by the formula $\nu(d\omega dt) = \operatorname{tr} a(\omega, t)\,\mu(d\omega dt)$. It follows from Lemma 5.3 (c) for $g := \operatorname{tr} a$ and $p = 1$ that $\nu((0, \sigma_n]\!]) \le n$, where $\sigma_n = n \wedge \tau_n(g)$; therefore ν is σ-finite. Hence, by Theorem 1.11, $\mathcal{L}_2(\mathcal{P}, \nu) = \mathcal{L}_2(\Pi, \nu)$. Note also that the sets of measure zero with respect to μ are naturally of measure zero with respect to ν. By Lemma 5.2 (c) this implies, in particular, that f is $\mathcal{P}^\nu$-measurable. In addition, $\mathcal{L}_{2,\mathrm{loc}}(\mathcal{P}, \nu)$ can be defined in the natural way, and it can be directly verified via Exercise 1.3 that, for our bounded f, we have $f^i \in \mathcal{L}_{2,\mathrm{loc}}(\mathcal{P}, \nu)$ and σ_n serves as a localizing subsequence. We can now repeat the proof of part (d) of Lemma 5.3, where we must take f^i, $\operatorname{tr} a\, dt$, f_n^j, 2 for g, dt, g_n, p; we can thus determine $f_n^j \in \mathcal{S}(\Pi)$ such that the right-hand side of (3) tends to zero in probability for any $T \in [0, \infty)$. It remains only to rewrite in an obvious way individual representations of f_n^j as sums of constants multiplied by indicators of sets in Π so that $f_n = (f_n^1, \dots, f_n^d)$ could be represented as in (1). □

Later we will need a theorem similar to Theorem 1.5; we state this theorem after the following remark.

2. REMARK. For $f, g \in E_d$, $t > 0$,

$$\big|(a(t)f, g)\big| \le \big(a(t)f, f\big)^{1/2}\big(a(t)g, g\big)^{1/2}.$$

Hence, by the Cauchy-Bunyakovsky inequality, if $f, g \in H_2(a)$ the integral in (4) (cf. below) exists, and is finite for all ω, s.

3. THEOREM. *For $f, g \in H_2(a)$, and real-valued processes α_s, β_s, γ_s, δ_s such that $\alpha_s - \delta_s$ is a local martingale and $\gamma_s - \beta_s$ is a local submartingale we write*

$$A_s(f, g) = \int_0^s \big(a(t)f(t), g(t)\big)\, dt, \qquad \alpha_s \sim \delta_s, \qquad \gamma_s \succ \beta_s. \tag{4}$$

(a) *Let $f \in H_2(a)$. Then there exists a local martingale α_s such that $\alpha_0 = 0$ and*

$$|\alpha_s|^2 \sim A_s(f, f), \qquad (\lambda, \xi_s)\alpha_s \sim A_s(\lambda, f) \tag{5}$$

for any constant vector $\lambda \in E_d$.

(b) *Let $f \in H_2(a)$, $\beta_s \sim 0$ and $(\lambda, \xi_s)\beta_s \sim A_s(\lambda, f)$ for any constant vector $\lambda \in E_d$. Then $|\beta_s|^2 \succ A_s(f, f)$.*

(c) *Under the assumptions of* (b), *let $|\beta_s|^2 \sim A_s(f, f)$, $g \in H_2(a)$, $\gamma_s \sim 0$ and*

$$(\lambda, \xi_s)\gamma_s \sim A_s(\lambda, g)$$

for any constant vector $\lambda \in E_d$. Then $\beta_s\gamma_s \sim A_s(f, g)$. If also $|\gamma_s|^2 \sim A_s(g, g)$, then

$$|\beta_s - \gamma_s|^2 = |\beta_s|^2 - 2\beta_s\gamma_s + |\gamma_s|^2 \sim A_s(f - g, f - g) \tag{6}$$

and if $(a(f - g), f - g) = 0$ (μ-a.e.), then $\beta_s - \gamma_s = \beta_0 - \gamma_0$ for all $s \in [0, \infty)$ at once a.s.

PROOF. (a) Let f_n be the functions of Lemma 1, and define

$$\alpha_s^n = \sum_{i=1}^{k(n)} \big(c_{in}, \xi_{s \wedge \tau(i,n)} - \xi_0\big).$$

Clearly, $\alpha_s^n \sim 0$ and $\alpha_0^n = 0$. In addition, by Theorem 4.9 (j) and Remark 9.4 for any $\lambda \in E_d$ and Markov times τ,

$$(7)\qquad \begin{aligned}(\lambda, \xi_{s\wedge\tau})\alpha_s^n &\sim (\lambda, \xi_{s\wedge\tau} - \xi_0)\alpha_s^n \\ &\sim \sum_{i=1}^{k(n)} A_s\big(\lambda I_{(0,\tau]]}, c_{in} I_{(0,\tau(i,n)]]}\big) = A_s\big(\lambda I_{(0,\tau]]}, f_n\big).\end{aligned}$$

Substituting here $\tau = \tau(i,n)$, $\lambda = c(i,n)$ and summing on i, we obtain

$$(8)\qquad |\alpha_s^n|^2 \sim A_s(f_n, f_n).$$

Since the difference $f_n - f_k$ has the same form as f_n and $\alpha^n - \alpha^k$ has the same form as α^n, it follows that $|\alpha_s^n - \alpha_s^k|^2 \sim A_s(f_n - f_k, f_n - f_k)$. Note also that

$$(9)\qquad A_s^{1/2}(f_n - f_k, f_n - f_k) \le A_s^{1/2}(f_n - f, f_n - f) + A_s^{1/2}(f_k - f, f_k - f),$$

$$(10)\qquad P\{\zeta + \eta \ge \varepsilon\} \le P\Big\{\Big(\zeta \ge \frac{\varepsilon}{2}\Big) \cup \Big(\eta \ge \frac{\varepsilon}{2}\Big)\Big\} \le P\Big\{\zeta \ge \frac{\varepsilon}{2}\Big\} + P\Big\{\eta \ge \frac{\varepsilon}{2}\Big\},$$

where ζ, η are arbitrary random variables, $\varepsilon \ge 0$. It follows from this and Lemma 1 that the first expression in (9) tends to zero in probability as $n, k \to \infty$. Considering the local martingale

$$m_s^{nk} := A_s(f_n - f_k, f_n - f_k) - |\alpha_s^n - \alpha_s^k|^2$$

we see that $m_s^{nk} \le A_s(f_n - f_k, f_n - f_k)$, and hence, by Theorem 6.12 (b) (cf. Exercise 6.16), for any $T \in [0, \infty)$,

$$\sup_{s \le T} |m_s^{nk}| \xrightarrow{P} 0$$

as $n, k \to \infty$. Similar assertion holds for $|\alpha_s^n - \alpha_s^k|^2$ (cf. (10)). In short, condition (6.20) holds and, by Lemma 6.18, there exists a continuous $\mathcal{F}_s$–adapted process α_s, such that relation (6.21) is true for any $T \in [0, \infty)$.

By Theorem 4.9 (k), α_s is a local martingale. Clearly, $\alpha_0 = 0$ (a.s.), since $\alpha_0^n = 0$. Redefining α, if necessary, outside the set of probability zero where $\alpha_0 \ne 0$, and defining $\alpha_s = 0$ there, we may assume that $\alpha_0 = 0$. In addition, as in (8) we have $\alpha_s^k \alpha_s^n \sim A_s(f_k, f_n)$, and for $s \le T$

$$(11)\qquad \begin{aligned}&\big|\alpha_s^k \alpha_s^n - A_s(f_k, f_n) - \big(\alpha_s^k \alpha_s - A_s(f_k, f)\big)\big| \\ &\qquad \le \sup_{s \le T} |\alpha_s^k| \sup_{s \le T} |\alpha_s^n - \alpha_s| + A_T^{1/2}(f_k, f_k) A_T^{1/2}(f_n - f, f_n - f).\end{aligned}$$

By previous arguments, letting $n \to \infty$ and using Theorem 4.9 (k) (and Lemma 6.13) we see that $\alpha_s^k \alpha_s \sim A_s(f_k, f)$. Hence we may replace α^k, f_k in (11) by α, f and obtain, letting $n \to \infty$, that $\alpha_s^2 \sim A_s(f, f)$. Finally, the second relation in (5) is obtained by taking (λ, ξ_s), λ in place of α_s^k, f_k in (11) and using (7). Thus part (a) is proved.

(b) By assumption and by Theorem 4.9, if τ is a Markov time, then

$$(\beta_s - \beta_{s\wedge\tau})(\lambda, \xi_{s\wedge\tau}) = (\beta_s - \beta_{s\wedge\tau})(\lambda, \xi_\tau) I_{\tau<\infty} \sim 0,$$
$$(\lambda, \xi_{s\wedge\tau} - \xi_0)\beta_s \sim (\lambda, \xi_{s\wedge\tau})\beta_s \sim (\lambda, \xi_{s\wedge\tau})\beta_{s\wedge\tau} \sim A_s\big(\lambda I_{(0,\tau]]}, f\big).$$

Taking f_n, α^n, α as in (a), we see from these relations that $\alpha_s^n \beta_s \sim A_s(f, f_n)$. Replacing α^k, f_k in (11) by β, f and again using Theorem 4.9 (k), we obtain $\beta_s \alpha_s \sim A_s(f, f)$. Finally, using Lemma II.8.5 (b), Theorem 4.9 (j) with $\tau = 0$ and a formula of the type $m_s^2 - m_0^2 = (m_s - m_0)^2 + 2(m_s - m_0)m_0$, we easily see that the square of a local martingale is a local submartingale, and hence

$$0 \prec |\beta_s - \alpha_s|^2 = \beta_s^2 - 2\beta_s\alpha_s + \alpha_s^2 \sim \beta_s^2 - 2A_s(f,f) + A_s(f,f) = \beta_s^2 - A_s(f,f).$$

(c) It follows from (b) that for any $\lambda \in (-\infty, \infty)$

$$\beta_s^2 + 2\lambda A_s(f,g) + \lambda^2 A_s(g,g) \sim A_s(f + \lambda g, f + \lambda g) \prec |\beta_s + \lambda\gamma_s|^2 = \beta_s^2 + 2\lambda\beta_s\gamma_s + \lambda^2\gamma_s^2.$$

Collecting like terms, dividing by λ and letting $\lambda \downarrow 0$, $\lambda \uparrow 0$, we obtain by Theorem 4.9 (k)

$$A_s(f,g) \prec \beta_s\gamma_s, \qquad A_s(f,g) \succ \beta_s\gamma_s, \qquad A_s(f,g) \sim \beta_s\gamma_s.$$

Formula (6) is now obvious. By Theorem 4.9 (j) we can replace β_s, γ_s in (6) by $\beta_s - \beta_0$, $\gamma_s - \gamma_0$. The last assertion in (c) follows from such a modification of (6) by Theorem 6.11 (a) and Fubini's Theorem, by which, if $h \in H_2(a)$ and $h = 0$ (μ-a.e.), then for almost any ω we have $A_s(h,h) = 0$ for all s, that is, $A_s(h,h)$ is a martingale. □

4. DEFINITION. Let f be a pseudopredictable function on $\Omega \times (0, \infty)$ with values in the set of $(d_1 \times d)$-matrices, such that $faf^* \in H_1$. Let η_s be a d_1-dimensional process defined for $\omega \in \Omega$, $s \in [0, \infty)$. We will call the process $\eta_s - \eta_0$ the *stochastic integral of $f(t)$ with respect to $d\xi_t$ with variable upper limit*, writing

$$\eta_s - \eta_0 = \int_0^s f(t)\, d\xi_t, \qquad d\eta_s = f(s)\, d\xi_s, \tag{12}$$

if the $(d + d_1)$-dimensional process (η_s, ξ_s) is an admissible local martingale and

$$d\langle \eta^i, \eta^j \rangle_s = (faf^*)^{ij}(s)\, ds, \qquad d\langle \eta^i, \xi^k \rangle_s = (fa)^{ik}(s)\, ds$$

for all $i, j = 1, \dots, d_1$, $k = 1, \dots, d$, that is, for any constants $\lambda \in E_{d_1}$, $\pi \in E_d$ the process $(\lambda, \eta_s) + (\pi, \xi_s)$ is an admissible local (one-dimensional) martingale and

$$d\langle (\lambda, \eta) + (\pi, \xi) \rangle_s = \big(a(f^*\lambda + \pi), f^*\lambda + \pi\big)(s)\, ds.$$

5. REMARK. Since $(faf^*)^{ij} = (af^i, f^j)$, $(fa)^{ik} = (af^i)^k$, where f^i is the ith column of the matrix f^*, it follows that $f^i \in H_2(a)$ and, by Theorem 3, one possible stochastic integral of f with respect to $d\xi$ with variable upper limit is the process η_s whose ith coordinate is taken from part (a) of Theorem 3 with f^i substituted for f. Therefore, under the assumptions of Definition 4, the stochastic integral of f with respect to $d\xi$ with variable upper limit always exists. Moreover, by part (c) of Theorem 3 the notation (12) is legitimate in the same sense in which all other similar notations are legitimate. In particular, two processes, each of which is a stochastic integral of f with respect to $d\xi$ with variable upper limit, are indistinguishable, that is, they coincide for all s at once (a.s.).

It must be said here that Definition 4 is also legitimate in the sense that, if ξ_s is a d-dimensional Wiener process with respect to $\{\mathcal{F}_s\}$, then the right-hand sides of (12) coincide with the corresponding expressions from Secs. 7, 8. This follows directly from the construction of α in the proof of Theorem 3 and from Theorem 6.6. In particular, Definition 4 also yields an alternative definition of stochastic integral with respect to a Wiener process.

Also observe that by Theorem 3 (c) modification of f on a set of zero measure μ does not change its stochastic integral with respect to $d\xi$, and obviously, indistinguishable processes are suitable only simultaneously as representations of the integral.

6. REMARK. If η is the process of Definition 4, one can also introduce stochastic integrals with respect to it. More rigorously stated: if g is a pseudopredictable function with values in the set of $(d_2 \times d_1)$-matrices such that $gfaf^*g^* = g(faf^*)g^* \in H_1$, then by Remark 5 there exists a process ζ_s such that $d\zeta_s = g\,d\eta_s$. It is very essential that then also $d\zeta_s = gf\,d\xi_s$.

To verify this, observe that $gfaf^*g^* = (gf)a(gf)^*$ and take a process $\varkappa_s$ such that $d\varkappa_s = gf\,d\xi_s$. Then, by Theorem 3 (c) and Definition 4,

$$d\langle \varkappa^i, \eta^k\rangle_s = (gfaf^*)^{ik}\,ds,$$
$$d\langle \varkappa^i, \varkappa^j\rangle_s = (gfaf^*g^*)^{ij}\,ds, \qquad i, j = 1, \dots, d_2,\ k = 1, \dots, d_1.$$

By Definition 4 we can replace $\varkappa$ in these expressions by ζ. Then by Theorem 3 (c) the process $\varkappa_s - \varkappa_0$ is indistinguishable from $\zeta_s - \zeta_0$, and (see the end of Remark 5) $d\zeta_s = d\varkappa_s = gf\,d\xi_s$.

In particular, it follows from these arguments that if $d\eta_s = f\,d\xi_s$ and ξ_s is as in Remark 9.3, that is, $d\xi_s = \sigma\,dw_s$, then $d\eta_s = f\sigma\,dw_s$.

Moreover, in the general case, if g is $\mathcal{F}_0$-measurable and independent of t and $faf^* \in H_1$, then obviously $gfaf^*g^* \in H_1$, and if also $d\zeta_s = g\,d\eta_s$, then it follows easily from Definition 4, Theorem 4.9 (j) and Theorem 3 (c) that $\zeta_s - \zeta_0 = g(\eta_s - \eta_0)$ for all s (a.s.). In addition, since $d\zeta_s = gf\,d\xi_s$, we have

$$\int_0^s gf(t)\,d\xi_t = g\int_0^s f(t)\,d\xi_t$$

for all s at once (a.s.). This shows, in particular (when g is independent of ω as well), that the stochastic integral with respect to $d\xi$ possesses linearity properties.

Now that we have defined a stochastic integral against admissible local martingales, we can write down the martingale version of Itô's formula in the usual form; it is readily seen that for $d\xi_s = \sigma(s)\,dw_s$, this coincides with (8.11).

7. THEOREM. *Under the assumptions and notation of Theorem* 9.5,

$$\alpha(s) = u(\zeta_0) + \int_0^s u_x(\zeta_t)\,d\xi_t$$

for all s at once (a.s.), where the row-vector u_x is the gradient of u.

PROOF. Fix a $k = 1, ..., d$ and define

$$\tilde{\xi}_s = (\xi_s^1, ..., \xi_s^d, \xi_s^k),\ \tilde{\eta}_s = (\eta_s^1, ..., \eta_s^d, 0),\ \tilde{\zeta}_0 = (\zeta_0^1, ..., \zeta_0^d, 0),\ \tilde{\zeta}_s = \tilde{\eta}_s + \tilde{\xi}_s.$$

Apply Theorem 9.5 to $v(\tilde{\zeta}_s)$, where $v(x^1, ..., x^{d+1}) = u(x^1, ..., x^d)x^{d+1}$, and then use Theorem 6.15. It follows that the processes

$$u(\zeta_s)\xi_s^k - \int_0^s \Big[\xi_t^k L_t^{a,b} u(\zeta_t) + \sum_{j=1}^d a^{kj}(t) u_{x^j}(\zeta_t)\Big]\, dt,$$

$$\alpha(s)\xi_s^k - \int_0^s \big(u_x(\zeta_t)a(t)\big)^k\, dt$$

are local martingales.

Furthermore, the same argument applied to the function $v(\zeta_s, \zeta_s^{d+1})$, where $v(x, x^{d+1}) = [u(x) + x^{d+1}]^2$ and

$$\zeta_s^{d+1} = -\int_0^s L_t^{a,b} u(\zeta_t)\, dt,$$

gives readily that the process

$$\gamma(s) := \alpha^2(s) - \int_0^s (a(t)u_x^*(\zeta_t), u_x^*(\zeta_t))\, dt$$

is a local martingale. The last fact can also be seen from the following simple manipulations:

$$\gamma(s) = \Big[u^2(\zeta_s) - \int_0^s L_t^{a,b}(u^2)(\zeta_t)\, dt\Big] - 2\int_0^s (\alpha(s) - \alpha(t)) L_t^{a,b} u(\zeta_t)\, dt$$

$$+ \Big[\Big(\int_0^s L_t^{a,b} u(\zeta_t)\, dt\Big)^2 - 2\int_0^s L_t^{a,b} u(\zeta_t) \int_t^s L_r^{a,b} u(\zeta_r)\, dr dt\Big].$$

Here the last term vanishes (cf., for example, (8.10)) and the first and second terms are local martingales (this follows from Theorem 9.5 and Theorem 6.15). By Definition 4, this proves the theorem. □

The *concept* of a stochastic integral with respect to a local martingale enabled us to derive Itô's formula easily from its martingale version. It also enables us to prove an important generalization of Lévy's Theorem, obtained by Doob [8] (1953) in the one-dimensional case and by Stroock and Varadhan [47] (1979) for several dimensions.

8. THEOREM. *The positive symmetric square root $\sqrt{a}$ of a matrix a is pseudopredictable, $\sqrt{a} \in H_2$, and if* $\det a > 0$ *(μ-a.e.), then for $t \geq 0$ there exists a d-dimensional Wiener process w_t with respect to $\{\mathcal{F}_t\}$ such that*

$$\xi_s - \xi_0 = \int_0^s \sqrt{a(t)}\, dw_t$$

for all $s \in [0, \infty)$ at once (a.s.).

PROOF. After reducing a to diagonal form and changing variables $u = \lambda v$, we easily see that for certain constant $c \in (0, \infty)$ and all ω, t

$$\sqrt{a} = c \int_0^\infty \frac{E - e^{-au}}{u^{3/2}} du, \tag{13}$$

where E is the identity matrix of order d. This formula, together with Fubini's Theorem, implies all the necessary measurability properties of $\sqrt{a}$. That $\sqrt{a} \in H_2$ follows immediately from the assumption $a \in H_1$ (cf. Definition 9.1). Now define $f = (\sqrt{a})^{-1}$ wherever $\det a > 0$ and $f = E$ at points where $\det a = 0$. By assumption, $\sqrt{a} f = E$, $faf^* = E$ (μ-a.e.). Consequently, we can define a local martingale w_s such that $w_0 = 0$, $dw_s = f(s)\, d\xi_s$, $d\langle w, w\rangle_s = faf^*\, ds = E\, ds$. By Lévy's Theorem w_s is a Wiener process with respect to $\{\mathcal{F}_s\}$ for $s \geq 0$. It remains only to use the fact that by Definition 4 and Remarks 5, 6 we have $d\xi_s = E\, d\xi_s = \sqrt{a} f\, d\xi_s = \sqrt{a}\, dw_s$. □

9. REMARK. The assumption $\det a > 0$ (μ-a.e.) of Theorem 8 is essential. To see this, consider the case in which Ω consists of only one point and $\xi_s \equiv 0$. Then there are no Wiener processes on Ω.

10. PROBLEM. Suppose that the assumption $\det a > 0$ (μ-a.e.) does not hold. Take a complete probability space $(\Omega', \mathcal{F}', P')$, and a d-dimensional Wiener process $(w', \mathcal{F}'_t)$ defined there. Assume that the σ-algebras $\mathcal{F}'_t$ are complete with respect to $\mathcal{F}', P'$. Let $\bar{\Omega} = \Omega \times \Omega'$, $\bar{P} = P \times P'$ and define $\bar{\mathcal{F}}$, $\bar{\mathcal{F}}_t$ as completions of $\mathcal{F} \otimes \mathcal{F}'$, $\mathcal{F}_t \otimes \mathcal{F}'_t$. Define also

$$g = \lim_{\varepsilon \downarrow 0} (\sqrt{a} + \varepsilon^2 E)(a + \varepsilon E)^{-1}, \qquad h = E - g\sqrt{a}.$$

Prove that $\xi_t(\omega)$ is an admissible local martingale and $w'_t(\omega')$ a d-dimensional Wiener process on $(\bar{\Omega}, \bar{\mathcal{F}}, \bar{P})$ with respect to $\{\bar{\mathcal{F}}_t\}$. Prove also that the process

$$\bar{w}_s(\omega, \omega') = \int_0^s g(\omega, t)\, d\xi_t(\omega) + \int_0^s h(\omega, t)\, dw'_t(\omega')$$

is well defined, is a Wiener process with respect to $\{\bar{\mathcal{F}}_t\}$ and

$$\xi_s(\omega) - \xi_0(\omega) = \int_0^s \sqrt{a(\omega, t)}\, d\bar{w}_t(\omega, \omega')$$

for all $s \in [0, \infty)$ at once $\bar{P}$-a.s.

11. EXERCISE. Let w_t be a one-dimensional Wiener process and $\xi_t = (w_t, -w_t)$ a two-dimensional process. Prove that *any* pseudopredictable function of the form $(f(t), f(t))$ is integrable with respect to $d\xi$ and its stochastic integral is indistinguishable from zero.

This exercise shows that, generally speaking, stochastic integral with respect to a vector-valued process ξ *cannot be reduced* to stochastic integrals with respect to its coordinates.

11. Regularly measurable processes

Let $(\Omega, \mathcal{F}, P)$ be a probability space and $\{\mathcal{F}_t\}$ a filtration of σ-algebras $\mathcal{F}_t$ defined on the space, $t \in [0, \infty)$. In Lemma 5.3 (c) the important question of describing the set $\mathcal{L}_{2,\text{loc}}(\mathcal{P}, \mu)$ was reduced to the problem of describing the set of pseudopredictable functions. In this section we will present a large class of such functions. The contents of this section will not be used later on.

All the previous arguments of this chapter may be applied (sometimes without any modifications; as far as Itô's formula is concerned, the necessary modifications are quite natural) to the construction and investigation of stochastic integrals with respect to arbitrary locally square integrable martingales, whose sample paths are right-continuous and have limits from the left. The only result that must be known *a priori* is that there exists the so-called Doléan's measure corresponding to this martingale which plays the role of μ. The existence of such a measure is easily proved using Carathéodory's theorem from measure theory, exposition of which, however, is not part of our plan. For the case of an arbitrary martingale (with the above properties), it is difficult to say anything about the set of $\mathcal{P}_\mu$-measurable functions except that it contains the predictable functions. Even for the case of continuous local martingales (and only this case is under consideration) $\mathcal{P}_\mu$ contains the sets $\{(\omega, t) \colon 0 < t < \tau(\omega)\}$ for any Markov time τ and the σ-algebra generated by all these sets, known as the *well-measurable* or *optional* σ-algebra (it turns to be larger than $\mathcal{P}$).

In our situation $\{(\omega, t) \colon 0 < t < \tau(\omega)\} \in \mathcal{P}_\mu$, since $I_{t<\tau} = I_{t\leq\tau}$ for almost all (ω, t) with respect to μ, for example, by Fubini's Theorem. Moreover, the indicator $I_{t<\tau}$ is right-continuous for every ω and is therefore a Borel function. For continuous *admissible* local martingales, e.g., Wiener processes, it is sometimes convenient to use a σ-algebra which is even larger than the optional one but is still a subalgebra of $\mathcal{P}_\mu$.

1. THEOREM. *Let $\mathcal{R}$ denote the set of all elements of $A \in \mathfrak{A}$ (see Sec. 2), for each of which $\{\omega \colon (\omega, t) \in A\} \in \mathcal{F}_t$ for all $t \in (0, \infty)$. Then*

(a) *$\mathcal{R}$ is a σ-algebra;*

(b) *a real-valued or vector function $f(\omega, t)$ is $\mathcal{R}$-measurable if and only if f is $\mathfrak{A}$-measurable and $\mathcal{F}_t$-measurable as a function of ω for any $t \in (0, \infty)$, that is, if and only if $f(t)$ is a $\mathcal{F}_t$-adapted $\mathfrak{A}$-measurable stochastic process;*

(c) *$\mathcal{R} \subset \mathcal{P}_\mu$.*

Before proving this theorem, we will discuss it and lay the ground.

2. DEFINITION. We will call the elements of $\mathcal{R}$ *regularly measurable* sets; $\mathcal{R}$-measurable functions will be called *regularly measurable processes.*

3. REMARK. It is not always desirable to consider regularly measurable processes rather than pseudopredictable functions. For example, in some situations one must define stochastic integrals with respect to dw_s of functions such as

$$\int_0^s f(\omega, s, t)\, dw_t, \tag{1}$$

when s also enters the integrand. Since $\mathcal{R} \subset \mathfrak{A}$, regularly measurable processes are $\mathfrak{A}$-measurable and by Fubini's Theorem they are Borel functions of t for any ω. It seems to us that this property of regularly measurable functions must be the reason why no *measurability and integrability* conditions on f can possibly make (1) regularly measurable by choosing different representations of the stochastic integral for different s. At the same time, we can *easily* make it pseudopredictable, for example, by replacing Ω, $w_t(\omega)$ by $\{(\omega, s) \colon \omega \in \Omega,\ s \in [0, \infty)\}$, $\bar{w}_t(\omega, s) := w_t(\omega)$; we shall not do this here, as we will not consider integrals of type (1).

Nevertheless, it should be noted that stochastic integrals of regularly measurable processes possess many useful properties, of which we will establish Theorem 11, Remark 12 and Corollary 13 below.

To prove Theorem 1 we will need two lemmas.

4. LEMMA. *Let $p \in [1, \infty)$, $f \in \mathcal{L}_p(\mathfrak{B}((0, \infty)), \ell)$. Set $f(t) = 0$ for $t \leq 0$. Then*

$$\lim_{r \to 0} \int_{-\infty}^{\infty} |f(t + r) - f(t)|^p dt = 0. \tag{2}$$

PROOF. We first note that for any nonnegative Borel function $g(t)$

$$\int_{-\infty}^{\infty} g(t + r)\, dt = \int_{-\infty}^{\infty} g(t)\, dt. \tag{3}$$

For continuous g this follows from the properties of the Riemann integral; for Borel functions that vanish outside some compact set it follows from Theorem I.5.4; and for arbitrary Borel functions $g \geq 0$ it follows from the Monotone Convergence Theorem.

Further, we may assume without loss of generality that f is a Borel function (cf. Lemma 5.2). Set $\Pi_1 := \{(0, s] \colon s \in [0, \infty)\}$. By Theorem 1.11 (cf. also Lemma I.1.7), $\mathcal{L}_p(\mathfrak{B}(0, \infty), \ell) = \mathcal{L}_p(\Pi_1, \ell)$. Consequently, for any $\varepsilon > 0$ there exists a function $f_\varepsilon \in S(\Pi_1)$ such that

$$\int_{-\infty}^{\infty} |f(t) - f_\varepsilon(t)|^p dt \leq \varepsilon^p.$$

It follows from this inequality, (3), and the triangle inequality that the $1/p$th power of the integral in (2) does not exceed

$$2\varepsilon + \left(\int_{-\infty}^{\infty} |f_\varepsilon(t + r) - f_\varepsilon(t)|^p dt \right)^{1/p}.$$

The last term vanishes as $r \to 0$. This is obvious if $f_\varepsilon = I_{(0,s]}$ and is easily proved using the triangle inequality for $f_\varepsilon \in S(\Pi_1)$. This clearly proves the lemma. $\square$

5. LEMMA. *Let $p \in [1, \infty)$, $f \in \mathcal{L}_p(\mathfrak{A}, \mu)$, and in addition assume f is a $\mathfrak{A}$-measurable function. Define $f(\omega, t) = 0$ for $t \leq 0$, $\varkappa_i(t) = k2^{-i}$ for $t \in [k2^{-i}, (k + 1)2^{-i})$, where $k = 0, \pm 1, \pm 2, \ldots$. Then there exists a sequence of integers $i(n) \to \infty$ such that for almost every $s \in [0, 1]$*

$$\lim_{n \to \infty} \mathbf{E} \int_{0}^{\infty} \left| f\left(\varkappa_{i(n)}(t + s) - s\right) - f(t) \right|^p dt = 0 \tag{4}$$

and $f\left(\varkappa_i(\cdot + s) - s\right) \in \mathcal{L}_p(\mathfrak{A}, \mu)$ for all i.

PROOF. As in Itô [21] (1951), we will use arguments due to Doob. Changing variables $t+s=u,\ t=v$ (to show that this is legitimate we can proceed as in the proof of (3), first considering $g(\omega,s,t)\geq 0$), using the fact that function $\varkappa_i(t)-t$ is periodic with period one and applying Fubini's Theorem, we obtain

$$\int_0^1 \mathbf{E}\int_{-\infty}^{\infty}\left|f\left(\varkappa_i(t+s)-s\right)-f(t)\right|^p dtds$$

$$= \mathbf{E}\int_{-\infty}^{\infty}\left(\int_v^{v+1}\left|f\left(\varkappa_i(u)-u+v\right)-f(v)\right|^p du\right)dv \tag{5}$$

$$= \int_0^1 du\, \mathbf{E}\int_{-\infty}^{\infty}\left|f\left(\varkappa_i(u)-u+v\right)-f(v)\right|^p dv.$$

It also follows from Fubini's Theorem that

$$f(\omega,\cdot)\in\mathcal{L}_p(\mathfrak{B}((0,\infty)),\ell)$$

for almost all ω. By the previous lemma the last integral with respect to dv tends to zero even uniformly in u, since $|u-\varkappa_i(u)|\leq 2^{-i}$. In addition, it is bounded by the function

$$2^{p-1}\int_{-\infty}^{\infty}\left(|f|^p\left(\varkappa_i(u)-u+v\right)+|f|^p(v)\right)dv = 2^p\int_{-\infty}^{\infty}|f|^p(v)\,dv,$$

which is integrable with respect to u,ω

By the Dominated Convergence Theorem, the last expression in (5) vanishes as $i\to\infty$. Consequently, the series whose generic term is the first expressions of (5) with $i=i(n)$, converges for an appropriate choice of the subsequence $i(n)$; the series of the expectations in (4) converges for almost all $s\in[0,1]$, and (4) is valid for almost all $s\in[0,1]$. Noting that the last and first expressions in (5) are finite, and, if necessary, making the set of "good" s-values slightly smaller, we verify the last assertion of the theorem. □

6. REMARK. Given a countable set of $\mathfrak{A}$-measurable functions $f_1,f_2,\dots\in\mathcal{L}_p(\mathfrak{A},\mu)$ and nonnegative numbers $\lambda_1,\lambda_2,\dots$ such that

$$\sum_{k=1}^{\infty}\lambda_k\mathbf{E}\int_0^{\infty}|f_k(t)|^p\,dt<\infty,$$

there exists a sequence of integers $i(n)\to\infty$, such that, for almost all $s\in[0,1]$,

$$\lim_{n\to\infty}\mathbf{E}\int_0^{\infty}\sum_{k=1}^{\infty}\lambda_k\left|f_k\left(\varkappa_{i(n)}(t+s)-s\right)-f_k(t)\right|^p dt=0.$$

Indeed, consider the sum with respect to k of the first expressions in (5), written for f_k and multiplied by λ_k, and then repeat the final part of the proof, also using Weierstrass's theorem on passage to the limit in a uniformly convergent series.

7. Proof of Theorem 1. Parts (a) and (b) follow immediately from the definitions. Let us prove (c). Let $A \in \mathcal{R}$. We want to show that $A \in \mathcal{P}_\mu$. Since $A = \cup_n \big(A \cap (0, n]\!]\big)$, and $(0, n]\!] \in \mathcal{R}$, we may assume without loss of generality that for $f := I_A$ we have $f \in \mathcal{L}_1(\mathfrak{A}, \mu)$. Define $f(t) = 0$ for $t \le 0$, take the sequence $i(n)$ of Lemma 5 and define $f_n(t) = f\big(\varkappa_{i(n)}(t+s) - s\big)$, where the number $s \in [0, 1]$ is fixed so that the assertions of Lemma 5 hold. Obviously, for $t > 0$,

$$\begin{aligned} f_n(t) &= \sum_{j=1}^{\infty} f\big(t(n,j)\big) I_{t(n,j) \le t < t(n,j+1)} \\ &= \sum_{j=1}^{\infty} \Big[f\big(t(n,j)\big) I_{t(n,j) \le t} - f\big(t(n,j)\big) I_{t(n,j+1) \le t} \Big], \end{aligned} \tag{6}$$

where $\{t(n,j)\}$ is the set $(0,\infty) \cap \{k2^{-i(n)} - s : k = 0, 1, 2, \dots\}$ arranged in increasing order. Since $f\big(t(n,j)\big)$ is $\mathcal{F}_{t(n,j)}$-measurable, we see by Lemma 2.2 (a) and by what was said just before Theorem 1 that the functions f_n are $\mathcal{P}_\mu$-measurable. It follows from Lemma 5 and Exercise 1.3 that $f_n \in \mathcal{L}_1(\mathcal{P}, \mu)$ for all n. By Lemma 5 and the completeness of $\mathcal{L}_1(\mathcal{P}, \mu)$ (Lemma 1.2) we conclude that $f \in \mathcal{L}_1(\mathcal{P}, \mu)$ and f is $\mathcal{P}^\mu$-measurable. Finally, since f is $\mathfrak{A}$-measurable, it follows by Fubini's Theorem that $f(t)$ is a Borel function of t for any ω and is therefore $\mathcal{P}_\mu$-measurable. □

8. Exercise. Prove that $\mathcal{P} \subset \mathcal{R}$ and deduce from this, using Theorem 1, that $\mathcal{P}^\mu = \mathcal{R}^\mu$.

In Lemma 5 we constructed an approximation of a more or less arbitrary $\mathfrak{A}$-measurable function not by simple functions, as usual, but by piecewise constant functions of t related to nonrandom partitions of $(0,\infty)$. It is natural to ask what this does for the theory of stochastic integral.

9. Lemma. *Let $p \in [1, \infty)$, f a d-dimensional $\mathfrak{A}$-measurable function, and*

$$\eta(r) := \int_0^r |f(t)|^p \, dt < \infty \quad (a.s.)$$

for any $r \in [0, \infty)$. Then there exists a sequence of integers $i(n) \to \infty$ such that, for almost any $s \in [0, 1]$,

$$\int_0^T \big| f(t) - f\big(\varkappa_{i(n)}(t+s) - s\big)\big|^p \, dt \xrightarrow{P} 0$$

for any $T \in [0, \infty)$, where f is assumed to vanish at negative values of $\varkappa_{i(n)}(t+s) - s$.

Proof. By Fubini's Theorem, $\eta(r)$ is $\mathcal{F}$-measurable. In addition, $\eta(r)$ is left-continuous in r for $r \in (0, \infty)$, by the Monotone Convergence Theorem. Therefore, η is $\mathfrak{A}$-measurable. Now let $f_k^j(t) = f^j I_{\eta(t) \le k}$, $f_k = (f_k^1, \dots, f_k^d)$. Then

$$\int_0^\infty |f_k^j(t)|^p \, dt \le k, \qquad \mathbf{E} \int_0^\infty |f_k^j(t)|^p \, dt \le k$$

and the constants λ_k in Remark 6 for f_k^j can be taken to be equal to 2^{-k}. By Remark 6 and Lemma 6.13, for suitable $i(n)$ and almost all $s \in [0,1]$,

$$I_{\eta(T)\le k}\int_0^T \left|f(t)-f\left(\varkappa_{i(n)}(t+s)-s\right)\right|^p dt$$

$$\le \int_0^T \left|f_k(t)-f_k\left(\varkappa_{i(n)}(t+s)-s\right)\right|^p dt \xrightarrow{P} 0.$$

It obviously remains to observe that $P\{\eta(T)>k\}\to 0$ as $k\to\infty$, since it is assumed that $\eta(T)<\infty$ (a.s.). □

We will now generalize Corollary 2.5.

10. LEMMA. (a) *Let the σ-algebras $\mathcal{F}, \mathcal{F}_t$ be complete with respect to $\mathcal{F}, P$, and let w_t be a d-dimensional Wiener process with respect to $\{\mathcal{F}_t\}$, defined on $(\Omega, \mathcal{F}, P)$ for $t\ge 0$.*

(b) *Let f be a function with values in E_d, defined on $\Omega\times(0,\infty)$, and suppose there exist Markov times $\tau_0,\tau_1,\dots$ such that $\tau_n\to\infty$ as $n\to\infty$, $0=\tau_0\le\tau_1\le\dots$, $f(t)=f(\tau_n)$ for $t\in[\tau_n,\tau_{n+1})$ for all $\omega\in\Omega$, $n=0,1,2,\dots$. Suppose, finally, that $f(\tau_n)$ is $\mathcal{F}_{\tau_n}$-measurable for $n=0,1,2,\dots$. Then f is regularly measurable, $f\in H_2$ and for all $s\in[0,\infty)$ at once (a.s.)*

$$\int_0^s f^*(t)\,dw_t=\sum_{n=0}^{\infty} f^*(\tau_n)\left(w_{s\wedge\tau_{n+1}}-w_{s\wedge\tau_n}\right). \tag{7}$$

PROOF. We are given that

$$f(t)=\sum_{n=0}^{\infty} f(\tau_n)I_{\tau_n\le t<\tau_{n+1}}=\sum_{n=0}^{\infty}\left[f(\tau_n)I_{\tau_n\le t}-f(\tau_n)I_{\tau_{n+1}\le t}\right].$$

Together with the definition of the σ-algebras $\mathcal{F}_\tau$ this easily implies that the process $f(t)$ is $\mathcal{F}_t$-adapted. In addition, it is right-continuous in t and is therefore $\mathfrak{A}$-measurable with respect to (ω,t). It obviously follows from Theorem 1 and the assumption $\tau_n\to\infty$ that $f\in H_2$. Furthermore, for any $m=1,2,\dots$,

$$f(t)I_{t\le\tau_m}=\sum_{n=0}^{m-1}\left[f(\tau_n)I_{\tau_n<t}-f(\tau_n)I_{\tau_{n+1}<t}\right]\quad(\mu\text{-a.e.}).$$

Hence, by Theorem 4.4 (b), Theorem 6.14 (in which we take $g=f(\tau_n)$ and $f\equiv 1$) and the equality (cf. (4.11))

$$\int_0^s I_{\tau_n<t}\,dw_t=\int_0^s (1-I_{t\le\tau_n})\,dw_t=w_s-w_{s\wedge\tau_n}\quad\text{(a.s.)}$$

we conclude that

$$\int_0^s f^*(t)I_{t\le\tau_m}\,dw_t=\sum_{n=0}^{m-1} f^*(\tau_n)\left(w_{s\wedge\tau_{n+1}}-w_{s\wedge\tau_n}\right) \tag{8}$$

for all s at once (a.s.). It remains to use Theorem 4.4 (a) and the fact that $\tau_m \to \infty$ and that when $s \le \tau_m$, the series in (7) is equal to the sum in (8). $\square$

The next theorem follows directly from Lemmas 9, 10, the relations $\varkappa_i(t+s)-s \le t$, $\mathcal{F}_u \subset \mathcal{F}_t$ for $u \le t$, and Theorem 6.6.

11. THEOREM. *Suppose that assumption* (a) *of Lemma* 10 *is valid. Let* $f(t)$ *be a regularly measurable process with values in* E_d *such that*

$$\int_0^T |f(t)|^2 dt < \infty$$

for all $\omega \in \Omega$, $T \in (0,\infty)$. *Then* $f \in H_2$ *and there exists a sequence of integers* $i(n) \to \infty$ *such that, for almost any* $s \in [0,1]$,

$$\sup_{r \le T} \left| \int_0^r f^*(t)\, dw_t - \sum_{j=1}^{\infty} f^*(t(n,j)) \left(w_{r\wedge t(n,j+1)} - w_{r\wedge t(n,j)}\right) \right| \xrightarrow{P} 0 \tag{9}$$

for any $T \in (0,\infty)$, *where* $\{t(n,j)\}$ *is the set* $(0,\infty) \cap \{k2^{-i(n)} - s : k = 0,1,2,\dots\}$ *arranged in the increasing order.*

12. REMARK. It is sometimes important to keep in mind that in this theorem the stochastic integral of f is approximated by sums involving values only of the function f itself. In a certain sense, Theorem 11, together with Lemma 6.13, contradicts what was said after Corollary II.3.7.

13. COROLLARY. *Under the assumptions of Theorem* 11, *let* $g(t)$ *be a process satisfying the same conditions as* f. *Let* $A \in \mathcal{F}$ *be a set,* S *a nonnegative function on* Ω *and* $f(\omega,t) = g(\omega,t)$ *for all* $\omega \in A$, $t \in (0,S)$. *Then*

$$\int_0^r f^*(t)\, dw_t = \int_0^r g^*(t)\, dw_t$$

for all $r \in [0,S)$ *at once a.s. on* A.

To see this it suffices to take $f - g$ for f in Theorem 11, use Lemma 6.13, and observe that the sums in (9) vanish for $\omega \in A$, $r < S$.

This corollary gives one more variant of the "locality" property of a stochastic integral, similar to that discussed at the beginning of Sec. 4.

14. EXERCISE. Let f be a real-valued function with finite Lebesgue integral over $(-\infty,\infty)$ (and independent of ω). Prove that there exists a sequence of integers $i(n) \to \infty$ such that, for almost all $s \in [0,1]$,

$$\int_{-\infty}^{\infty} f(t)\, dt = \lim_{n\to\infty} \sum_{k=-\infty}^{\infty} 2^{-i(n)} f(k2^{-i(n)} - s).$$

This equality demonstrates one more relation between the Lebesgue and Riemann integrals.

CHAPTER IV

Some Applications of Itô's Formula

Itô's formula is the main instrument for investigating a great number of problems in the theory of diffusion processes. It was this formula that gave rise to a number of new methods for the analysis of random processes, now known as methods of stochastic analysis.

In this chapter we discuss only the most elementary examples of the application of Itô's formula. In Sec. 1 several particular cases of the formula will be discussed. It turns out that the basic difference between Itô's formula and the formula for the ordinary differential of a composite function, namely the presence of the terms with $d\xi_t^i d\xi_t^j$ in the expression for $du(\xi_t)$, easily explains the connection between diffusion processes and parabolic and elliptic operators. Incidentally, Itô's formula sometimes looks exactly like the formula for the usual differential of a composite function: $du(z_t) - u_z(z_t)\,dz_t$, if $u(z)$ is an *analytic* function of one complex variable z and $z_t = w_t^1 + iw_t^2$ is what is called a *complex Wiener process*, that is, (w_t^1, w_t^2) is a two-dimensional Wiener process.

In order to understand the subsequent material, the reader must be familiar with the results presented in Sec. 1 and the Burkholder-Davis-Gundy inequalities from Sec. 4. The extremely important methods of random time change (Sec. 2) and change of probability measure (Sec. 3), which make it possible to transform stochastic integrals of one kind into others, will be used only in Chap. VI and partly in Sec. V.2, which is devoted to some examples and can be omitted. These explanations may be useful for the reader who wants to gain a more rapid acquaintance with Itô's stochastic equations.

1. Transformation of Itô's formula; particular cases

Let $(\Omega, \mathcal{F}, P)$ be a complete probability space, $d, d_1 \geq 1$ integers, $(w_t, \mathcal{F}_t)$ a d_1-dimensional Wiener process on $(\Omega, \mathcal{F}, P)$ for $t \geq 0$, where the σ-algebras $\mathcal{F}_t$ are complete with respect to $\mathcal{F}, P$. Let us also take a d-dimensional process ξ_t that has a stochastic differential with respect to w and set $d\xi_t = \sigma(t)\,dw_t + b(t)\,dt$.

In Secs. III.8 and III.10 we derived two versions of Itô's formula. Naturally, for our process ξ_t these formulas give the same result, which can be obtained from (III.8.11) by performing all necessary operations, using the multiplication table (III.8.2), and combining like terms. We get

$$du(\xi_t) = L_t^{a,b} u(\xi_t)\,dt + u_x(\xi_t)\sigma(t)\,dw_t, \tag{1}$$

where $u = u(x)$ is a twice continuously differentiable function, $u_x(x)$ the row-vector with coordinates $u_{x^i}(x)$, $a = a(t) := \sigma(t)\sigma^*(t)$ a matrix,

$$L_t^{a,b} u(x) := \frac{1}{2} \sum_{i,j=1}^{d} a^{ij}(t) u_{x^i x^j}(x) + \sum_{i=1}^{d} b^i(t) u_{x^i}(x). \tag{2}$$

Expression (2) defines a second-order differential operator for all ω and t. It turns out that this operator is always elliptic, possibly degenerate, because for arbitrary $\lambda \in E_d$:

$$(a\lambda, \lambda) = (\sigma(t)\sigma^*(t)\lambda, \lambda) = (\sigma^*(t)\lambda, \sigma^*(t)\lambda) = |\sigma^*(t)\lambda|^2 \geq 0.$$

A few remarks concerning the structure of operator (2) will simplify its writing down in some cases. The variable x in this operator takes values in the same space as the process ξ_t; $b(t)$ is, so to speak, the velocity of the deterministic component of ξ_t at time t; the matrix $a(t)$ describes the diffusion component of the process. It is symmetric: $a^* = (\sigma\sigma^*)^* = \sigma\sigma^* = a$, and as we have already seen, it is positive semidefinite. Moreover,

$$a^{ii} = \sum_{j=1}^{d_1} (\sigma^{ij})^2,$$

so that, for example, $a^{11} \equiv 0$ if and only if $\sigma^{1j} \equiv 0$ for all $j = 1, \ldots, d_1$, that is, when ξ_t^1 has no diffusion component. When this happens, also $a^{1i} = a^{i1} \equiv 0$ and operator (2) involves no *second* derivatives of u in which one of the differentiations is with respect to x^1. In the general case a^{ij} can be determined from the equation $d\xi_t^i d\xi_t^j = a^{ij}\, dt$.

Sometimes it may be worthwhile to remember that if a *Wiener* process w_t is "built up" from two Wiener processes w_t^1 and w_t^2, that is,

$$w_t = (w_t^{(1)1}, \ldots, w_t^{(1)k}, w_t^{(2)1}, \ldots, w_t^{(2)m}),$$

$k + m = d_1$ and $\sigma(t)\, dw_t = \sigma^{(1)}(t)\, dw_t^{(1)} + \sigma^{(2)}(t)\, dw_t^{(2)}$, where $\sigma^{(i)}$ are matrices of appropriate dimensions, then

$$a(t) = a^{(1)}(t) + a^{(2)}(t),$$

where $a^{(i)} := \sigma^{(i)}\sigma^{(i)*}$.

It is also possible to derive Itô's formula in the form (1) with an at first sight more general operator than (2).

1. THEOREM. *Let $u(t,x)$ be a real function defined on $[0,\infty) \times E_d$, continuous in (t,x) together with its derivatives $\frac{\partial u}{\partial t}, u_x, u_{xx}$. For a real function c of class H_1 define*

$$\varphi_t = \int_0^t c(s)\, ds,$$

Then the process $u(t,\xi_t)e^{-\varphi_t}$ has a stochastic differential and

$$\begin{aligned} d\left[u(t,\xi_t)e^{-\varphi_t}\right] &= e^{-\varphi_t} u_x(t,\xi_t)\sigma(t)\, dw_t \\ &\quad + e^{-\varphi_t}\left[\frac{\partial u}{\partial t}(t,\xi_t) + L_t^{a,b} u(t,\xi_t) - c(t)u(t,\xi_t)\right] dt. \end{aligned} \tag{3}$$

In addition, if $t \geq 0$ is fixed and η_t is a d-dimensional, $\mathcal{F}_t$-measurable vector such that for $s > t$

$$\eta_s := \eta_t + \int_t^s \sigma(r)\, dw_r + \int_t^s b(r)\, dr$$

then a.s. for all $s > t$ at once

$$(4)\qquad \begin{aligned} u(s,\eta_s)e^{-\varphi_s} &= u(t,\eta_t)e^{-\varphi_t} + \int_t^s e^{-\varphi_r} u_x(r,\eta_r)\sigma(r)\,dw_r \\ &+ \int_t^s e^{-\varphi_r}\left[\frac{\partial u}{\partial r}(r,\eta_r) + L_r^{a,b}u(r,\eta_r) - c(r)u(r,\eta_r)\right]dr. \end{aligned}$$

PROOF. By definition, $d\varphi_t = c(t)\,dt$ and (e.g., by Itô's formula) $d\exp(-\varphi_t) = -c(t)\exp(-\varphi_t)\,dt$. Hence Lemma III.8.5 easily implies that we need to prove Itô's formula (3) only for $c \equiv 0$. Let us consider the $(d+1)$-dimensional process $\tilde{\xi}_t := (\xi_t^0, \xi_t)$, where $\xi_t^0 \equiv t$. Then, applying Itô's formula (1) to $u(\tilde{\xi}_t)$, we could formally obtain (3) for the case $c \equiv 0$, provided we assume in addition that $\partial^2 u/\partial t^2$ and $\partial u_x/\partial t$ are continuous with respect to (t,x). Now, in the proof of Itô's formula (III.8.11) we needed the continuity of the derivatives of u only when effecting the passage to the limit from polynomials to u. But since the derivatives of u entering Itô's formula (3) are continuous by the assumption and, by Remark III.8.10, the approximating polynomials u_n can be selected in such a way that their derivatives $\partial u_n/\partial t, u_{nx}, u_{nxx}$ converge uniformly on compact sets to the corresponding derivatives of u, it follows that the proof of (III.8.11) carries over almost literally to our case. This proves (3).

It follows from (3) for $u(s, x+y)$ with arbitrary fixed $y \in E_d$ and $t \geq 0$ that for every $s \geq t$ (a.s.)

$$(5)\qquad \begin{aligned} u(s,\xi_s+y)e^{-\varphi_s} &= u(t,\xi_t+y)e^{-\varphi_t} + \int_t^s e^{-\varphi_t} u_x(r,\xi_r+y)\sigma(r)\,dw_r \\ &+ \int_t^s e^{-\varphi_r}\left[L_r^{a,b}u(r,\xi_r+y) + \frac{\partial u}{\partial r}(r,\xi_r+y) - c(r)u(r,\xi_r+y)\right]dr. \end{aligned}$$

Now let $y_n = y_n(\omega) = \varkappa_n(\eta_t - \xi_t)$, where $\varkappa_n(x) = (\varkappa_n(x^1), \dots, \varkappa_n(x^d))$ and $\varkappa_n(p) = 2^{-n}[2^n p]$, $p \in (-\infty, \infty)$. Multiply (5) by the indicator of the $\mathcal{F}_t$-measurable event $\{\omega : y_n(\omega) = y\}$. By Theorem III.6.14, this indicator can be brought through the integral signs. Once this is done, we can replace y by y_n in the integrands and take the indicator outside again. Then we see that on every set $\{\omega : y_n(\omega) = y\}$ equality (5) is true (a.s.) with y_n substituted for y. Since the set of all possible values of y_n is countable, the union of the sets $\{\omega : y_n(\omega) = y\}$ over a certain countable set of all values of $y_n(\omega)$ is Ω and hence (5) is valid with y_n substituted for y (a.s.). To prove (4) it now remains to let $n \to \infty$, noting that $\xi_t + y_n \to \xi_t + \eta_t - \xi_t = \eta_t$ and applying the known rules for limits of stochastic and usual integrals. □

We can now confirm our remark, made just before Theorem 1, that Itô's formula (3) is a particular case of Itô's formula (1). Indeed, up to the less stringent smoothness conditions on u, formula (3) can be obtained from (1) if we apply the latter to $\tilde{u}(\tilde{\xi}_t, \varphi_t)$, where $\tilde{u}(t,x,\varphi) = u(t,x)\exp(-\varphi)$ and $\tilde{\xi}_t$ is the same process as in the previous proof.

2. COROLLARY. *For fixed $t \in [0, \infty)$, let us write $d\eta_s = \sigma(s)\,dw_s + b(s)\,ds$, $s \geq t$, if a.s. for all $s \geq t$ at once*

$$\eta_s = \eta_t + \int_t^s \sigma(r)\,dw_r + \int_t^s b(r)\,dr.$$

We will also say in that case that η_t has a stochastic differential after time t. *Theorem 1 shows that formula* (III.8.11) *can be used to calculate $du(\eta_s)$ for $s \geq t$, for a twice continuously differentiable function $u(x)$, just as in the case $t = 0$. In particular, if the scalar processes η_s, ξ_s have stochastic differentials after time t, then $d(\eta_s \xi_s) = \eta_s\,d\xi_s + \xi_s\,d\eta_s + d\eta_s d\xi_s$, $s \geq t$.*

Let us apply Theorem 1 to the case when $d_1 = d$, σ is the identity matrix, b is a constant vector and thus $\xi_t = \xi_0 + w_t + bt$ for all $t \geq 0$. Moreover, let $c(t) = c(t, \xi_t)$ for some function $c(t, x)$ defined on $[0, \infty) \times E_d$. In this situation,

$$d[u(t, \xi_t)e^{-\varphi_t}] = e^{-\varphi_t} u_x(t, \xi_t)\,dw_t - e^{-\varphi_t} g(t, \xi_t)\,dt, \tag{6}$$

where

$$-g(t, x) := \frac{\partial u}{\partial t}(t, x) + \frac{1}{2}\Delta u(t, x) + u_{(b)}(t, x) - c(t, x)u(t, x).$$

Theorem II.8.6 can be derived from Itô's formula (6) or from its martingale version (Theorem III.9.5). The reader can convince himself that this new proof of the theorem does not use its assertion. Indeed, let us assume that all the conditions of the theorem are fulfilled except those concerning c, which we replace by the requirement that c be a bounded Borel function. Then, using the functions u_n and the Markov times $\tau(n)$ from the proof of Theorem II.8.6, we apply Itô's formula (6) to u_n instead of u, concluding that the process $\varkappa_{s \wedge \tau(n)}$ from the proof of Theorem II.8.6 is a local martingale. Since by assumption it is bounded, it follows that $\varkappa_{s \wedge \tau(n)}$ is a martingale, and we can finish the proof as before.

The same arguments work in a much more general situation, when ξ_t is a solution of a *stochastic equation* $d\xi_t = \sigma(t, \xi_t)\,dw_t + b(t, \xi_t)\,dt$, $\sigma(t, x)$ and $b(t, x)$ being nonrandom functions on $[0, \infty) \times E_d$. Indeed, if in addition to this $c(\omega, t) = c(t, \xi_t)$ for a nonrandom function $c(t, x)$, formula (3) gives

$$d\left[u(t, \xi_t)e^{-\varphi_t}\right] = e^{-\varphi_t} u_x(t, \xi_t)\sigma(t, \xi_t)\,dw_t - e^{-\varphi_t} g(t, \xi_t)\,dt,$$

where for $a(t, x) := \sigma(t, x)\sigma^*(t, x)$, we put

$$-g(t, x) := \frac{\partial u}{\partial t}(t, x) + \frac{1}{2}\sum_{i,j=1}^{d} a^{ij}(t, x)u_{x^i x^j}(t, x)$$
$$+ \sum_{i=1}^{d} b^i(t, x)u_{x^i}(t, x) - c(t, x)u(t, x).$$

In complete analogy to the proof of Theorem II.8.6, we can now prove that *if $T \in (0, \infty)$, a domain $G \subset (-\infty, T) \times E_d$, $Q := ([0, T) \times E_d) \cap G \neq \emptyset$, the function u is continuous and bounded in $\bar{Q}$ and its derivatives $\frac{\partial u}{\partial t}, u_x, u_{xx}$ are continuous in Q, the* *functions* $c(t,x)$ *and* $g(t,x)$ *are bounded in* Q, ξ_0 *does not depend on* ω *and* $(0,\xi_0)\in Q$, *then*

$$u(0,\xi_0)=\mathbf{E}\,e^{-\varphi_\tau}u(\tau,\xi_\tau)+\mathbf{E}\int_0^\tau g(t,\xi_t)e^{-\varphi_t}\,dt,$$

where $\tau=\inf\{t\geq 0:(t,\xi_t)\notin Q\}$. This equality gives a probabilistic representation of the solution of a parabolic equation if we regard the definition of g as an equation in u. We can make the same remarks concerning this formula as after Remark II.8.7 and use it for the same purposes as in Secs. II.9–10.

Sometimes it is helpful to keep in mind that Itô's formula can be used to calculate certain stochastic integrals, as well as to give meaning to usual integrals of random functions.

For example, if w_t is a one-dimensional Wiener process, then $d(w_t{}^2)=2w_t\,dw_t+(dw_t)^2=2w_t\,dw_t+dt$ (Lemma III.8.5), that is, formula (III.2.7) is valid a.s.:

$$\int_0^s w_t\,dw_t=\frac{1}{2}(w_s^2-s).$$

For the same process, we can make the integral

$$\int_0^s \frac{dv}{dx}(w_t)\,dt \tag{7}$$

meaningful for any bounded (possibly discontinuous) Borel function $v(x)$. Indeed, define

$$u(x)=2\int_0^x v(y)\,dy,\qquad \int_0^s\frac{dv}{dx}(w_t)\,dt=u(w_s)-2\int_0^s v(w_t)\,dw_t \tag{8}$$

and note that by Itô's formula this definition is legitimate for continuously differentiable functions. Definition (8) also has the following natural property: if $v_n(x)$ are uniformly bounded Borel (for example, smooth) functions and $v_n(x)\to v(x)$ a.e. as $n\to\infty$, then

$$\int_0^s\frac{dv_n}{dx}(w_t)\,dt\xrightarrow{P}\int_0^s\frac{dv}{dx}(w_t)\,dt. \tag{9}$$

Indeed, by standard theorems on passage to the limit for usual and stochastic integrals, it is sufficient to observe that

$$\mathbf{E}\int_0^s|v(w_t)-v_n(w_t)|^2\,dt=\int_0^s\frac{1}{\sqrt{2\pi t}}\int_{-\infty}^{\infty}|v(x)-v_n(x)|^2e^{-x^2/2t}\,dt\,dx\to 0.$$

If $v(x)=\frac{1}{2}\operatorname{sgn}x$, the process (7) is called a *local time* of w at zero, denoted by l_s. It follows from the arguments about (9) with smooth v_n that we can choose l_s so that this process is increasing and continuous. By definition, $|w_s|=\bar{w}_s+l_s$ (a.s.), where

$$\bar{w}_s=\int_0^s\operatorname{sgn}w_t\,dw_t, \tag{10}$$

and moreover, by Lévy's Theorem, $\bar{w}_s$ is a Wiener process. Thus, local time enables us to express the Wiener process *reflected at the origin* $|w_s|$ as the sum of $\bar{w}_s$ and l_s. Local time plays an important role in consideration of other types of interaction between diffusion processes and domain boundaries.

3. PROBLEM. Let $u(x)$ be twice continuously differentiable on $[0,\infty)$, $\eta_s = |w_s|$. Considering the function $u(|x|) - u'(0)|x|$, prove that

$$u(\eta_s) = u(0) + \int_0^s u'(\eta_t)\, d\bar{w}_t + \frac{1}{2}\int_0^s u''(\eta_t)\, dt + u'(0)l_s \qquad \text{(a.s.)}.$$

4. PROBLEM. Approximating $\operatorname{sgn} x$ by smooth odd functions in x, prove that l_s and $\bar{w}_s = |w_s| - l_s$ are measurable with respect to the completion of the σ-algebra $\mathcal{F}_s^{|w|}$. Prove also that $\operatorname{sgn} w_s$ and $\mathcal{F}_s^{|w|}$ are independent, w_s is not $\mathcal{F}_\infty^{|w|}$-measurable, and consequently *is not determined* by $\bar{w}$, although according to (10) it is a solution of the equation $dw_t = \operatorname{sgn} w_t\, d\bar{w}_t$. (This argument belongs to Tanaka.)

We finish this section with a discussion of the concept of quadratic variation, introduced in Sec. III.5.

5. THEOREM. *As everywhere in this section, let ξ_t be a d-dimensional process with stochastic differential $d\xi_t = \sigma(t)\, dw_t + b(t)\, dt$. Let $0 = t(0,n) < t(1,n) < \cdots$ be a sequence of nonrandom partitions of $[0,\infty)$ such that*

$$\sup\{t(i+1,n) - t(i,n) : i = 0,1,\cdots\} \to 0 \quad \textit{as } n \to \infty.$$

Then, for any $T \in (0,\infty)$,

$$\sup_{s\le T}\left|\zeta_s(n) - \int_0^s a(t)\, dt\right| \xrightarrow{P} 0, \tag{11}$$

where $a = \sigma\sigma^*$ and

$$\zeta_s(n) = \sum_{i=0}^{\infty} (\xi_{s\wedge t(i+1,n)} - \xi_{s\wedge t(i,n)})(\xi_{s\wedge t(i+1,n)} - \xi_{s\wedge t(i,n)})^*.$$

PROOF. It is easy to see, using the formula $(a-b)(c-d) = ac - bd - (a-b)d - (c-d)b$, that for $k,j = 1,\dots,d$

$$\zeta_s^{kj}(n) = \xi_s^k\xi_s^j - \xi_0^k\xi_0^j - \sum_{i=0}^{\infty} \xi_{t(i,n)}^j(\xi_{s\wedge t(i+1,n)}^k - \xi_{s\wedge t(i,n)}^k)$$
$$- \sum_{i=0}^{\infty} \xi_{t(i,n)}^k(\xi_{s\wedge t(i+1,n)}^j - \xi_{s\wedge t(i,n)}^j).$$

Proceeding exactly as in the proof of Lemma III.8.4, we obtain assertion (11), but with the integral of a replaced by the matrix with entries

$$\xi_s^k\xi_s^j - \xi_0^k\xi_0^j - \sum_{r=1}^{d_1}\int_0^s (\xi_t^j\sigma^{kr}(t) + \xi_t^k\sigma^{jr}(t))\, dw_t^r - \int_0^s (\xi_t^j b^k(t) + \xi_t^k b^j(t))\, dt.$$

It remains only to observe that there is no need for this substitution, because by Itô's formula, both matrix processes coincide for all $s \in [0, \infty)$ (a.s.). □

2. Random time change in stochastic integrals

Let $(\Omega, \mathcal{F}, P)$ be a complete probability space and $(w_t, \mathcal{F}_t)$ a one-dimensional Wiener process on it, defined for $t \geq 0$, where the σ-algebras $\mathcal{F}_t$ are complete with respect to $\mathcal{F}, P$. In Definition III.5.5 we introduced the classes of functions H_p, which obviously depend on the filtration of σ-algebras $\{\mathcal{F}_t\}$. We will write $H_p(\mathcal{F}_t)$ for H_p, in order to make clear which filtration of σ-algebras underlies the construction of H_p, because we will be varying it in this section. Similarly, we write $\mathcal{P}(\mathcal{F}_t)$ for $\mathcal{P}$, etc.

Let us fix a function $\beta \in H_1(\mathcal{F}_t)$ and assume that for all ω, t

$$(1) \qquad \beta(\omega, t) > 0, \qquad \int_0^\infty \beta(\omega, r)\, dr = \infty.$$

Let

$$(2) \qquad \varphi(s) = \int_0^s \beta(t)\, dt, \qquad \psi(s) = \inf\{t \geq 0 : \varphi(t) \geq s\}.$$

It follows from (1) that $\varphi(s)$ is continuous and strictly increasing from 0 to ∞, $\psi(s)$ is finite and varies continuously from 0 to ∞ for $s \in [0, \infty)$. Moreover, we know that $\psi(s)$ are Markov times, the σ-algebras $\mathcal{F}_{\psi(s)}$ are defined and form a filtration of σ-algebras. These σ-algebras are complete, because $\mathcal{F}_{\psi(s)} \supset \mathcal{F}_0$. We also note that

$$\varphi(\psi(s)) = \psi(\varphi(s)) = s, \qquad \varphi(s) = \inf\{t \geq 0 : \psi(t) \geq s\},$$

and by the last formula, $\varphi(s)$ is a Markov time for $\{\mathcal{F}_{\psi(t)}\}$ for any $s \in [0, \infty)$ (see Exercise II.4.16).

1. Exercise. Let $G_t = \mathcal{F}_{\psi(t)}$ and consider $G_{\varphi(t)}$. Naturally, we would expect that $G_{\varphi(t)} = \mathcal{F}_{\psi(\varphi(t))} = \mathcal{F}_t$. However, this needs not be true, and the reader is asked to verify this, starting with the case in which $\Omega = [1, 2]^2$, $P = \ell$, $\mathcal{F}_t = \mathcal{F} = \mathfrak{B}^\ell(\Omega)$ for $t > 1$, $\mathcal{F}_t$ is the completion of $\mathfrak{B}([1, 2]) \otimes \{\emptyset, [1, 2]\}$ for $t \leq 1$, $\beta(\omega_1, \omega_2, t) = \omega_1$. Prove that in this case $G_{\varphi(1)} = \mathcal{F}_2$.

As a rule, the general theory of stochastic analysis deals with a filtration of σ-algebras $\mathcal{F}_t$ which is right-continuous in t, passing if necessary from $\mathcal{F}_t$ to $\mathcal{F}_{t+} := \bigcap_{s>t} \mathcal{F}_s$. It is possible to show that the assumption $\mathcal{F}_t = \mathcal{F}_{t+}$ for all $t \geq 0$ guarantees equality $G_{\varphi(t)} = \mathcal{F}_t$, but we will need neither this assumption nor this equality.

2. Lemma. (a) *If τ is a Markov time for $\{\mathcal{F}_t\}$, then $\varphi(\tau)$ is a Markov time for $\{\mathcal{F}_{\psi(t)}\}$.*

(b) *A function $f(t)$ is $\{\mathcal{F}_t\}$-predictable if and only if $f(\psi(t))$ is $\{\mathcal{F}_{\psi(t)}\}$-predictable.*

(c) *A continuous process ξ_t is $\{\mathcal{F}_t\}$-adapted if and only if $\xi_{\psi(t)}$ is $\{\mathcal{F}_{\psi(t)}\}$-adapted.*

(d) $\beta^{1/p} f(\cdot) \in H_p(\mathcal{F}_t)$ *if and only if* $f(\psi(\cdot)) \in H_p(\mathcal{F}_{\psi(t)})$, *and if* $\beta f \in H_1(\mathcal{F}_t)$ *then for all* $\omega \in \Omega$, $s \in [0, \infty)$

$$(3) \qquad \int_0^{\varphi(s)} f(\psi(t))\,dt = \int_0^s f(t)\beta(t)\,dt, \qquad \int_0^s f(\psi(t))\,dt = \int_0^{\psi(s)} f(t)\beta(t)\,dt;$$

in particular (for $f = \beta^{-1}$),

$$\psi(s) = \int_0^s \beta^{-1}(\psi(t))\,dt.$$

Proof. Assertion (a) follows from the fact that $\{\varphi(\tau) \le t\} = \{\tau \le \psi(t)\} \in \mathcal{F}_{\psi(t)}$ (see Exercise II.4.16). By (a), if $f(t) = I_{t \le \tau}$, where τ is a Markov time, then $f(\psi(t)) = I_{t \le \varphi(\tau)}$ is $\{\mathcal{F}_{\psi(t)}\}$-predictable. Hence, by the lemma on π- and λ-systems, we easily conclude that for any $\{\mathcal{F}_t\}$-predictable set A and $f = I_A$, the function $f(\psi(t))$ is $\{\mathcal{F}_{\psi(t)}\}$-predictable. Standard approximation of measurable functions by simple functions now proves the "only if" part of assertion (b). Conversely, if τ is a Markov time relative to $\{\mathcal{F}_{\psi(t)}\}$, then $\{\tau < s\} \in \mathcal{F}_{\psi(s)}$ and

$$(4) \qquad \begin{gathered} \{\tau < s \le \varphi(t)\} = \{\tau < s, \psi(s) \le t\} \in \mathcal{F}_t, \\ \{\tau < \varphi(t)\} = \bigcup_{s \in \rho} \{\tau < s \le \varphi(t)\} \in \mathcal{F}_t, \quad \{\varphi(t) \le \tau\} \in \mathcal{F}_t, \end{gathered}$$

where ρ is the set of all rational points in $[0, \infty)$. Hence, putting $g(t) := I_{t \le \tau}$, we see that $g(\varphi(t))$ is $\mathcal{F}_t$-adapted, and, since it is also left-continuous, it follows that $g(\varphi(t))$ is $\{\mathcal{F}_t\}$-predictable (see Lemma III.2.2). As above, the transition to an arbitrary $\{\mathcal{F}_{\psi(t)}\}$-predictable function g is carried out by standard means and this proves (b).

Assertion (c) is a direct consequence of (b), Lemma III.2.2 , and the fact that $f(t)$ is $\mathcal{F}_t$-measurable for any $t > 0$ if f is an $\{\mathcal{F}_t\}$-predictable function.

Let us prove (d). We observe, that equalities (3) hold for any fixed ω and s and any *Borel* function $f(t) \ge 0$ (which is independent of ω). Indeed, the second equality follows from the first one by the substitution $s \to \psi(s)$. The first equality is true for $f = I_{(0,u]}$ by the definitions in (2), and standard arguments immediately extend it to all Borel functions $f \ge 0$.

In addition, if f is still independent of ω, nonnegative and $\mathfrak{B}^\ell((0,\infty))$-measurable, then there exist Borel functions $h, g \ge 0$ such that $g(t) = 0$ (a.e.), $|f(t) - h(t)| \le g(t)$ and $g(t)\beta(t) = 0$ (a.e.). By (3), $g(\psi(t)) = 0$, $f(\psi(t)) = h(\psi(t))$, $f(t)\beta(t) = h(t)\beta(t)$ (a.e.), and as equalities (3) are true for h, they are also true for f. We have also proved that $f(\psi(\cdot))$ is $\mathfrak{B}^\ell((0,\infty))$-measurable for any ω. This is naturally true not only for nonnegative but for any $\mathfrak{B}^\ell((0,\infty))$-measurable function f.

On the other hand, if $f(\psi(\cdot))$ is $\mathfrak{B}^\ell((0,\infty))$-measurable for some ω, then there exist Borel functions g, h such that $g(t) = 0$ (a.e.) and $|f(\psi(t)) - h(t)| \le g(t)$. Then $h(\varphi), g(\varphi)$ are Borel measurable, $g(\varphi(t))\beta(t) = 0$ (a.e.) by (3), $g(\varphi(t)) = 0$ (a.e.) by (1), $f(t) = h(\varphi(t))$ (a.e.) and $f(t)$ is $\mathfrak{B}^\ell((0,\infty))$-measurable.

Definitions III.5.1 and III.5.5 now show that, to prove (d), it is sufficient to establish that f is $\mathcal{P}^\mu(\mathcal{F}_t)$-measurable if and only if $f(\psi)$ is $\mathcal{P}^\mu(\mathcal{F}_{\psi(t)})$-measurable. In order to do this, it is enough to repeat the argument of the two previous paragraphs with slight changes: take g, h to be not Borel functions, but predictable, use assertion (b) instead of the measurability of a superposition of Borel functions, and use Fubini's Theorem together with (3). □

3. THEOREM. (a) *The process*

$$\tilde{w}_s := \int_0^{\psi(s)} \beta^{1/2}(t)\, dw_t, \quad s \in [0, \infty), \tag{5}$$

is a Wiener process with respect to $\{\mathcal{F}_{\psi(s)}\}$.

(b) *If* $\beta^{\frac{1}{2}} f \in H_2(\mathcal{F}_t)$, *then* $f(\psi) \in H_2(\mathcal{F}_{\psi(t)})$ *and*

$$\int_0^{\psi(s)} f(t)\beta^{1/2}\, dw_t = \int_0^s f(\psi(t))\, d\tilde{w}_t \tag{6}$$

a.s. for all $s \in [0, \infty)$ *at once. In particular, for* $f = \beta^{-1/2}$,

$$w_{\psi(s)} = \int_0^s \beta^{-1/2}(\psi(t))\, d\tilde{w}_t. \tag{7}$$

PROOF. First of all let us note that by Lemma 2 the process $\tilde{w}_s$ is $\{\mathcal{F}_{\psi(s)}\}$-adapted. If τ is a bounded Markov time for $\{\mathcal{F}_{\psi(s)}\}$, then although in general $\psi(\tau)$ is not a Markov time for $\{\mathcal{F}_t\}$, nevertheless $\{\psi(\tau) < t\} \in \mathcal{F}_t$ by (4), for any $\varepsilon > 0$, $\tau + \varepsilon$ is also a Markov time for $\{\mathcal{F}_{\psi(s)}\}$,

$$\{\psi(\tau + \varepsilon) \le t\} = \bigcap_{n=1}^{\infty} \left\{\psi\left(\tau + \varepsilon - \frac{1}{n}\right) < t\right\} \in \mathcal{F}_t$$

and $\psi(\tau + \varepsilon)$ *is* a Markov time for $\{\mathcal{F}_t\}$. The following formulas are also valid:

$$\int_0^{\infty} \left|I_{t \le \psi(\tau)} \beta^{1/2}(t)\right|^2 dt = \varphi(\psi(\tau)) = \tau,$$

$$\int_0^{\infty} \left|I_{t \le \psi(\tau+\varepsilon)} - I_{t \le \psi(\tau)}\right|^2 \beta(t)\, dt = \int_{\psi(\tau)}^{\psi(\tau+\varepsilon)} \beta(t)\, dt = \varepsilon,$$

from which we see that

$$I_{(0, \psi(\tau+\varepsilon)]\!]} \beta^{1/2},\ I_{(0, \psi(\tau)]\!]} \beta^{1/2} \in \mathcal{L}_2(\mathcal{P}(\mathcal{F}_t), \mu),$$

$$\tilde{w}_\tau = \lim_{n \to \infty} \tilde{w}_{\tau + \frac{1}{n}} = \lim_{n \to \infty} \int_0^{\infty} I_{t \le \psi(\tau + \frac{1}{n})} \beta^{1/2}(t)\, dw_t$$

$$= \int_0^{\infty} I_{t \le \psi(\tau)} \beta^{1/2}(t)\, dw_t \qquad \text{(a.s.)}.$$

It follows from these equalities and the Wald identities (see (III.2.4), (III.2.8), and (III.5.5)) that $\tilde{w}_s$ is an admissible local martingale with respect to $\{\mathcal{F}_{\psi(s)}\}$ and $\langle\tilde{w}\rangle_s = s$. By Lévy's Theorem, $\{\tilde{w}_s, \mathcal{F}_{\psi(s)}\}$ is a Wiener process.

Let us denote the left-hand side of (6) by ξ_s. Arguments analogous to the previous ones show that ξ_s is an $\mathcal{F}_{\psi(s)}$-adapted process, and if f is *bounded* and τ is a bounded Markov time for $\{\mathcal{F}_{\psi(s)}\}$, then $I_{(0,\psi(\tau)]\!]} f\beta^{1/2} \in \mathcal{L}_2(\mathcal{P}(\mathcal{F}_t), \mu)$,

$$\xi_\tau = \int_0^\infty I_{t\le\psi(\tau)} f(t)\beta^{1/2}(t)\, dw_t \qquad \text{(a.s.)}.$$

Together with the Wald identities, this shows that $\xi_s + \lambda\tilde{w}_s$ (for any λ) and the pair $(\xi_s, \tilde{w}_s)$ are admissible local martingales with respect to $\{\mathcal{F}_{\psi(s)}\}$, and by (3)

$$\langle\xi\rangle_s = \int_0^{\psi(s)} f^2(t)\beta(t)\, dt = \int_0^s f^2(\psi(t))\, dt,$$

$$\langle\xi, \tilde{w}\rangle_s = \int_0^{\psi(s)} f(t)\beta(t)\, dt = \int_0^s f(\psi(t))\, dt.$$

By Definition III.10.4, these two equalities imply (6). Finally, we can prove (6) for arbitrary f by putting $f_n = (-n)\vee(f\wedge n)$, letting $n \to \infty$ and using (3) and Theorem III.6.6. □

4. Remark. Theorem 3 remains valid if w is a multidimensional Wiener process and f a function with values in the set of matrices of appropriate dimension. For brevity, we will not prove this elementary generalization.

There are many sometimes rather unexpected applications of Theorem 3.

5. Corollary. *Let $\sigma(x)$ be a real bounded Borel function defined on E_1 such that*

$$\int_{-n}^{n} \sigma^{-2}(x)\, dx < \infty \qquad (0^{-2} := \infty)$$

for all $n < \infty$. Then there exist a Wiener process $(\bar{w}_t, G_t)$ on $(\Omega, \mathcal{F}, P)$ and a G_t-adapted process ξ_t such that $\xi_0 = 0$, $d\xi_t = \sigma(\xi_t)\, d\bar{w}_t$, $t \ge 0$.

To verify this, we use Remark II.1.3, from which it follows that for $n = 1, 2, \ldots$ and $\tau_n := \inf\{t \ge 0\colon |w_t| \ge n\}$

$$\mathbf{E}\int_0^{n\wedge\tau_n} \sigma^{-2}(w_t)\, dt \le \int_0^n \mathbf{E}\sigma^{-2}(w_t) I_{|w_t|\le n}\, dt \le \int_{-n}^{n} \sigma^{-2}(x)\, dx \int_0^n \frac{dt}{\sqrt{t}} < \infty.$$

Hence, for

$$\Omega' := \left\{\omega : \int_0^s \sigma^{-2}(w_t)\,dt < \infty \quad \text{for all } s \in [0,\infty)\right\}$$
$$= \bigcap_n \left\{\omega : \int_0^{n\wedge\tau_n} \sigma^{-2}(w_t)\,dt < \infty\right\}$$

we obtain $P(\Omega') = 1$. Let $w_t^0(\omega) = w_t(\omega)$ for $\omega \in \Omega', t \geq 0$; $w_t^0(\omega) = w_t(\omega_0)$ for $\omega \notin \Omega', t \geq 0$, where ω_0 is an arbitrary fixed point in Ω'. Then it is clear that $(w_t^0, \mathcal{F}_t)$ is a Wiener process, and for the predictable process $\beta_t = \sigma^{-2}(w_t^0)$ conditions (1) are satisfied and $\beta \in H_1(\mathcal{F}_t)$. Thus, passing, if necessary, from w to w^0, we may — and therefore will — assume that the function $\beta(t) := \sigma^{-2}(w_t)$ satisfies conditions (1) and $\beta \in H_1(\mathcal{F}_t)$.

Now let $\xi_t = w_{\psi(t)}$, $G_t = \mathcal{F}_{\psi(t)}$. By (7) we have $d\xi_t = |\sigma(\xi_t)|\,d\tilde{w}_t$. Defining a process $\bar{w}_t$ so that $\bar{w}_0 = 0$, $d\bar{w}_t = \operatorname{sgn}\sigma(\xi_t)\,d\tilde{w}_t$ ($\operatorname{sgn} 0 := 1$) and noticing that by Lévy's Theorem $(\bar{w}_t, G_t)$ is a Wiener process, we finally obtain $d\xi_t = \sigma(\xi_t)\,d\bar{w}_t$.

6. Remark. If $\alpha \in (0, \frac{1}{2})$ we can take $|x|^\alpha \wedge 1$ as σ and the solution constructed above for the *equation* $d\xi_t = \sigma(\xi_t)\,d\bar{w}_t$ with initial value $\xi_0 = 0$ is *distinct from zero*. At the same time, this equation with initial value zero has also a trivial solution. We also note that if $\alpha \geq \frac{1}{2}$ and $\sigma(x) = |x|^\alpha \wedge 1$ then, for any Wiener process $(w_t, \mathcal{F}_t)$, the $\mathcal{F}_t$-adapted solution of the equation $d\xi_t = \sigma(\xi_t)\,dw_t$ with initial value zero is indistinguishable from zero.

Indeed, applying Itô's formula to any twice continuously differentiable function u, we see that the process

$$u(\xi_s) - \frac{1}{2}\int_0^s u''(\xi_t)\sigma^2(\xi_t)\,dt$$

is a local martingale. If we substitute here $u(x) = u_\varepsilon(x) = (\varepsilon + x^2)^{1/2}$, where $\varepsilon \in (0,1)$, and use the easily verified fact that $|u''(x)\sigma^2(x)|$ is uniformly bounded with respect to ε and x and tends to zero for every x as $\varepsilon \downarrow 0$, then by the Dominated Convergence Theorem and Theorem III.4.9 (k) we obtain that $|\xi_s|$ is a local martingale. Finally, our assertion follows from Theorem III.6.11 (a) by virtue of the assumption $\xi_0 = 0$.

We mention one more corollary of Theorem 3 and Lemma 2.

7. Corollary. *Let $f \in H_2(\mathcal{F}_t)$. Put $\beta = f^2$,*

$$\xi_s = \int_0^s f(t)\,dw_t \qquad \langle\xi\rangle_s = \int_0^s f^2(t)\,dt \tag{8}$$

and suppose that conditions (1) *are satisfied for all ω and t. Then there exists a Wiener process $(\tilde{w}_t, G_t)$ on $(\Omega, \mathcal{F}, P)$, defined for $t \geq 0$, such that (a.s.)*

$$\xi_s = \tilde{w}_{\langle\xi\rangle_s} \tag{9}$$

for all s and in addition, if τ is any Markov time for $\{\mathcal{F}_t\}$, then $\langle\xi\rangle_\tau$ is a Markov time for $\{G_t\}$.

Indeed, it is sufficient to define $\bar{w}_s$ so that $\bar{w}_0 = 0$, $d\bar{w}_s = \operatorname{sgn} f(s)\, dw_s$, to observe that by Lévy's Theorem, $\{\bar{w}_s, \mathcal{F}_s\}$ is a Wiener process, $f(s)\, dw_s = \beta^{\frac{1}{2}}(s)\, d\bar{w}_s$ and then to replace w in (5) by $\bar{w}$.

This corollary has a natural generalization to the case when conditions (1) are not satisfied. We will not formulate it, in order not to burden the reader; it should nevertheless be mentioned that together with properties of Wiener processes known from Chapter II, the generalization illuminates many properties of stochastic integrals.

8. Problem. Use notation (8) for $f \in H_2(\mathcal{F}_t)$. Prove that the *Burkholder-Davis-Gundy inequalities*

$$\mathbf{E} \sup_s |\xi_{s\wedge\tau}|^p \le N\, \mathbf{E}\, \langle \xi \rangle_\tau^{p/2}, \qquad \mathbf{E}\, \langle \xi \rangle_\tau^{p/2} \le N\, \mathbf{E} \sup_s |\xi_{s\wedge\tau}|^p$$

hold with the same constants N as in (II.7.1) for any $p \in (0, \infty)$ and any Markov time τ for $\{\mathcal{F}_t\}$.

Below, in Sec. 4, we will prove these inequalities without using Corollary 7, in a slightly more general case (with other values of N).

9. Corollary. *Under the assumptions of Corollary 7, let $a, b \in (0, \infty]$, $K, T \in (0, \infty)$ be constants, and let $|f| \le K$ for all ω and t. Let τ (respectively, γ) denote the first exit time of ξ_s (resp., Kw_s) from $(-a, b)$. Then*

$$P\{\tau \le T\} \le P\{\gamma \le T\}, \tag{10}$$

$$P\Big\{\sup_{s\le T} \xi_s \ge b\Big\} \le P\Big\{\sup_{s\le T} Kw_s \ge b\Big\} \le \left(1 \wedge \sqrt{\frac{2T}{\pi c^2}}\right) e^{-\frac{1}{2T}c^2}, \tag{11}$$

where $c = b/K$.

Indeed, let γ' be the first exit time of $\tilde{w}_{K^2 s}$ from $(-a, b)$, where $\tilde{w}$ is the same as in Corollary 7. Then it obviously follows from the inequality $\langle \xi \rangle_s \le K^2 s$ and from (9) that $\{\tau \le T\} \subset \{\gamma' \le T\}$. To prove (10) it now remains only to notice that $\tilde{w}_{K^2 s} = Kw'_s$, where w' is a Wiener process and, as a consequence, γ' is the first exit time of Kw'_s from $(-a, b)$ (see also Remark II.5.8). The first inequality in (11) is another, perhaps more expressive form of (10) when $a = \infty$. Since the middle term in (11) is known from (II.3.8), the second inequality in (11) follows from Remark II.3.14.

The inequality between the first and last terms in (11), in the special case that the coefficient of the exponential function is 1 is known as *McKean's estimate* (see also Corollary 3.3 below).

The following assertion can be proved by an obvious modification of the previous arguments.

10. Corollary. *Under the assumptions of Corollary 7, let $a, b \in (0, \infty]$, $\varepsilon, T \in (0, \infty)$ be constants, and let $\varepsilon \le |f|$ for all ω and t. Let τ (respectively, γ) be the first exit time of ξ_s (respectively, εw_s) from $(-a, b)$. Then $P\{\tau \ge T\} \le P\{\gamma \ge T\}$ and, for example, for $\alpha > 0$*

$$\mathbf{E}\tau^\alpha = \int_0^\infty P\{\tau^\alpha \ge T\} dT \le \int_0^\infty P\{\gamma^\alpha \ge T\} dT = \mathbf{E}\gamma^\alpha.$$

Moreover, if $\alpha \in (0, \frac{1}{2})$ *and* $a = \infty$, *then* (*see Corollary* II. 3.13)

$$\mathbf{E}\tau^\alpha \le \mathbf{E}\gamma^\alpha = \frac{b}{\varepsilon\sqrt{2\pi}}\int_0^\infty u^{-\frac{3}{2}+\alpha}\exp\Big(-\frac{b^2}{2u\varepsilon^2}\Big)du = Nb^{2\alpha},$$

where $N = N(\varepsilon, \alpha) < \infty$ *and the last equality is obtained by substituting* $u = b^2 v$.

11. Problem. Instead of ξ_s, consider the process

$$\int_0^s (f(t) + \varepsilon \operatorname{sgn} f(t))\, dw_t,$$

where $\varepsilon > 0$, $\operatorname{sgn} 0 := 1$. Prove that Corollary 7 remains valid without the assumption that $\beta := f^2$ satisfies conditions (1).

3. Girsanov's theorem

Let $(\Omega, \mathcal{F}, P)$ be a complete probability space, $(w_t, \mathcal{F}_t)$ a d-dimensional Wiener process for $t \ge 0$ with complete σ-algebras $\mathcal{F}_t$, and let the one-dimensional processes ξ_t, η_t have stochastic differentials: $d\xi_t = f^*(t)\, dw_t$, $d\eta_t = b(t)\, dt$. Assume that $\xi_0 = 0$, $\eta_0 = 0$. Define $\zeta_t = \xi_t + \eta_t$ and consider the stochastic equation $d\varkappa_t = \varkappa_t\, d\zeta_t$ with initial condition $\varkappa_0 = 1$.

Such equations are often encountered, and it is therefore useful to keep in mind that for a very wide class of functions ζ_t, even discontinuous ones, for which it is natural to write $\varkappa_{t-}\, d\zeta_t$ instead of $\varkappa_t\, d\zeta_t$, the solution to this equation can be obtained from the following purely symbolic formula:

$$\varkappa_s = \exp \int_0^s \ln(1 + d\zeta_t).$$

This formula involves an integral to which it is generally easy to assign a meaning. In our case, for example, the multiplication table (III.8.2) shows that in Taylor's formula for $\ln(1 + d\zeta_t)$ we have to take only two terms and therefore, so to speak, $\ln(1 + d\zeta_t) = d\zeta_t - \frac{1}{2}(d\zeta_t)^2$. This gives

$$\varkappa_s = \exp\Big(\zeta_s - \frac{1}{2}\langle\xi\rangle_s\Big) \tag{1}$$

which for $d = 1, \zeta_t = w_t$, is in good agreement with (III.2.11).

1. Lemma. *The process* $\varkappa_s$ *defined by* (1) *satisfies the equation* $d\varkappa_s = \varkappa_s\, d\zeta_s$ *for* $s \ge 0$. *In particular, for the process*

$$\rho_s := \exp\Big(\xi_s - \frac{1}{2}\langle\xi\rangle_s\Big) = \exp\Big(\int_0^s f^*(t)\, dw_t - \frac{1}{2}\int_0^s |f(t)|^2\, dt\Big), \tag{2}$$

which is called the exponential martingale, *we have* $d\rho_s = \rho_s\, d\xi_s$; *that is, for all* $s \in [0, \infty)$ *at once* (*a.s.*)

$$\rho_s = 1 + \int_0^s \rho_t f^*(t)\, dw_t. \tag{3}$$

Since the stochastic differential of the exponent in (2) is known, this lemma immediately follows from Itô's formula.

Now let

$$\rho_\infty = \varlimsup_{t\to\infty} \rho_t,$$

and note that by Theorem III.6.11 (b), by the obvious inequality $-\rho_t \le 0$ and relation (3), $\rho_t \to \rho_\infty$ as $t \to \infty$ (a.s.).

2. LEMMA. *Let τ, γ be Markov times, $\tau \le \gamma$, and let α be an $\mathcal{F}_\tau$-measurable nonnegative random variable. Then*

$$1 \ge \mathbf{E}\,\rho_\tau \ge \mathbf{E}\,\rho_\gamma, \;\mathbf{E}\,\alpha\rho_\tau \ge \mathbf{E}\,\alpha\rho_\gamma \quad (0\cdot\infty := 0); \tag{4}$$

in particular, $(\rho_t, \mathcal{F}_t)$ is a supermartingale. In addition, if $\mathbf{E}\,\rho_\gamma = 1$, all inequalities in (4) become equalities, and this is true not only for nonnegative α but also for any $\mathcal{F}_\tau$-measurable α, such that at least one of the expressions $\mathbf{E}\,\alpha\rho_\tau$ or $\mathbf{E}\,\alpha\rho_\gamma$ is meaningful. Also, in this case $(\rho_{t\wedge\gamma}, \mathcal{F}_t)$ is a martingale.

PROOF. It follows from (3) that ρ_s is a local martingale and a local supermartingale. The same is true of the process identically equal to n, where n is a constant. Hence, by Lemma II.8.5 (a), it clearly follows that $n \wedge \rho_s$ is a local supermartingale. Since it is bounded, it is a supermartingale. In particular, for $n \ge 1$ the first set of inequalities in (4) are true with $n\wedge\rho, m\wedge\tau, m\wedge\gamma$ substituted for ρ, τ, γ. Applying this substitution and then letting first $m \to \infty$ and then $n \to \infty$, we obtain the first inequalities in (4) by the Dominated Convergence Theorem. The second inequality in (4) is obtained for bounded $\alpha \ge 0$ by repeating the proof of Lemma II.8.5 (a). The Monotone Convergence Theorem now enables us to go to the limit from $\alpha\wedge n$ to α.

If $\mathbf{E}\,\rho_\gamma = 1$, the first inequalities in (4) of course become equalities. They show that $\rho_{t\wedge\gamma}$ is a martingale. As in the proof of Lemma II.8.5 (a), this makes the second relation in (4) an equality for bounded $\alpha \ge 0$. As before, this equality extends to all $\mathcal{F}_\tau$-measurable $\alpha \ge 0$. It now remains to note that $(\rho_\tau\alpha)_\pm = \rho_\tau\alpha_\pm$, $(\rho_\gamma\alpha)_\pm = \rho_\gamma\alpha_\pm$. □

This lemma yields yet another derivation of McKean's estimate.

3. COROLLARY. *If $\langle\xi\rangle_\infty \le K$ (a.s.) for some constant K, then for any constant $c \ge 0$ we have*

$$P\Big\{\sup_s \xi_s \ge c\Big\} \le e^{-\frac{1}{2K}c^2}. \tag{5}$$

Indeed, take $\lambda = c/K$ and note that the left-hand side of (5) is obviously bounded above by

$$P\Big\{\sup_s\Big(\xi_s - \frac{\lambda}{2}\langle\xi\rangle_s\Big) \ge c - \frac{\lambda}{2}K\Big\} = P\Big\{\sup_s e^{\lambda\xi_s - \frac{\lambda^2}{2}\langle\xi\rangle_s} \ge e^{\lambda c - \frac{\lambda^2}{2}K}\Big\}.$$

It now remains to apply Theorem III.6.11 (a).

4. THEOREM (Girsanov). *Suppose that $\mathbf{E}\,\rho_\infty = 1$ and define a new measure on $(\Omega, \mathcal{F})$ by $\bar P(d\omega) = \rho_\infty(\omega)P(d\omega)$. Then $\bar P$ is a probability measure on $(\Omega, \mathcal{F})$. A*

process β_t is a local martingale with respect to $\{\mathcal{F}_t\}$ on $(\Omega, \mathcal{F}, \bar{P})$ if and only if $\beta_t \rho_t$ is a local martingale with respect to $\{\mathcal{F}_t\}$ on $(\Omega, \mathcal{F}, P)$. For $s \geq 0$,

$$\bar{w}_s := w_s - \int_0^s f(t)\,dt$$

is a d-dimensional Wiener process with respect to $\{\mathcal{F}_s\}$ on the probability space $(\Omega, \mathcal{F}, \bar{P})$. If the function $\sigma(t) = \big(\sigma^{ij}(t)\big)$ on $\Omega \times (0, \infty)$ with values in the set of $(d_1 \times d)$-matrices is an element of H_2 with respect to P, then it is in H_2, with respect to $\bar{P}$, and for all $s \in [0, \infty)$ at once, both P-a.s. and $\bar{P}$-a.s.,

$$\int_0^s \sigma(t)\,d\bar{w}_t = \int_0^s \sigma(t)\,dw_t - \int_0^s \sigma(t) f(t)\,dt. \tag{6}$$

PROOF. By definition, $\bar{P}(\Omega) = \mathbf{E}\rho_\infty = 1$, hence $\bar{P}$ is a probability measure. Let $\bar{\mathbf{E}}$ denote integrals with respect to $\bar{P}$.

If β_t is a local martingale with respect to $\bar{P}$, then for $\bar{\beta}_t = \beta_t - \beta_0$ and

$$\tau(n) = n \wedge \inf\{t \geq 0 : |\beta_t - \beta_0| + \rho_t \geq n\}$$

and for any Markov time τ, it follows by Lemma 2 that

$$\begin{aligned} 0 = \bar{\mathbf{E}}\bar{\beta}_{\tau\wedge\tau(n)} &= \int_\Omega \bar{\beta}_{\tau\wedge\tau(n)} \bar{P}(d\omega) = \int_\Omega \bar{\beta}_{\tau\wedge\tau(n)} \rho_\infty P(d\omega) \\ &= \mathbf{E}\bar{\beta}_{\tau\wedge\tau(n)}\rho_\infty = \mathbf{E}\bar{\beta}_{\tau\wedge\tau(n)}\rho_{\tau\wedge\tau(n)}. \end{aligned} \tag{7}$$

Thus $\bar{\beta}_t\rho_t$ is a local martingale with respect to P and so is $\beta_t\rho_t = \bar{\beta}_t\rho_t + \beta_0\rho_t$ (see Theorem III.4.9 (j) with $\tau = 0$). Conversely, if $\beta_t\rho_t$ is a local martingale with respect to P, then $\bar{\beta}_t\rho_t = \beta_t\rho_t - \beta_0\rho_t$ is a local martingale, and the fact that β_t is a local martingale with respect to $\bar{P}$ also follows immediately by inverting (7).

Further, it follows easily from Itô's formula or from Lemma III.8.5 that the processes $\bar{w}_t\rho_t, \beta_t\rho_t$, where $\beta_t = \bar{w}_t\bar{w}_t^* - t(\delta^{ij})$, are local martingales with respect to P. The foregoing, together with Lévy's Theorem, now implies our assertion about $\bar{w}_t$.

Our assertion about σ follows from the fact, that by Fubini's Theorem, $P \times \ell$-null sets are simultaneously $\bar{P} \times \ell$-null sets. Hence, σ is measurable with respect to the completion $\mathcal{P}$ with respect to $\bar{P} \times \ell$, and the integral on the left of (6) is meaningful, as a stochastic integral against a Wiener process on $(\Omega, \mathcal{F}, \bar{P})$. In addition, the last integral in (6) is meaningful, since $|\sigma f|^2 = (f, \sigma^*\sigma f) \leq |f|^2 \operatorname{tr}\sigma^*\sigma$, $|\sigma f| \leq |f|^2 + \operatorname{tr}\sigma^*\sigma$.

Although formula (6) is quite natural due to the definition of $\bar{w}_s$, we need yet more arguments to prove it. Letting γ_s denote the right-hand side of (6), we easily deduce from Itô's formula that the processes

$$\gamma_s\rho_s, \quad \Big(\gamma_s\bar{w}_s^* - \int_0^s \sigma(t)\,dt\Big)\rho_s, \quad \Big(\gamma_s\gamma_s^* - \int_0^s a(t)\,dt\Big)\rho_s,$$

where $a = \sigma\sigma^*$, are local martingales on $(\Omega, \mathcal{F}, P)$. By what we have already proved, we can omit ρ if we replace P with $\bar{P}$. Hence, by Remark III.10.5, equality (6) is true $\bar{P}$-a.s. for all $s \in [0, \infty)$ at once. In particular, the squared difference of the left- and right-hand sides of (6), multiplied by ρ_s^{-1} vanishes ($\bar{P}$-a.s.) and is a local martingale with respect to $\bar{P}$. Then the squared difference itself is a local martingale with respect to P, and by Theorem III.6.11 (a), it vanishes for all s at once P-a.s. □

Girsanov's Theorem raises the question of when $\mathbf{E}\,\rho_\infty = 1$, a partial answer to which is contained in the following theorem (see also Problems 12, 13).

5. THEOREM. (a) *Let γ be a Markov time, $\varepsilon > 0$,*

$$\mathbf{E}\exp\frac{1+\varepsilon}{2}\langle\xi\rangle_\gamma < \infty.$$

Then $\mathbf{E}\,\rho_\gamma = 1$.

(b) *A more general assertion is also true: if*

$$\lim_{\varepsilon\downarrow 0}\varepsilon\ln\mathbf{E}\exp\frac{1-\varepsilon}{2}\langle\xi\rangle_\gamma = 0,$$

in particular, if $\mathbf{E}\exp\frac{1}{2}\langle\xi\rangle_\gamma < \infty$, *then* $\mathbf{E}\,\rho_\gamma = 1$.

(c) *If* $|f(\omega,t)|^2 \le g(t, w_t(\omega))$, *where* $g(t,x)$ *is a Borel function such that for some* $T \in (0,\infty)$, $q > 1$, $r > q\,d/(2q-2)$,

$$(8) \qquad \int_{E_d}\left(\int_0^T g^q(t,x)\,dt\right)^{r/q} dx < \infty,$$

then $\mathbf{E}\exp K\langle\xi\rangle_T < \infty$ *for any constant* K, *and* $\mathbf{E}\,\rho_T = 1$.

PROOF. (a) Let us denote the process ρ in (2) by $\rho(f)$. Then for any $p, r > 1$ and Markov times $\tau \le \gamma$, we easily obtain by the Hölder inequality

$$\mathbf{E}\,\rho_\tau^p \le \Big(\mathbf{E}\,\rho_\tau(prf)\Big)^{1/r}\left(\mathbf{E}\exp\left(\frac{rp-1}{2}p\frac{r}{r-1}\langle\xi\rangle_\tau\right)\right)^{(r-1)/r}.$$

By (4), the first factor on the right is at most 1. The second factor can be estimated uniformly with respect to $\tau \le \gamma$, since for $p = 1+\delta$, $r = 1+\sqrt{\delta}$ and sufficiently small $\delta > 0$, it is not difficult to see that the coefficient of $\langle\xi\rangle_\tau$ can be made smaller than $(1+\varepsilon)/2$. Fix δ, p with these properties.

For a suitable sequence of bounded Markov times $\tau(n) \uparrow \infty$, the process $\rho_{t\wedge\tau(n)}$ is a martingale, the process $\rho^p_{t\wedge\tau(n)}$ is a submartingale, and by Doob's inequality for moments of a submartingale $\mathbf{E}\sup_t \rho^p_{t\wedge\tau(n)\wedge\gamma}$ is bounded by a quantity independent of n. By the Monotone Convergence Theorem, this is true even if $\tau(n)$ is replaced by infinity. By the Hölder inequality $\mathbf{E}\sup_t \rho_{t\wedge\gamma} < \infty$. Thus, using the Dominated Convergence Theorem and the equality $\mathbf{E}\,\rho_{\tau(n)\wedge\gamma} = 1$, we see, by letting $n \to \infty$, that $\mathbf{E}\,\rho_\gamma = 1$.

(b) Since $\mathbf{E}\exp\frac{1}{2}(1+\varepsilon)\langle(1-2\varepsilon)\xi\rangle_\gamma < \infty$ for $\varepsilon \in (0,1/2)$, it follows from (a) and the Hölder inequality that

$$1 = \mathbf{E}\,\rho_\gamma((1-2\varepsilon)f) \le \big[\mathbf{E}\,\rho_\gamma\big]^{1-2\varepsilon}\left[\mathbf{E}\exp\frac{1-2\varepsilon}{2}\langle\xi\rangle_\gamma\right]^{2\varepsilon},$$

$$1 \le \big[\mathbf{E}\,\rho_\gamma\big]^{1-2\varepsilon}\left[\mathbf{E}\exp\frac{1-2\varepsilon}{2}\langle\xi\rangle_\gamma\right]^{2\varepsilon}.$$

Letting here $\varepsilon \to 0$, we obtain $\mathbf{E}\,\rho_\gamma \ge 1$. Together with (4) this yields $\mathbf{E}\,\rho_\gamma = 1$.

(c) By the Dominated Convergence Theorem, the function

$$A(s,t) := \Big(\int_{E_d}\Big(\int_0^T I_{s<u<t}\, g^q(u,x)\,du\Big)^{r/q}dx\Big)^{1/r}$$

is continuous with respect to (s,t) in the square $[0,T]^2$. Since $A(s,s)=0$, it follows that for any constant $\delta>0$ one can find $h>0$ such that $A(s,t)\le\delta$ whenever $|t-s|\le h$. Clearly, part (b) will be proved for $K=1$ if we prove that for sufficiently small h

$$\text{(9)} \qquad \mathbf{E}\exp\langle\xi\rangle_t \le 2\mathbf{E}\exp\langle\xi\rangle_s$$

provided that $s\le t\le s+h,\ s,\ t\in[0,T]$. To do this, we observe that for any $x\in E_d$, by the Hölder inequality,

$$\begin{aligned}F(x,s,t) :&= \mathbf{E}\int_s^t g(u,x+w_{u-s})\,du\\ &= \int_{E_d}dy\int_s^t g(u,x+y)\big(2\pi(u-s)\big)^{-d/2}e^{-|y|^2/(2u-2s)}du\\ &\le \int_{E_d}dy\Big(\int_s^t g^q(u,x+y)\,du\Big)^{1/q}\Big(\int_0^T(2\pi u)^{-pd/2}e^{-p|y|^2/(2u)}\,du\Big)^{1/p}\\ &\le A(s,t)\Big(\int_{E_d}\Big(\int_0^T(2\pi u)^{-pd/2}e^{-p|y|^2/(2u)}du\Big)^{r/(pr-p)}dy\Big)^{(r-1)/r},\end{aligned}$$

where $p=q/(q-1)$. In the last part obviously the inner integral with respect to u decreases faster than $e^{-|y|}$ as $|y|\to\infty$. For small $|y|$ it does not exceed the integral from 0 to ∞, which, after the substitution $u=|y|^2v$, turns out to be equal to $N|y|^{2-pd}$, where $N<\infty$ if $pd>2$. If $pd<2$, then for small $|y|$ the integral with respect to u does not exceed its value at $|y|=0$, which is finite. Finally, if $pd=2$, when automatically $d=1$, it equals

$$\int_0^{T|y|^{-2}}(2\pi u)^{-1}e^{-p/(2u)}\,du \sim (2\pi)^{-1}\ln\frac{T}{|y|^2},\quad |y|\to 0.$$

Hence, it follows that the coefficient of $A(s,t)$ is finite and $F(x,s,t)\le 1/2$, if h is sufficiently small. Now consider the simple equality (which follows, for example, from Itô's formula)

$$e^{\langle\xi\rangle_t}-e^{\langle\xi\rangle_s} = e^{\langle\xi\rangle_s}(\langle\xi\rangle_t-\langle\xi\rangle_s)+\int_s^t(\langle\xi\rangle_t-\langle\xi\rangle_u)|f(u)|^2e^{\langle\xi\rangle_u}du,$$

take the expectation on the left- and the right-hand sides and use the Markov property of w. Then we obtain

$$\mathbf{E}\, e^{\langle\xi\rangle_t} \le \mathbf{E}\, e^{\langle\xi\rangle_s} + \mathbf{E}\, e^{\langle\xi\rangle_s} F(w_s, s, t) + \int_s^t \mathbf{E}\, |f(u)|^2 e^{\langle\xi\rangle_u} F(w_u, u, t)\, du$$

$$\le \frac{3}{2} \mathbf{E}\, e^{\langle\xi\rangle_s} + \frac{1}{2} \int_s^t \mathbf{E}\, |f(u)|^2 e^{\langle\xi\rangle_u} du = \frac{3}{2} \mathbf{E}\, e^{\langle\xi\rangle_s} + \frac{1}{2} \mathbf{E}\left(e^{\langle\xi\rangle_t} - e^{\langle\xi\rangle_s}\right).$$

Comparing the first expression in this chain with the last two and canceling like terms, we obtain (9), subject to the assumption, that the left-hand side of (9) is finite. However, this assumption is easily avoided if $|f|$ is replaced by $|f| \wedge n$; we then obtain (9) for the same s, t, h, and can apply the Monotone Convergence Theorem. This proves part (c) for $K = 1$. The case of arbitrary $K > 0$ can be reduced to $K = 1$ by replacing $\sqrt{K} f$ with f. □

Neither Girsanov's Theorem nor Theorem 5 says anything about differential equations. Hence the following result is all the more surprising.

6. COROLLARY (Girsanov). *Let $T \in (0, \infty)$ be a constant, $f(t, x)$ a Borel function with values in E_d, defined on $[0, T] \times E_d$, which is either bounded or satisfies condition (8) for $g = f^2$ and some $q > 1$, $r > qd/(2q - 2)$. Then for every $\varepsilon > 0$ there exist a probability space $(\bar{\Omega}, \bar{\mathcal{F}}, \bar{P})$, a d-dimensional Wiener process $\{\bar{w}_t, \bar{\mathcal{F}}_t\}$ on that space, and a d-dimensional $\bar{\mathcal{F}}_t$-adapted continuous process ξ_t, such that for all $s \in [0, T]$ at once ($\bar{P}$-a.s.)*

$$\xi_s = \int_0^s f(t, \xi_t + \varepsilon \bar{w}_t)\, dt. \tag{10}$$

Indeed, take our original probability space $(\Omega, \mathcal{F}, P)$ and define

$$(\bar{\Omega}, \bar{\mathcal{F}}, \bar{\mathcal{F}}_t) = (\Omega, \mathcal{F}, \mathcal{F}_t),$$

$$\rho = \exp\left\{\frac{1}{\varepsilon} \int_0^T f^*(t, \varepsilon w_t) dw_t - \frac{1}{2\varepsilon^2} \int_0^T |f(t, \varepsilon w_t)|^2 dt\right\},$$

$$\bar{P}(d\omega) = \rho P(d\omega), \quad \bar{w}_t = w_s - \frac{1}{\varepsilon} \int_0^s f(t, \varepsilon w_t) I_{t \le T}\, dt, \quad \xi_s = \varepsilon(w_s - \bar{w}_s).$$

Then (10) is obvious for $s \le T$, and $\{\bar{w}_s, \bar{\mathcal{F}}_s\}$ is a Wiener process by Theorems 4, 5.

Note that Corollary 6 means, in particular, that for almost any (in the sense of Wiener measure) function $\bar{w}_\cdot \in C([0, \infty), E_d)$ equation (10), considered as an equation in ξ_s, has a solution in $[0, T]$.

Girsanov's Theorem plays an important role in many areas of the theory of diffusion processes, in particular, in the statistics of these processes, since it enables one to write $C([0, \infty), E_d)$-integrals with respect to distributions of different processes in terms of integrals against Wiener measure. In particular:

7. COROLLARY. *Let $T \in (0,\infty)$ be a constant, $g(x)$ a nonnegative $\mathfrak{A}_T(C)$-measurable function on C (see the definition of $\mathfrak{A}_T(C)$ in Remark* II.5.8). *Let $\bar{w}_t = w_t + bt$, where $b \in E_d$. Then*

$$(11) \qquad \mathbf{E}\, g(\bar{w}.) = \mathbf{E}\, g(w.) \exp\left((b, w_T) - \frac{1}{2}|b|^2 T\right).$$

Indeed, by Girsanov's Theorem for $f = bI_{[0,T]}$, the right-hand side is $\bar{\mathbf{E}}\, g(w.)$, and in addition $\tilde{w}_t := w_t - b(t \wedge T)$ is a Wiener process with respect to $\bar{P}$. Moreover, since the distributions of Wiener processes on C coincide with the same Wiener measure (Exercise II.5.4), it follows from Theorem I.4.3 that $\mathbf{E}\, g(\bar{w}.) = \bar{\mathbf{E}}\, g(\xi.)$, where $\xi_t = \tilde{w}_t + bt$. It remains to note that $g(\xi.) = g(w.)$ for all ω since $\xi_t = w_t$ for $t \le T$ and $g(x.)$ is $\mathfrak{A}_T(C)$-measurable.

8. EXERCISE. Using (11) and Theorem II.3.12, find another derivation of the result formulated in Exercise II.9.4. Prove that if $b \in E_d$, then also

$$\lim_{\varepsilon \downarrow 0} P\Big\{\max_{t \le T} |w_t + bt| \le \varepsilon\Big\}\Big(P\Big\{\max_{t\le T}|w_t| \le \varepsilon\Big\}\Big)^{-1} = e^{-\frac{1}{2}|b|^2 T}.$$

9. EXERCISE. Use Theorem 5 to prove that if $a \in H_2$, $d\zeta_t = a(t)dw_t$, $\zeta_0 = 0$ and for some Markov time τ the matrix $\langle \zeta, \zeta\rangle_\tau$ does not depend on ω, then $\zeta_\tau \sim \mathcal{N}(0, \langle\zeta,\zeta\rangle_\tau)$.

10. PROBLEM. Prove (11) by using independence of increments of w, first considering $g(x) = g_1(x_{t(2)} - x_{t(1)}) \cdot \ldots \cdot g_n(x_{t(n)} - x_{t(n-1)})$, where $0 \le t(1) \le \ldots \le t(n) \le T$ and g_i are bounded Borel functions, and then using the lemma on π- and λ-systems.

11. PROBLEM. Let $\alpha(t)$ be a continuous real-valued function on $[0,\infty)$, w_t a one-dimensional Wiener process, $n = 1, 2, \ldots$,

$$\sigma(n) = \inf\{t \ge 0 \colon w_t + \alpha(t) \le -n\}, \quad \tau(n) = \inf\{t \ge 0\colon w_t + \alpha(t) - t \le -n\}.$$

Use (11) to prove that for any constant $T \in (0,\infty)$

$$(12) \qquad \mathbf{E}\, e^{w_T - T/2} I_{T \le \tau(n)} = P\{\sigma(n) \ge T\}.$$

Assuming that (cf. Problems II.9.9, II.9.10)

$$(13) \qquad \overline{\lim_{t\to\infty}}\,(w_t - \alpha(t)) = \infty \quad \text{(a.s.)}$$

prove that $\sigma(n) < \infty$ (a.s.), and derive that for some sequence $T = T(n) \to \infty$ as $n \to \infty$ the left-hand side of (12) tends to zero.

12. PROBLEM. Under the assumptions of the previous problem let τ be a Markov time and assume that

$$(14) \qquad \mathbf{E}\, e^{\tau/2 - \alpha(\tau)} < \infty$$

and that $\frac{1}{2}t - \alpha(t)$ is an increasing function of t for all sufficiently large t. Noting that if $\alpha(0) > -n$ and $\tau(n) < \infty$, then $w_{\tau(n)} = -n + \tau(n) - \alpha(\tau(n))$, prove that

$$(15) \qquad \lim_{n\to\infty} \mathbf{E}\, e^{w_{\tau(n)} - \tau(n)/2} I_{\tau(n) < \tau} \le \lim_{n\to\infty} e^{-n}\mathbf{E}\, e^{\tau/2 - \alpha(\tau)} = 0.$$

Assuming that (13) is true and taking $T(n)$ from the previous problem, prove further that if $\gamma(n) := T(n) \wedge \tau(n)$, then

$$I_n := \mathbf{E}\, e^{w_{\gamma(n)} - \gamma(n)/2} I_{\gamma(n) < \tau} \to 0$$

as $n \to \infty$. Combining these facts with the formula

$$1 = \mathbf{E}\, e^{w_{\gamma(n)\wedge\tau} - (\gamma(n)\wedge\tau)/2} = \mathbf{E}\, e^{w_\tau - \tau/2} I_{\tau \le \gamma(n)} + I_n$$

deduce the following result due to Novikov: If conditions (13) and (14) are satisfied and $\frac{1}{2}t - \alpha(t)$ increases for large t, then

$$\mathbf{E}\, e^{w_\tau - \tau/2} = 1. \tag{16}$$

Next, observe that if $\alpha(t) \equiv 0$ the expression after the limit sign on the left of (15) equals

$$e^{-n/2}\mathbf{E}\, e^{w_{\tau(n)}/2} I_{\tau(n)<\tau} \le e^{-n/2}\mathbf{E}\, e^{w_{\tau(n)\wedge\tau}/2}.$$

Hence deduce the following result of Kazamaki: equality (16) is valid if

$$\sup_{t\ge 0} \mathbf{E}\, e^{w_{t\wedge\tau}/2} < \infty. \tag{17}$$

It turns out that condition (17) is always weaker than (14) for $\alpha(t) \equiv 0$; sometimes it is very easy to verify. For example, if τ is the first exit time of w_t from $(-\infty, 1)$, then $w_{t\wedge\tau} \le 1$ and condition (17) is satisfied. For the same τ, as we know, even $\mathbf{E}\sqrt{\tau} = \infty$, so that the left-hand side of (14) is also infinite when $\alpha \equiv 0$.

13. PROBLEM. By modifying the proof of part (b) in Theorem 5 show that if

$$\lim_{\varepsilon\downarrow 0} \varepsilon \sup_t \ln \mathbf{E} \exp \frac{1-\varepsilon}{2}\xi_{t\wedge\gamma} = 0,$$

then $\mathbf{E}\,\rho_\gamma = 1$. Also show that here the equality "$= 0$" can be replaces by "$< \infty$" and the same is true for assertion (b) in Theorem 5. (Hint for assertion (b):

$$\rho_\gamma((1-2\varepsilon)f) = \rho_\gamma^{1-2\varepsilon}\Big(I_{\langle\xi\rangle_\gamma<N} \exp\frac{1-2\varepsilon}{2}\langle\xi\rangle_\gamma\Big)^{2\varepsilon} + \big(I_{\langle\xi\rangle_\gamma\ge N}\rho_\gamma\big)^{1-2\varepsilon}\Big(\exp\frac{1-2\varepsilon}{2}\langle\xi\rangle_\gamma\Big)^{2\varepsilon}.$$

14. PROBLEM. Let τ be the first exit time of the process (t, w_t) from some domain $G \subset E_2$, σ the first exit time of $(t, w_t + t)$ from the *same* domain. Use a formula analogous to (12) to prove that equality (16) is true if and only if $\sigma < \infty$ (a.s.). Hence show once again (see Problem II.9.12) that (II.6.9) is valid for $x > 0, b \ge 0, \nu \ge -b^2/2$, and if $\nu < -b^2/2$, then the left-hand side is infinite. Using this, show by an example that one cannot, generally speaking, take $\varepsilon < 0$ in the part (a) of Theorem 5.

4. Burkholder-Davis-Gundy inequalities for multidimensional random processes

In Sec. II.7 we have proved the Burkholder-Davis-Gundy inequalities for one-dimensional Wiener processes; in Problem 2.8 we have generalized them to one-dimensional stochastic integrals. This generalization is easily extended to the case of multidimensional stochastic integrals, by considering each coordinate separately. However, the constants in the inequalities will then depend on the dimension, which is sometimes extremely undesirable. Partly because of this and partly because that it is always desirable to have several proofs of the most important facts, we will now present a derivation of the Burkholder-Davis-Gundy inequalities that avoids these shortcomings; as the reader will see, we will not use the results of Sec. II.7 at all.

Let $(\Omega, \mathcal{F}, P)$ be a complete probability space, $\{\mathcal{F}_t\}$ a filtration of σ-algebras $\mathcal{F}_t \subset \mathcal{F}$ defined for $t \geq 0$, and complete with respect to $\mathcal{F}$ and P; let ξ_t be a d-dimensional admissible local martingale with respect to $\{\mathcal{F}_t\}$, defined for $t \geq 0$ such that $\xi_0 = 0$. Define $d\langle \xi, \xi \rangle_t = a(t)\,dt$, $c(t) = \operatorname{tr} a(t)$.

1. THEOREM. *For any $p \in (0, \infty)$ there exists a constant $N = N(p) < \infty$ such that for any Markov time τ*

$$\mathbf{E}\sup_t |\xi_{t\wedge\tau}|^p \leq N\,\mathbf{E}\langle \xi \rangle_\tau^{p/2}, \qquad \mathbf{E}\langle \xi \rangle_\tau^{p/2} \leq N\,\mathbf{E}\sup_t |\xi_{t\wedge\tau}|^p. \tag{1}$$

PROOF. It is not difficult to verify that if σ is a Markov time then the process $\eta_t = \xi_{t\wedge\sigma}$ is an admissible local martingale and $\langle \eta \rangle_t = \langle \xi \rangle_{t\wedge\sigma}$ for all t at once (a.s.). Choosing a sequence $\sigma(n) \uparrow \infty$ such that $\xi_{t\wedge\sigma(n)}$ and $\langle \xi \rangle_{t\wedge\sigma(n)}$ are bounded for every n, we see, in view of the Monotone Convergence Theorem, that it will suffice to consider only the case in which ξ_t and $\langle \xi \rangle_t$ are bounded with respect to (ω, t).

In that case Itô's formula shows that for any $n \geq 2$

$$\begin{aligned} d|\xi_t|^2 &= 2\xi_t^*\,d\xi_t + c(t)\,dt, \\ d|\xi_t|^{2n} &= 2n|\xi_t|^{2n-2}\xi_t^*\,d\xi_t + \Big[n|\xi_t|^{2n-2}c(t) + 2n(n-1)|\xi_t|^{2n-4}\big(a(t)\xi_t, \xi_t\big)\Big]dt. \end{aligned} \tag{2}$$

Consequently, the process

$$|\xi_s|^{2n} - \int_0^s \Big[n|\xi_t|^{2n-2}c(t) + 2n(n-1)|\xi_t|^{2n-4}\big(a(t)\xi_t, \xi_t\big)\Big]dt \tag{3}$$

is a local martingale. Here $\big(a(t)\xi_t, \xi_t\big) \leq c(t)|\xi_t|^2$, which, together with our assumptions, implies that the process (3) is bounded and is a martingale. Replacing s by $\gamma \in \mathfrak{M}$, using the above inequality, and calculating the expectation, we obtain

$$\mathbf{E}|\xi_\gamma|^{2n} \leq \mathbf{E}\int_0^\gamma n(2n-1)|\xi_t|^{2n-2}c(t)\,dt.$$

Hence, by Theorem III.6.8 and the Hölder inequality, for any $\alpha \in (0, 1)$

$$\begin{aligned} \mathbf{E}\sup_t |\xi_t|^{2n\alpha} &\leq N\,\mathbf{E}\Big(\int_0^\infty |\xi_t|^{2n-2}c(t)\,dt\Big)^\alpha \\ &\leq N\,\mathbf{E}\langle \xi \rangle_\infty^\alpha \sup_t |\xi_t|^{(2n-2)\alpha} \leq N\Big(\mathbf{E}\langle \xi \rangle_\infty^{n\alpha}\Big)^{1/n}\Big(\mathbf{E}\sup_t |\xi_t|^{2n\alpha}\Big)^{(n-1)/n}, \end{aligned} \tag{4}$$

where $N = (2-\alpha)(2n^2+2n)^\alpha/(1-\alpha)$. Canceling like terms in the inequality between the first and the last terms in (4) we obtain the first inequality in (1) for $p = 2n\alpha$, $\tau = \infty$. But any $p \in (0, \infty)$ can be represented (in more than one way) as $2n\alpha$, where $n \geq 2$, $\alpha \in (0, 1)$, while the case of arbitrary τ can be reduced to the case $\tau = \infty$ by considering $\xi_{t\wedge\tau}$ instead of ξ_t. Hence the first inequality in (1) is proved.

We will now derive the second inequality from the first one, assuming, as before, that ξ and $\langle \xi \rangle$ are bounded. To that end we note that, by (2), $\zeta_t := |\xi_t|^2 - \langle \xi \rangle_t$

is an admissible local martingale and $d\langle\zeta\rangle_t = 4\big(a(t)\xi_t,\xi_t\big)\,dt$. Using this, the first inequality in (1) and the inequality $(a\xi,\xi)\le c|\xi|^2$, we obtain

$$\mathbf{E}\sup_t\Big||\xi_t|^2-\langle\xi\rangle_t\Big|^{p/2}\le N\,\mathbf{E}\Big(\int_0^\infty|\xi_t|^2c(t)\,dt\Big)^{p/4}$$
$$\le N\,\mathbf{E}\sup_t|\xi_t|^{p/2}\langle\xi\rangle_\infty^{p/4}\le N\Big(\mathbf{E}\sup_t|\xi_t|^p\Big)^{1/2}\Big(\mathbf{E}\langle\xi\rangle_\infty^{p/2}\Big)^{1/2}.$$

To obtain a lower bound for the first expression in these inequalities, we use the simple inequality $|v|^\alpha\le\big(|u|+|u-v|\big)^\alpha\le 2^\alpha|u|^\alpha+2^\alpha|u-v|^\alpha$, where $\alpha\ge0$. Then, putting

$$\lambda=\Big(\mathbf{E}\sup_t|\xi_t|^p\Big)^{1/2},\quad \mu=\Big(\mathbf{E}\langle\xi\rangle_\infty^{p/2}\Big)^{1/2}$$

we obtain $\mu^2\le N(\lambda^2+\lambda\mu)$. Since μ is finite, it is clear that $\mu\le N\lambda$ and this gives us the second inequality in (1) for the case $\tau=\infty$. The case of arbitrary τ can be reduced to that of $\tau=\infty$ as before. □

2. REMARK. It is sometimes useful to know that in the case of the Davis inequalities for $p=1$ one can put $N=3$ in (1). A proof of this assertion for bounded ξ and $\langle\xi\rangle$ can be obtained, using Theorem III.6.8, from the Wald identities $\mathbf{E}\,|\xi_\gamma|^2=\mathbf{E}\,\langle\xi\rangle_\gamma$, which hold for any $\gamma\in\mathfrak{M}$. The general case is dealt with in the same way as in the proof of the theorem.

Theorem 1 can be generalized as follows.

3. THEOREM. *Under the assumptions of Theorem* 1, *let b be a function on $\Omega\times(0,\infty)$ with values in E_d, $b\in H_1$. Let η_t be an $\mathcal{F}_t$-adapted process such that $d\eta_t=b(t)\,dt$. Define $\zeta_t=\xi_t+\eta_t$.*

(a) *If $\big(\zeta_t,b(t)\big)\le K\operatorname{tr}a(t)$ for all (ω,t) for some constant $K\in[0,\infty)$, then*

$$\mathbf{E}\sup_t|\zeta_{t\wedge\tau}|^p\le N(p,K)\Big[\mathbf{E}\,|\zeta_0|^p+\mathbf{E}\,\langle\xi\rangle_\tau^{p/2}\Big]\tag{5}$$

for any $p\in(0,\infty)$ and Markov times τ;

(b) *if $2\big(\zeta_t,b(t)\big)\ge-\varepsilon\operatorname{tr}a(t)$ for all (ω,t) for some constant $\varepsilon\in(0,1)$, then*

$$\mathbf{E}\,\langle\xi\rangle_\tau^{p/2}\le N(p,\varepsilon)\,\mathbf{E}\sup_t|\zeta_{t\wedge\tau}|^p$$

for any $p\in(0,\infty)$ and Markov times τ.

PROOF. Instead of ζ_t we can consider the product of ζ_t and the indicator of the set $\big\{|\eta_0|\le R\big\}\in\mathcal{F}_0$ and then reasoning as at the beginning of the proof of Theorem 1, we can reduce everything to the case when the processes ζ_t and $\langle\xi\rangle_t$ are bounded on $\Omega\times[0,\infty)$.

Now, by Itô's formula,

$$|\zeta_s|^2=|\zeta_0|^2+\int_0^s\Big(c(t)+2\big(\zeta_t,b(t)\big)\Big)\,dt+2\int_0^s\zeta_t^*\,d\xi_t.\tag{6}$$

We now raise this equality to the power $\alpha = p/2$ and use the inequality $|u+v+w|^\alpha \le 3^\alpha|u|^\alpha + 3^\alpha|v|^\alpha + 3^\alpha|w|^\alpha$ and Theorem 1. Then, in case (a),

$$\mathbf{E}\sup_s |\zeta_{s\wedge\tau}|^{2\alpha} \le N\Big[\mathbf{E}|\zeta_0|^{2\alpha} + \mathbf{E}\langle\xi\rangle_\tau^\alpha + \mathbf{E}\Big(\int_0^\tau (a(t)\zeta_t,\zeta_t)\,dt\Big)^{\alpha/2}\Big]$$
$$\le N\Big[\mathbf{E}|\zeta_0|^{2\alpha} + \mathbf{E}\langle\xi\rangle_\tau^\alpha\Big] + N\,\mathbf{E}\sup_s |\zeta_{t\wedge\tau}|^\alpha \langle\xi\rangle_\tau^{\alpha/2}.$$

Applying the Cauchy-Bunyakovsky inequality to the last expression, as at the end of the proof of Theorem 1, we obtain (5).

In case (b) from (6) and Theorem 1 we have

$$(1-\varepsilon)\langle\xi\rangle_s \le 2\sup_{t\le s}|\zeta_t|^2 - 2\int_0^s \zeta_t^*\,d\xi_t,$$

$$\mathbf{E}\langle\xi\rangle_\tau^{p/2} \le N\,\mathbf{E}\sup_s|\zeta_{s\wedge\tau}|^p + N\,\mathbf{E}\Big(\int_0^\tau (a(t)\zeta_t,\zeta_t)\,dt\Big)^{p/4}.$$

The rest of the proof follows already familiar lines. □

4. EXERCISE. Prove Liouville's Theorem II.9.1 by observing that under its assumptions $u(t,w_t) - u(0,0)$ is a bounded stochastic integral and that $\lim_{t\to\infty} u(t,w_t) - u(0,0)$ is a constant (a.s.), which by Wald's identities imply that the constant is zero and the stochastic integral vanishes.

CHAPTER V

Itô Stochastic Equations

We have already met some Itô stochastic equations. Besides the simplest equations $d\xi_s = \sqrt{2}\,dw_s - \xi_s\,ds$, $d\xi_s = \xi_s\,d\zeta_s$, considered in Problem II.9.11 and Lemma IV.3.1, we considered equation $d\xi_s = \operatorname{sgn}\xi_s\,dw_s$ in Problem IV.1.4; in Corollary IV.2.5 we encountered a more general equation $d\xi_s = \sigma(\xi_s)\,dw_s$ and in Corollary IV.3.6, the equation $d\xi_s = \varepsilon\,dw_s + f(s,\xi_s)\,ds$. The assertions concerning the last three equations were rather specific: it was stated that they have solutions on a *certain* probability space with a *certain* Wiener process w_s. Surprisingly, as far as the equations $d\xi_s = \operatorname{sgn}\xi_s\,dw_s$ and $d\xi_s = \sigma(\xi_s)\,dw_s$ are concerned, this assertion cannot be strengthened: there exist probability spaces and Wiener processes w_s for which they are unsolvable. For the first equation this fact, due to Tanaka, is easily derived using Problem IV.1.4. For the second one, Barlow [1] (1982) gave a counterexample which even had a continuous, strictly positive, bounded σ. The equation $d\xi_s = \varepsilon\,dw_s + f(s,\xi_s)\,ds$, for any $\varepsilon \neq 0$ and any bounded Borel function $f(s,x)$, has a solution on any probability space with any Wiener process, even in the vector case. This nontrivial result was obtained by Veretennikov [48] (1980).

We do not plan to consider all these subtle points. The main content of the present chapter is a proof that stochastic equations are solvable in "good" cases (Sec. V.1), followed by a detailed analysis of equations whose coefficients do not depend explicitly on ω. We will prove that solutions of these equations have the strong Markov property (Sec. V.5). After that, our main effort will be directed toward deriving Kolmogorov's equations for certain functionals of the solutions; these functionals are usually called probabilistic solutions of differential equations. The main method to be used is a systematic application of Euler's approximation method. This will enable us, for example, to avoid proving theorems about the differentiability of solutions of stochastic equations with respect to initial values when we investigate the smoothness of probabilistic solutions of differential equations. Finally, in Secs. V.9–10, we will study probabilistic solutions that are not even continuous. The results of Secs. V.9–10 are not necessary for understanding Chapter VI.

The following assumptions will be adopted *throughout this chapter*.

Let $(\Omega, \mathcal{F}, P)$ be a complete probability space, $(w_t, \mathcal{F}_t)$ a d_1-dimensional Wiener process on it for $t \geq 0$, where the σ-algebras $\mathcal{F}_t$ are complete with respect to $\mathcal{F}, P$. Let $\sigma(t,x) \in H_2$ and $b(t,x) \in H_1$ be functions, defined on $\Omega \times (0,\infty)$ for every $x \in E_d$, with values in the set of $(d \times d_1)$-matrices and in E_d, respectively. We will assume that σ and b are continuous in x for all (ω, t), and that for all $T, R \in [0,\infty)$, $\omega \in \Omega$,

$$(1) \qquad \int_0^T \sup_{|x| \leq R} \left[\|\sigma(t,x)\|^2 + |b(t,x)|\right] dt < \infty,$$

where $\|\sigma\|^2 := \operatorname{tr}\sigma\sigma^* = \sum_{i,j}(\sigma^{ij})^2$. Note that the supremum in (1) may be replaced by the supremum over the set of all points with rational coordinates, which is dense in $\{|x| \le R\}$. Hence it is $\mathfrak{B}^\ell((0,\infty))$-measurable and condition (1) is meaningful.

In this chapter we will investigate the equation

$$d\xi_s = \sigma(s,\xi_s)\,dw_s + b(s,\xi_s)\,ds, \qquad s \ge t,$$

for fixed $t \ge 0$, with initial value $\xi_t = \xi_t^0$, where ξ_t^0 is an $\mathcal{F}_t$-measurable d-dimensional vector. In other words, we will analyze the equation

$$\xi_s = \xi_t^0 + \int_t^s \sigma(r,\xi_r)\,dw_r + \int_t^s b(r,\xi_r)\,dr, \qquad s \ge t. \tag{2}$$

(Recall that by Definition III.4.10

$$\int_t^s \sigma(r,\xi_r)\,dw_r := \int_0^s I_{t<r}\sigma(r,\xi_r)\,dw_r.)$$

By a "solution of equation (2)" we, naturally, mean a d-dimensional process $\xi_s = \xi_s(\omega)$, $\mathcal{F}_s$-measurable with respect to ω for any $s \ge t$, continuous in s, defined for $\omega \in \Omega$, $s \in [0,\infty)$, and satisfying equation (2) for all $s \in [t,\infty)$ at once a.s.

Note that since, for example, $\sigma(r,x)$ and $b(r,x)$ are continuous functions of x, they are also $\mathcal{P}_\mu \otimes \mathfrak{B}(E_d)$-measurable with respect to (ω,r,x) (cf. Lemma I.5.7). In addition, for any process ξ_r that is continuous in r, $\mathcal{F}_r$-measurable with respect to ω for any $r \ge t$, and defined on $\Omega \times [t,\infty)$, the process $\xi_r I_{t<r}$ is left-continuous in r and therefore predictable. Consequently, the mappings

$$(\omega,r) \longrightarrow (\omega,r), \qquad (\omega,r) \longrightarrow \xi_r I_{t<r}, \qquad (\omega,r) \longrightarrow (\omega,r,\xi_r I_{t<r})$$

are $\mathcal{P}_\mu$-measurable as mappings of $\Omega \times (0,\infty)$ to $(\Omega \times (0,\infty), \mathcal{P}_\mu)$, $(E_d, \mathfrak{B}(E_d))$ and $(\Omega \times (0,\infty) \times E_d, \mathcal{P}_\mu \otimes \mathfrak{B}(E_d))$, respectively. Hence, $I_{t<r}\sigma(r,\xi_r)$, $I_{t<r}b(r,\xi_r)$ are pseudopredictable. Since the continuous function ξ_r is bounded on every interval $[t,T]$ and condition (1) holds, we see that for any ω

$$\int_0^T \left[\left\|I_{t<r}\sigma(r,\xi_r)\right\|^2 + \left|I_{t<r}b(r,\xi_r)\right|\right] dr < \infty.$$

Thus, the right-hand side of (2) is meaningful for $s \ge t$ for *any* process ξ_r with the properties listed above.

1. Existence and uniqueness of solutions of Itô stochastic equations

As already stated, equation (0.2) may have no solution if no additional assumptions are made. Some sufficient conditions for the solvability of equation (0.2) are stated in the following theorem.

1. THEOREM. *Assume that*

(a) (monotonicity condition) *for any $R > 0$, there exists a function $K_r(R) > 0$, belonging to H_1 as a function of (ω,r), such that for all $|x|, |y| \le R$, $r > 0$, $\omega \in \Omega$,*

$$2(x-y, b(r,x)-b(r,y)) + \left\|\sigma(r,x)-\sigma(r,y)\right\|^2 \le K_r(R)|x-y|^2; \tag{1}$$

(b) (growth condition) *for all* $x \in E_d$, $r > 0$, $\omega \in \Omega$,

$$2\big(x, b(r,x)\big) + \|\sigma(r,x)\|^2 \leq K_r(1)\big(1 + |x|^2\big). \tag{2}$$

Then stochastic equation (0.2) *has a solution and any two solutions are indistinguishable.*

2. REMARK. The simplest conditions that imply inequalities (1) and (2) are as follows: for $|x|, |y| \leq n$, $z \in E_d$ and all ω, r,

$$\big|b(r,x) - b(r,y)\big| + \big\|\sigma(r,x) - \sigma(r,y)\big\| \leq K(n)|x - y|, \tag{3}$$

$$\big|b(r,z)\big| + \big\|\sigma(r,z)\big\| \leq K(1)\big(1 + |z|\big), \tag{4}$$

where $K(n)$ are constants, $n = 1, 2, \ldots$. To verify this, just use the inequalities $(a,b) \leq |a| \cdot |b|$, $|a| \leq 1 + |a|^2$.

The statement of Theorem 1 with conditions (3) and (4) is known as *Itô's Theorem*. We will describe its proof, omitting inessential details and assuming additionally that $\mathbf{E}\,|\xi_t^0|^2 < \infty$ and that $K(n)$ is independent of n. In this simple case we will not even need Itô's formula.

Evidently, it suffices to prove the existence and uniqueness of a solution of (0.2) for an arbitrary finite time interval $[t, T]$. Fix $T \in (t, \infty)$ and let V denote the space of all processes ξ_s with values in E_d such that $\xi_s I_{t<s\leq T}$ is a predictable function and

$$\|\xi.\| := \left(\mathbf{E} \int_t^T |\xi_s|^2 ds\right)^{1/2} < \infty.$$

Define

$$I\xi_s = \int_t^s \sigma(r, \xi_r)\, dw_r + \int_t^s b(r, \xi_r)\, dr, \qquad s \in [t, T]. \tag{5}$$

This formula defines an operator I on V that assigns to every element $\xi. \in V$ a process $I\xi_s$ in accordance with formula (5). It is easy to see, using the inequality $(a+b)^2 \leq 2a^2 + 2b^2$, the isometric property of the stochastic integration operator, the Cauchy-Bunyakovsky inequality and conditions (3) and (4), that I maps V into V and, for some constant α and all $\xi., \eta. \in V$,

$$\mathbf{E}\,|I\xi_s - I\eta_s|^2 \leq \alpha \mathbf{E} \int_t^s |\xi_r - \eta_r|^2\, dr. \tag{6}$$

Now set $\xi_s^{(0)} \equiv 0$, $\xi_s^{(n+1)} = \xi_s^0 + I\xi_s^{(n)}$, $n = 0, 1, 2, \ldots$. It follows from (6) that

$$\mathbf{E}\,|\xi_s^{(n+1)} - \xi_s^{(n)}|^2 \leq \alpha \int_t^s \mathbf{E}\,|\xi_r^{(n)} - \xi_r^{(n-1)}|^2\, dr.$$

Iterating this inequality, we obtain

$$\|\xi.^{(n+1)} - \xi.^{(n)}\|^2 \leq \frac{T^n \alpha^n}{n!} \|\xi.^{(1)}\|^2. \tag{7}$$

Since the series $\sum (T\alpha)^{n/2}(n!)^{-1/2}$ is convergent, it follows that the function series $\sum (\xi_\cdot^{(n+1)} - \xi_\cdot^{(n)})$ is convergent in V. In other words, the functions $\xi_\cdot^{(n+1)}$ converge in V and there exists a function $\hat{\xi}_\cdot \in V$ such that $\|\xi_\cdot^{(n)} - \hat{\xi}_\cdot\| \to 0$ as $n \to \infty$. In addition, integrating (6), we obtain

$$\|I\xi_\cdot - I\eta_\cdot\|^2 \le \alpha T \|\xi_\cdot - \eta_\cdot\|^2. \tag{8}$$

In particular, the operator I is continuous in V. Letting $n \to \infty$ in the equality $\|\xi_\cdot^{(n+1)} - (\xi_t^0 + I\xi_\cdot^{(n)})\| = 0$, we conclude that $\|\hat{\xi}_\cdot - (\xi_t^0 + I\hat{\xi}_\cdot)\| = 0$. By (8), it follows that $I\hat{\xi}_s = I(\xi_t^0 + I\hat{\xi}_\cdot)_s$ for almost all ω, s. But the two sides of this equality are continuous in s, hence they coincide for all $s \in [t, T]$ at once a.s. Setting $\xi_s = \xi_t^0 + I\hat{\xi}_s$, we obtain $\xi_s = \xi_t^0 + I(\xi_t^0 + I\hat{\xi}_\cdot)_s = \xi_t^0 + I\xi_s$ for all $s \in [t, T]$ at once (a.s.). This implies the existence of a solution of (0.2) on $[t, T]$. Uniqueness follows immediately from (6).

In applications, the Lipschitz condition (3) is usually not very restrictive. On the other hand, the condition of linear growth is often undesirable in applications to the theory of stochastic partial differential equations. Even in our case of ordinary Itô stochastic equations, it is much stronger than (2), as is demonstrated for $d = 1$ by the equation $d\xi_s = b(s, \xi_s)\, ds$ with *any* function $b(s, x)$ (independent of ω) that is continuous in (s, x) and decreasing in x.

In the proof of Theorem 1 and for some other purposes we will need the following lemma.

3. LEMMA. *Under the assumptions of Theorem* 1, *let* $t, t^n \ge 0$, $t^n \to t$. *Let* ξ_s^n, $n = 1, 2, \dots$, *be given* d*-dimensional continuous processes on* $\Omega \times [t^n, \infty)$, $\mathcal{F}_s$*-measurable with respect to* ω *for any* $s \ge t^n$, *such that*

$$d\xi_s^n = \sigma(s, \xi_s^n + p_s^n)\, dw_s + b(s, \xi_s^n + p_s^n)\, ds, \qquad s \ge t^n, \tag{9}$$

where p_s^n *are functions, defined on* $\Omega \times (t^n, \infty)$, *such that the functions* $I_{t^n < s}|p_s^n|$ *are at least pseudopredictable. Let* $\tau^n(R) \ge t^n$ *be Markov times such that*

$$|\xi_s^n| + |p_s^n| \le R$$

for $t^n < s \le \tau^n(R)$, *for all* $n \ge 1$, $R > 0$. *Assume moreover that* $\xi_{t^n}^n \xrightarrow{P} \eta$, *where* η *is a random vector, and that for all* $T \in (t+1, \infty)$, $R \in (0, \infty)$,

$$\lim_{n\to\infty} \mathbf{E} \int_{t^n}^{T\wedge\tau^n(R)} |p_r^n|\, dr = 0. \tag{10}$$

Finally, let $\rho(R)$ *be a function such that* $\rho(R) \to \infty$ *as* $R \to \infty$ *and*

$$\lim_{R\to\infty} \overline{\lim_{n\to\infty}} P\Big\{\tau^n(R) \le T, \sup_{t^n \le s \le \tau^n(R)} |\xi_s^n| \le \rho(R)\Big\} = 0 \tag{11}$$

for any $T \in (0, \infty)$. *Then for any* $T \in (0, \infty)$

$$\sup_{0 \le s \le T} \big|\xi_{s\vee t^n}^n - \xi_{s\vee t^m}^m\big| \xrightarrow{P} 0 \tag{12}$$

as $n, m \to \infty$.

PROOF. It suffices to prove (12) for $T > \sup t^n$. Take such a T and for $s, R > 0$, $n, m = 1, 2, \ldots$, denote

$$\psi_s(R) = \exp\left(-2\int_0^s K_r(R)\,dr\right), \qquad u^{nm} = t^n \vee t^m,$$
$$\gamma^{nm}(R) = \left(\tau^n(R) \wedge \tau^m(R)\right) \vee u^{nm}.$$

Let us fix n, m, R temporarily, omitting the superscripts n, m and the argument R, and define $\xi_s = \xi_s^n$, $p_s = p_s^n$, $\eta_s = \xi_s^m$ and $q_s = p_s^m$. Then we use Itô's formula (Corollary IV.1.2) to find $d|\xi_s - \eta_s|^2$, $d\left(|\xi_s - \eta_s|^2\psi_s\right)$ for $s \geq u$ to show that (a.s.)

$$\begin{aligned}(13)\qquad &\left|\xi_{s\wedge\gamma} - \eta_{s\wedge\gamma}\right|^2 \psi_{s\wedge\gamma} \\ &= |\xi_u - \eta_u|^2\psi_u + \int_0^{s\wedge\gamma} I_{u<r}\Big[2\big(\xi_r - \eta_r, b(r, \xi_r + p_r) - b(r, \eta_r + q_r)\big) \\ &\quad + \left\|\sigma(r, \xi_r + p_r) - \sigma(r, \eta_r + q_r)\right\|^2 - 2K_r|\xi_r - \eta_r|^2\Big]\psi_r\,dr \\ &\quad + 2\int_0^{s\wedge\gamma} I_{u<r}(\xi_r - \eta_r)^*\left[\sigma(r, \xi_r + p_r) - \sigma(r, \eta_r + q_r)\right]\psi_r\,dw_r.\end{aligned}$$

To obtain an upper bound for the integral with respect to dr, we use the monotonicity condition and the fact that if $u < r \leq \gamma$, then $|\xi_r + p_r|$, $|\eta_r + q_r| \leq R$. In addition, we rewrite $\xi_r - \eta_r$ in the first scalar product on the right of (13) as $\left[(\xi_r + p_r) - (\eta_r + q_r)\right] + q_r - p_r$ and observe that for $u < r \leq \gamma$

$$2|\xi_r - \eta_r|^2 \geq \left|(\xi_r + p_r) - (\eta_r + q_r)\right|^2 - 2|p_r - q_r|^2,$$
$$|p_r - q_r|^2 \leq 2R\left(|p_r| + |q_r|\right).$$

Then for *all* $s \geq 0$ (a.s.) it follows from (13) that

$$(14)\qquad I_{u\leq s}\left|\xi^n_{s\wedge\gamma} - \xi^m_{s\wedge\gamma}\right|^2\psi_{s\wedge\gamma}(R) \leq \left|\xi^n_u - \xi^m_u\right|^2 + 4\beta^n_s(R) + 4\beta^m_s(R) + \alpha^{nm}_s(R),$$

where $u = u^{nm}$, $\gamma = \gamma^{nm}(R)$, $\alpha^{nm}_s(R)$ is a local martingale equal to the last term in (13), $\alpha^{nm}_0(R) = 0$,

$$\beta^k_s(R) = \int_0^{s\wedge\tau^k(R)} I_{t^k<r}|p^k_r|K'_r(R)\,dr, \qquad k = n, m,$$
$$K'_r(R) = \sup_{|x|\leq R}\left|b(r, x)\right| + RK_r(R).$$

It is evident that $K'(R) \in H_1$, and by condition (10), Chebyshev's inequality, and Lemma III.6.19,

$$(15)\qquad \beta^k_s(R) \xrightarrow{P} 0$$

as $k \to \infty$ for any s.

Suppose now that as $n, m \to \infty$,

$$\sup_{0 \le s \le u^{nm}} |\xi^n_{s \vee t^n} - \xi^n_{t^n}| = \sup_{t^n \le s \le u^{nm}} |\xi^n_s - \xi^n_{t^n}| \xrightarrow{P} 0. \tag{16}$$

Then the first term on the right in (14) tends to zero in probability, since $|\xi^n_u - \xi^m_u| \le |\xi^n_u - \xi^n_{t^n}| + |\xi^m_u - \xi^m_{t^n}| + |\xi^n_{t^n} - \xi^m_{t^n}|$ and $\xi^k_{t^k} \xrightarrow{P} \eta$ as $k \to \infty$. Since the left-hand side of (14) is nonnegative, this, together with (15), implies by Theorem III.6.12 (b) that the supremum over $s \in [t, T]$ of the right-hand side of (14) tends to zero in probability as $n, m \to \infty$. Since $\psi_{s \wedge \gamma}(R) \ge \psi_T(R)$ for $s \le T$ and the latter is independent of n, m, it follows that our conclusion is also valid for the right-hand side of (14) multiplied by $\psi^{-1}_{s \wedge \gamma}(R)$. Thus, for all $R, \varepsilon > 0$,

$$\begin{aligned} &\lim_{n,m \to \infty} P\Big\{\tau^n(R) \ge T,\ \tau^m(R) \ge T,\ \sup_{u^{nm} \le s \le T} |\xi^n_s - \xi^m_s| \ge \varepsilon\Big\} \\ &\qquad \le \lim_{n,m \to \infty} P\Big\{\sup_{s \le T} I_{u^{nm} \le s} \big|\xi^n_{s \wedge \gamma^{nm}(R)} - \xi^m_{s \wedge \gamma^{nm}(R)}\big|^2 \ge \varepsilon^2\Big\} = 0. \end{aligned} \tag{17}$$

Hence, it is clear, that in order to prove (12) we need only prove (16) and the fact that

$$\lim_{R \to \infty} \varlimsup_{n \to \infty} P\{\tau^n(R) \le T\} = 0. \tag{18}$$

Recalling once again that $|\xi^n_r + p^n_r| \le R$ for $t^n < r \le \tau^n(R)$, we find from (0.1) by the Dominated Convergence Theorem that for any R,

$$\int_0^T I_{t^n < r \le u^{nm} \wedge \tau^n(R)} \Big(\big\|\sigma(r, \xi^n_r + p^n_r)\big\|^2 + \big|b(r, \xi^n_r + p^n_r)\big|\Big)\, dr \longrightarrow 0$$

as $n, m \to \infty$. In addition, for $t^n \le s \le u^{nm}$ (a.s.),

$$\begin{aligned} \xi^n_{s \wedge \tau^n(R)} - \xi^n_{t^n} &= \int_0^s I_{t^n < r \le u^{nm} \wedge \tau^n(R)} \sigma(r, \xi^n_r + p^n_r)\, dw_r \\ &\quad + \int_0^s I_{t^n < r \le u^{nm} \wedge \tau^n(R)} b(r, \xi^n_r + p^n_r)\, dr. \end{aligned}$$

Consequently, Theorem III.6.6 implies (16) with $u^{nm} \wedge \tau^n(R)$ substituted for u^{nm}. Together with inequalities similar to (17), this shows that it suffices to prove only (18).

Applying Itô's formula to $(|\xi^n_s|^2 + 1)\psi_s$ for $\psi_s = \psi_s(1)$ and using the growth condition, we obtain as before

$$I_{t^n \le s} (|\xi^n_{s \wedge \tau}|^2 + 1) \psi_{s \wedge \tau} \le |\xi^n_{t^n}|^2 + 1 + 2\beta^n_s(R) + \delta^n_s(R), \tag{19}$$

where $\tau = \tau^n(R)$, $\delta^n_s(R)$ is a certain local martingale, $\delta^n_0(R) = 0$. Since the left-hand side of (19) is positive, and by (15) and Exercise III.6.17 $\xi^n_{t^n}$ and $\beta^n_s(R)$ are bounded in probability, it follows by Theorem III.6.12 (c) that the right-hand side of (19) and, therefore, its left-hand side are bounded in probability on $[0, T]$.

Consequently, for any $\varepsilon > 0$

$$\lim_{R\to\infty}\overline{\lim_{n\to\infty}}\, P\Big\{\tau^n(R) \le T, \sup_{t^n\le s\le\tau^n(R)} |\xi_s^n| \ge \rho(R)\Big\}$$

$$\le \lim_{R\to\infty}\overline{\lim_{n\to\infty}}\, P\Big\{\sup_{s\le T} I_{t^n\le s}\big|\xi^n_{s\wedge\tau^n(R)}\big|^2\psi_{s\wedge\tau^n(R)} \ge \rho^2(R)\varepsilon, \psi_T \ge \varepsilon\Big\}$$

$$+ P\{\psi_T \le \varepsilon\} = P\{\psi_T \le \varepsilon\}.$$

Letting $\varepsilon \downarrow 0$, we see that the first expression in these inequalities vanishes and, together with condition (11), this implies (18). □

4. PROOF OF THEOREM 1. We will use Euler's method, putting $\xi_t^n = \xi_t^0$,

$$\xi_s^n = \xi^n_{t(k,n)} + \int_{t(k,n)}^s \sigma\big(r, \xi^n_{t(k,n)}\big)\, dw_r + \int_{t(k,n)}^s b\big(r, \xi^n_{t(k,n)}\big)\, dr \tag{20}$$

for $s \in \big(t(k,n), t(k+1,n)\big]$, $k = 0, 1, 2, \dots$, where $t(k,n) = t + kn^{-1}$. It is easy to see that ξ_s^n satisfies the following equation for all $s \ge t$ at once with probability one:

$$\xi_s^n = \xi_t^0 + \int_t^s \sigma\big(r, \xi^n_{\varkappa(n,r)}\big)\, dw_r + \int_t^s b\big(r, \xi^n_{\varkappa(n,r)}\big)\, dr, \tag{21}$$

where $\varkappa(n,r) = t(k,n)$ for $r \in \big(t(k,n), t(k+1,n)\big]$. Our further arguments will be based on Lemma 3 and on the fact that ξ_s^n obviously satisfies equation (9), where $p_s^n = \big(\xi^n_{\varkappa(n,s)} - \xi_s^n\big)I_{t<s}$ is a predictable left-continuous process.

Fix $s \ge t$, $R > 0$, set $t^n = t$, $\tau^n = \tau^n(R) = \inf\{r \ge t \colon |\xi_r^n| \ge R/3\}$, and observe that (a.s.)

$$\begin{aligned} -p_s^n I_{s<\tau^n} &= I_{s<\tau^n}\int_t^s I_{\varkappa(n,s)<r\le\tau^n}\sigma\big(r, \xi^n_{\varkappa(n,r)}\big)\, dw_r \\ &\quad + I_{s<\tau^n}\int_t^s I_{\varkappa(n,s)<r\le\tau^n} b\big(r, \xi^n_{\varkappa(n,r)}\big)\, dr. \end{aligned} \tag{22}$$

Since $\big|\xi^n_{\varkappa(n,r)}\big| \le R$ for all $t < r \le \tau^n$, it follows by (0.1) that

$$\int_t^s I_{\varkappa(n,s)<r\le\tau^n}\big\|\sigma\big(r, \xi^n_{\varkappa(n,r)}\big)\big\|^2 dr \le \int_t^s I_{\varkappa(n,s)<r\le s}\sup_{|x|\le R}\|\sigma(r,x)\|^2 dr \to 0 \tag{23}$$

as $n \to \infty$ for all ω, s. It follows from Theorem III.6.6 that the first term on the right of (22) tends to zero in probability. Inequality (0.1) implies that the same is true of the second term.

By Fubini's Theorem and Lemma III.6.13 (f) ($|p_r^n| \le R$ for $r < \tau^n$), we obtain that the left-hand side of (10) equals

$$\lim_{n\to\infty}\int_t^T \mathbf{E}\,|p_r^n| I_{r<\tau^n}\, dr = 0.$$

Thus condition (10) is satisfied. Obviously, if $\rho(R) = R/4$, the event under the probability sign in condition (11) is empty. Hence, condition (11) is also satisfied, (12) holds and by Lemma III.6.18 there exists a continuous $\mathcal{F}_s$-adapted process ξ_s on $[t, \infty)$ such that

$$\sup_{t<s\leq T} \left|\xi_s^n - \xi_s\right| \xrightarrow{P} 0 \tag{24}$$

as $n \to \infty$ for any $T \in (t, \infty)$. The argument s of ξ_s^n and ξ_s can obviously be replaced by $\varkappa(n, s)$. Moreover, this substitution may be done only for ξ_s^n, since $\xi_s - \xi_{\varkappa(n,s)} \to 0$ uniformly in $s \in (t, T)$ for all ω, by the continuity of ξ_s.

As in the proof of Lemma III.6.18, it now follows that $\xi_s^n \to \xi_s$, $\xi_{\varkappa(n,s)}^n \to \xi_s$ (a.s.) for some subsequence $n = n(k)$, uniformly on any interval (t, T). By condition (0.1), the *continuity* of $\sigma(r, x)$ and $b(r, x)$ in x, Theorem III.6.6 and the Dominated Convergence Theorem, we may therefore pass to the limit in (21) on the subsequence $n = n(k)$, concluding that ξ satisfies (0.2).

Finally, suppose that there exists another solution of equation (0.2), say, $\tilde{\xi}_s$. In Lemma 3 put $p_s^n \equiv 0$ and $\xi_s^n = \xi_s$ for even n and $\xi_s^n = \tilde{\xi}_s$ for odd n. Then Lemma 3 yields $\xi_s = \tilde{\xi}_s$ for all $s \in [t, \infty)$ at once a.s. □

In many situations it is useful to have a theorem about passing to limits in Itô stochastic equations.

5. Theorem. *Let the functions $\sigma_n(s, x)$, $b_n(s, x)$, $n = 1, 2, \ldots$, satisfy the conditions for σ, b, formulated in the introduction to this chapter. Let σ, b satisfy the assumptions of Theorem 1. Assume also that $t, t^n \geq 0$, $t^n \to t$. Assume that on $\Omega \times [t^n, \infty)$ for $n = 1, 2, \ldots$ we are given the solutions η_s^n of the equations*

$$d\eta_s^n = \sigma_n(s, \eta_s^n)\, dw_s + b_n(s, \eta_s^n)\, ds, \qquad s \geq t^n,$$

and suppose that for all $T \in (t + 1, \infty)$, $R \in (0, \infty)$,

$$\eta_{t^n}^n \xrightarrow{P} \xi_t^0, \qquad \int_{t^n}^{T\wedge\gamma^n(R)} \sup_{|x|\leq R} \left[\|\sigma_n - \sigma\|^2 + |b_n - b|\right](r, x)\, dr \xrightarrow{P} 0 \tag{25}$$

as $n \to \infty$, where $\gamma^n(R) = \inf\left\{s \geq t^n : |\eta_s^n| \geq R\right\}$. Then for all $T \in (0, \infty)$

$$\sup_{0\leq s\leq T} \left|\eta_{s\vee t^n}^n - \xi_{s\vee t}\right| \xrightarrow{P} 0 \tag{26}$$

as $n \to \infty$, where ξ_s is the solution of equation (0.2).

Proof. For even $n = 2k$, $s \geq t^n$, denote

$$p_s^n = \int_{t^k}^{s} (\sigma_k - \sigma)(r, \eta_r^k)\, dw_r + \int_{t^k}^{s} (b_k - b)(r, \eta_r^k)\, dr,$$

$\bar{t}^n = t^k$, $\xi_s^n = \eta_s^k - p_s^n$; for odd n, let $p_s^n = 0$, $\xi_s^n = \xi_s$, $\bar{t}^n = t$. Since ξ_s^n obviously satisfies (9), we will apply Lemma 3. To this end, set $\rho(R) = R/2$, $\tau^n(R) = \inf\left\{s \geq \bar{t}^n : |\xi_s^n| + |p_s^n| \geq R\right\}$.

Observe that $|\eta_s^k| \le R$ for $t^k < s \le \gamma^k(R)$, and therefore it follows from (25) and Theorem III.6.6 that for any $T \in (0, \infty)$

$$(27) \qquad P\text{-}\lim_{k\to\infty} \sup_{t^k \le s \le \gamma^k(R)\wedge T} |p_s^{2k}| = 0 \quad \text{(a.s.)}.$$

Since $\tau^{2k}(R) \le \gamma^k(R)$ and $|p_s^{2k}| \le R$ for $s \le \tau^{2k}(R)$, Lemma III.6.13 (f) implies that condition (10) is fulfilled. In addition, if the event contained in (11) occurs, then n is even, and the expression whose limit appears in (27) is greater than $R/2$. Therefore, condition (11) is satisfied by virtue of (27). By Lemma 3 this implies (26) with ξ_s^{2n} substituted for η_s^n.

This, in turn, together with (27) (and Lemma III.6.13) proves (26) without replacing η_s^n, but with $\gamma^n(R)\wedge T$ substituted for T. Thus, for any $R, \varepsilon > 0$,

$$\begin{aligned}\varlimsup_{n\to\infty} P\Big\{\sup_{s\le T} |\eta^n_{s\vee t^n} - \xi_{s\vee t}| \ge \varepsilon\Big\} &\le \varlimsup_{n\to\infty} P\{\gamma^n(R)\le T\}\\ &\le \varlimsup_{n\to\infty} P\Big\{\gamma^n(R)\le T,\ \sup_{s\le\gamma^n(R)} |\eta^n_{s\vee t^n} - \xi_{s\vee t}| \le 1\Big\}\\ &\quad + \varlimsup_{n\to\infty} P\Big\{\sup_{s\le\gamma^n(R)\wedge T} |\eta^n_{s\vee t^n} - \xi_{s\vee t}| > 1\Big\} \le P\Big\{\sup_{s\le T}|\xi_{s\vee t}| \ge R-1\Big\}.\end{aligned}$$

Since the last expression vanishes as $R\to\infty$, the theorem is proved. □

6. REMARK. We did not use the continuity of σ, b in x in Lemma 3. Continuity is not needed in Theorem 5 either, provided that we assume that there exists a solution of equation (0.2). However, it can be shown that condition (1) always implies that σ is continuous in x.

2. Two examples of application of Itô Stochastic equations

Some applications of the theory of diffusion processes to the theory of differential equations were seen in Secs. II.8–10, where we considered differential operators in which the principal part is the Laplace operator and the coefficients of the first derivatives are constants. Of course, by a linear transformation these applications carry over from the Laplace operator to more general elliptic operators, but still with constant coefficients. We will now show how the theory of Itô stochastic equations makes it possible to consider elliptic operators with variable coefficients as well: through only "qualitative" arguments the theory will enable us to obtain certain results for equations with such operators. This possibility was touched upon in Sec. IV.1.

1. EXAMPLE. Let $a(x)$ be a function on E_d with values in the set of $(d\times d)$-matrices, $b(x)$ a function with values in E_d and $c(x)$ a real-valued function. Let us assume that a and b satisfy a Lipschitz condition on E_d, c is bounded and $(a(x)\lambda,\lambda)\ge \delta|\lambda|^2$ for all $x,\lambda\in E_d$, where $\delta>0$ is a constant, that is, a is uniformly nonsingular. Let $u(x)$ be a twice continuously differentiable function on E_d. Denote $D = D_\delta = \{x\in E_d\colon 0<x^1<\delta,\ |x|<\delta\}$, fix constants $\alpha\in(0,1)$, $K\ge 0$, and suppose that $|u|\le K$ in D, $|u(0)-u(x)|\le K|x|^\alpha$ for $x\in\partial D$, $|Lu|\le K$ in D, where

$$Lu(x) := \frac{1}{2}\sum_{i,j=1}^d a^{ij}(x)u_{x^ix^j}(x) + \sum_{i=1}^d b^i(x)u_{x^i}(x) + c(x)u(x).$$

We claim that under these assumptions

$$|u(0) - u(x)| \le N|x|^{\alpha} \tag{1}$$

for all $x \in D$, where N is a constant depending *only* on δ, α, K and the suprema of $\operatorname{tr} a$, $|b|$, c over D. It is evident that since u is smooth, inequality (1) holds with some constant even for $\alpha = 1$. The essence of our assertion is that the constant N of (1) can be made dependent only on the above-mentioned constants.

To prove this we first note that $|Lu - cu| \le |Lu| + K|c|$; hence we may assume without loss of generality that $c \equiv 0$. Then, replacing u by $u - u(0)$, we may assume that $u(0) = 0$. Moreover, it clearly suffices to prove (1) in any domain D_ε, where ε depends only on the parameters on which we allow N to depend. Since $|(x, b(x))| \le \varepsilon|b(x)|$, $\operatorname{tr} a \ge \delta$ in D_ε, it follows that $2(x, b(x)) \ge -\operatorname{tr} a/2$ in D_ε for suitable ε. The reader can easily verify that ε may also be chosen so that in D_ε

$$\tilde{b}^1 := v^{-1}(b^1 v + a^{11} v_{x^1}) \le 0, \qquad v \ge 1,$$

where $v(x) := 3 - \exp(x^1 \varepsilon^{-1/2})$. Letting $\tilde{u} = v^{-1}u$, $\tilde{c} = v^{-1}Lv$ and $\tilde{b}^i = b^i + a^{1i} v_{x^1}$ for $i \ge 2$, we obtain

$$\frac{1}{2}\sum_{i,j=1}^{d} a^{ij}\tilde{u}_{x^i x^j} + \sum_{i=1}^{d} \tilde{b}^i \tilde{u}_{x^i} + \tilde{c}\tilde{u} = v^{-1}Lu$$

in D_ε, and this expression is bounded in D_ε by a constant like K. As $\tilde{c}$ can easily be eliminated and it evidently suffices to prove (1) for $\tilde{u}$, we see, passing if necessary from δ to ε, that no loss of generality is incurred by assuming that

$$c \equiv 0, \qquad u(0) = 0, \qquad 2(x, b(x)) \ge -\operatorname{tr} a(x)/2, \qquad b^1 \le 0. \tag{2}$$

After this purely technical reduction, we proceed to the use of Itô stochastic equations. Let $(\Omega, \mathcal{F}, P)$ be a probability space and $(w_t, \mathcal{F}_t)$ a d-dimensional Wiener process defined there for $t \ge 0$, where the $\mathcal{F}_t$ are, as usual, σ-algebras complete with respect to $\mathcal{F}$, P. Put $\sigma = \sqrt{a}$ (see, for example, (III.10.13)) and for every $x \in E_d$ consider the Itô stochastic equation

$$\xi_s = x + \int_0^s \sigma(\xi_t)\,dw_t + \int_0^s b(\xi_t)\,dt. \tag{3}$$

We claim that σ and b satisfy Lipschitz conditions. This is true for b by assumption and can be proved for σ as follows. If A and B are symmetric $(d \times d)$-matrices, $\alpha, \beta > 0$ constants and $(B\lambda, \lambda) \ge \beta|\lambda|^2$, $|(BA\lambda, \lambda)| \le \alpha|\lambda|^2$ for all $\lambda \in E_d$, then for any eigenvector ν of A

$$\beta|A\nu|^2 \le |\nu|^{-1}|A\nu| \cdot |(BA\nu, \nu)| \le \alpha|A\nu| \cdot |\nu|$$

and $|A| \le \alpha\beta^{-1}$. If $y, z \in E_d$, $A = \sigma(y) - \sigma(z)$, $B = \sigma(y) + \sigma(z)$, then it is not difficult to see that $(BA\lambda, \lambda) = ((a(y) - a(z))\lambda, \lambda)$. Since a satisfies a Lipschitz condition and $(B\lambda, \lambda) \ge 2\sqrt{\delta}|\lambda|^2$, it follows that σ also satisfies a Lipschitz condition.

By Theorem 1.1, there exists a unique solution of equation (3), which we denote by $\xi_s(x)$. Let $\tau(x)$ be the first exit time of $\xi_s(x)$ from D. Then by Itô's formula

$$u(\xi_s(x)) - \int_0^s Lu(\xi_t(x))\,dt$$

is a local martingale. It remains a local martingale if s is replaced by $s\wedge\tau(x)$. But since after this substitution the expression is bounded on every finite time interval $[0, T]$ (u, Lu are bounded in D and $\xi_t(x) \in D$ for $t < \tau(x)$), we obtain a martingale. Evaluating its expectation, letting $s \to \infty$ and using the Dominated Convergence Theorem and the fact that $\mathbf{E}\,\tau < \infty$ by conditions (2) and Theorem IV.4.3 (b), we obtain

$$u(x) = \mathbf{E}\,u(\xi_{\tau(x)}(x)) - \mathbf{E}\int_0^{\tau(x)} Lu(\xi_t(x))\,dt. \tag{4}$$

We now use the conditions: $|u(x)| \le K|x|^\alpha$ on ∂D (observe that $\xi_{\tau(x)}(x) \in \partial D$ for $x \in D$ at points where $\tau(x) < \infty$, that is, almost surely), $|Lu| \le K$ in D and Theorem IV.4.3, by which

$$\mathbf{E}\sup_{t\le\tau(x)}|\xi_t(x)|^\alpha \le N(|x|^\alpha + \mathbf{E}\,\tau^{\alpha/2}(x)),$$

$$\mathbf{E}\,\tau(x) \le N_1\mathbf{E}\sup_{t\le\tau(x)}|\xi_t(x)|^2 \le N_1\delta^{2-\alpha}\mathbf{E}\sup_{t\le\tau(x)}|\xi_t(x)|^\alpha.$$

Thus $|u(x)| \le N|x|^\alpha + N\mathbf{E}\,\tau^{\alpha/2}(x)$ for $x \in D$ and to prove (1) it will suffice to estimate the last term for $x \in D$.

We will consider two processes: $\xi_s^1(x)$ and

$$\eta_s(x) = \xi_s^1(x) - \int_0^s b^1(\xi_t(x))\,dt = x^1 + \sum_{j=1}^d \int_0^s \sigma^{1j}(\xi_t(x))\,dw_t^j.$$

By condition (2), $b^1(\xi_t(x)) \le 0$ for $t < \tau(x)$, hence $\xi_s^1(x) \le \eta_s(x)$ for $s \le \tau(x)$. It follows that $\tau(x) \le \gamma(x)$, where $\gamma(x)$ is the time at which the process $\eta_s(x)$ first vanishes. It remains to notice that $a^{11} \ge \delta$, apply Theorem III.10.8, by which $d\eta_t(x) = (a^{11}(\xi_t(x)))^{1/2}d\tilde{w}_t$, where $\tilde{w}_t$ is some one-dimensional Wiener process, and use Corollary IV.2.10, by which $\mathbf{E}\,\gamma^{\alpha/2}(x) \le N|x^1|^\alpha \le N|x|^\alpha$.

Our next example is based on the application of a formula similar to (4) in the case when $D = E_1$, the process $\xi_t(x)$ leaves D in a finite time $\tau(x)$ and $\xi_{\tau(x)}(x)$ is defined as the limit of ξ_t as $t \uparrow \tau(x)$.

2. Example. Let $x \in (-\infty, \infty)$ and consider the ordinary differential equation

$$\frac{1}{2}u_{xx} + x^3u_x - u = 0. \tag{5}$$

We will see that this equation has a positive, bounded, twice continuously differentiable solution and, moreover, that any twice continuously differentiable solution is bounded.

Take $(\Omega, \mathcal{F}, P, w_t, \mathcal{F}_t)$ as in the previous example with $d = 1$, and for every fixed $x \in (-\infty, \infty)$ consider the Itô stochastic equation

$$(6) \qquad \tilde{\xi}_s = x + w_s + \int_0^s \tilde{\xi}_t^3 \, dt.$$

It turns out that this equation has a solution on some time interval $[0, \tau(x))$, where $\tau(x)$ is a Markov time, known as the *explosion time* of $\tilde{\xi}_s(x)$, that is, $|\tilde{\xi}_s(x)| \to \infty$ as $s \uparrow \tau(x)$ (a.s.). In addition, $P\{\tau(x) < \infty\} = 1$ and if we define

$$(7) \qquad \tilde{u}(x) = \mathbf{E}\, e^{-\tau(x)} I_{\lim_{s \to \tau(x)} \tilde{\xi}_s(x) = \infty},$$

then $\tilde{u}(x) > 0$, $\tilde{u}$ is twice continuously differentiable with respect to x and satisfies equation (5), $\tilde{u}(x) \to 1$ as $x \to \infty$ and $\tilde{u}(x) \to 0$ as $x \to -\infty$.

These assertions are quite natural. Indeed, if there exists a solution $\tilde{u}$ of equation (5) with boundary values $\tilde{u}(\infty) = 1$, $\tilde{u}(-\infty) = 0$, then, applying Itô's formula to $\tilde{u}(\tilde{\xi}_s(x))e^{-s}$ as in the derivation of (4), we conclude that $\tilde{u}(x)$ must have the form of (7). Moreover, the change of variables $\eta_s = \tilde{\xi}_s - w_s$ turns equation (6) into an ordinary equation $\dot{\eta}_s = (\eta_s + w_s)^3$. Since $(y + w_s)^3$ increases as $y \to \infty$ and decreases as $y \to -\infty$ with higher rate than linear functions, it is quite natural in this context of the theory of ordinary differential equations that η_s should "explode" at some time $\tau(x)$, that is, $|\eta_s| \to \infty$ as $s \uparrow \tau(x)$, where $\tau(x) < \infty$ (a.s.). The same result is true for $\tilde{\xi}_s$. Hence our assertion about the right-hand side of (7) are quite plausible.

The previous arguments are perhaps too heuristic. A rigorous justification would require consideration of Itô stochastic equations over random time intervals and generalization of Itô's formula to such processes. Such a theory would be useful, but since we are concerned here with equation (5) as an example only, we will not develop it. Instead of (6) we will consider an equation obtained from it by a random time change. Note that the study of processes, that move "too quickly", in their "natural" time in which large displacements of the process lead to large increments of the natural time, may be useful in many other cases. By the way, in the context of the theory of differential equations, the use of random time change is equivalent to the completely harmless (and useless) multiplication of equation (5) by some function $f(x)$ (we will take $f(x) = (|x|^3 + 1)^{-1}$).

Set $\sigma(x) = (|x|^3 + 1)^{-1/2}$, $b(x) = x^3(|x|^3 + 1)^{-1}$ and for every $x \in (-\infty, \infty)$ let $\xi_s(x)$ denote the solution of equation (3). Since σ, b and their first derivatives are bounded, it follows from Theorem 1.1 that there exists a unique solution of equation (3) for all $s \geq 0$. Define

$$\varphi_s(x) = \int_0^s \left(|\xi_t(x)|^3 + 1\right)^{-1} dt, \qquad \tau_n(x) = \inf\{t \geq 0 \colon |\xi_t(x)| \geq n\},$$

$$\xi^n(x) = \xi_{\tau_n(x)}(x), \qquad \varphi^n(x) = \varphi_{\tau_n(x)}(x), \qquad n = 1, 2, \ldots.$$

By Itô's formula applied to $v(\xi_s(x))$, where v is a solution of the equation $\frac{1}{2}v_{xx} + x^3 v_x = 0$, for example,

$$v(x) = \int_0^x e^{-\frac{1}{2}r^4} dr,$$

we see that $v(\xi_s(x)) - v(x)$ is a stochastic integral. Since v is a bounded function, this stochastic integral is also bounded. By Theorem III.5.6 (c) it has a limit as $s \to \infty$ (a.s.), and since v is a strictly increasing function, $\xi_s(x)$ has a limit as $s \to \infty$ (a.s.).

Furthermore, the function

$$w(x) := 2\int_0^\infty \left(e^{\frac{1}{2}p^4} \int_{p\vee|x|}^\infty e^{-\frac{1}{2}r^4}\,dr \right) dp \tag{8}$$

does not exceed its value at zero. That value is finite, since it can be proved by l'Hôpital's rule that the expression in parenthesis in (8) is equivalent to $(2p^3)^{-1}$ as $p \to \infty$. Inverting the order of integration in (8), we can easily calculate w_x and w_{xx} and prove that $\frac{1}{2}w_{xx} + x^3 w_x = -1$. Hence, applying Itô's formula to $w(\xi_s(x))$, we conclude that $w(\xi_s(x)) + \varphi_s(x)$ is a local martingale. It is obviously nonnegative, hence, by Theorem III.6.11(a), almost all its sample paths are bounded and $\varphi_\infty(x) < \infty$ (a.s.). Together with what was proved previously and the definition of $\varphi_s(x)$, this implies that $|\xi_s(x)| \to \infty$ (a.s.) as $s \to \infty$, the definitions

$$u_{(\pm)}(x) = \mathbf{E}\, e^{-\varphi_\infty(x)} I_{\lim_{s\to\infty} \xi_s(x) = \pm\infty}$$

are meaningful, and $u_{(+)} + u_{(-)} > 0$.

Let us now recall some facts from the theory of ordinary differential equations. The coefficient of u_{xx} in (5) is a constant, and all the other coefficients are continuous. Hence equation (5) has two linearly independent solutions u_1, u_2 on the whole real line. It easily follows from the maximum principle that given $x < y$, the only solution of (5) on $[x, y]$ which vanishes at x and y is identically zero. This clearly implies that the vectors $(u_1(x), u_2(x))$ and $(u_1(y), u_2(y))$ are not proportional for $x \neq y$. Consequently, for any $n = 1, 2, \dots$ equation (5) (on $(-\infty, \infty)$) has a solution $u^n_{(+)}$ such that $u^n_{(+)}(-n) = 0$, $u^n_{(+)}(n) = 1$. It is not difficult to see that $u^n_{(-)}(x) := u^n_{(+)}(-x)$ also satisfies equation (5) and the conditions $u^n_{(-)}(-n) = 1$, $u^n_{(-)}(n) = 0$.

It is readily proved by Itô's formula that the processes $u^n_{(\pm)}(\xi_s(x)) \exp(-\varphi_s(x))$ are local martingales. Arguing just as before (4), one proves that for $|x| \le n$

$$u^n_{(+)}(x) = \mathbf{E}\, e^{-\varphi^n(x)} I_{\xi^n(x) = n}, \qquad u^n_{(-)}(x) = \mathbf{E}\, e^{-\varphi^n(x)} I_{\xi^n(x) = -n}. \tag{9}$$

Hence, by the Dominated Convergence Theorem, $u^n_{(\pm)} \to u_{(\pm)}$ as $n \to \infty$. Since $u^n_{(\pm)}$ is a linear combination of u_1 and u_2 and $u^n_{(\pm)}(0) \to u_{(\pm)}(0)$ and $u^n_{(\pm)}(1) \to u_{(\pm)}(1)$, we conclude that $u_{(\pm)}$ are solutions of equation (5). In addition, it can be proved similarly to (4) and (9) that for $-n \le x \le y \le n$ and $\tau = \inf\{t \colon \xi_t(x) \notin (-n, y)\}$

$$u^n_{(+)}(x) = \mathbf{E}\, e^{-\varphi_\tau(x)} u^n_{(+)}(\xi_\tau(x)) = u^n_{(+)}(y) \mathbf{E}\, e^{-\varphi_\tau(x)} I_{\xi_\tau(x) = y} \le u^n_{(+)}(y).$$

Thus $u^n_{(+)}$ increases on $[-n, n]$. Hence, $u_{(+)}$ increases on $(-\infty, \infty)$, $u_{(-)}$ decreases, both functions are nonnegative, do not vanish identically ($u_{(+)} + u_{(-)} > 0$), cannot be proportional, and form a fundamental system of solutions of (5). In addition, they are bounded, hence any solution of (5) is bounded. Finally, $u_{(+)} + u_{(-)}$ is a positive solution of (5). Thus we have proved all the assertions we made after equation (5).

3. REMARK. As stated, our assertions about (5) are based essentially on the fact that any solution of the ordinary differential equation $\dot{\eta}_s = \eta_s^3$ goes to infinity in a finite time, and that the behavior of the solutions of the very similar equation (6) is similar. It must be noted, however, that in some cases addition of a diffusion component alters dramatically the behavior of solutions of ordinary equations.

For example, solutions of equation $\dot{\eta}_s = \eta_s$ go to infinity as $s \to \infty$ if $\eta_0 \neq 0$. However, the solutions $\xi_0 \exp(2w_s - s)$ of the equation $d\xi_s = 2\xi_s\,dw_s + \xi_s\,ds$ tend exponentially to zero as $s \to \infty$ (a.s.). It can be shown that solutions of the equation $d\xi_s = (\xi_s^3 + \xi_s)\,dw_s + \xi_s^3\,ds$, unlike those of the equation $\dot{\eta}_s = \eta_s^3$, not only exist for all $s \geq 0$ but also tend to zero as $s \to \infty$ (a.s.).

3. Equations solvable by Euler's method

For given $t \geq 0$, integer n, and a $\mathcal{F}_t$-measurable d-dimensional vector ξ_t^0 denote by $\xi_s^n(t, \xi_t^0)$, $s \geq t$, Euler's approximation process, which is defined by formula (1.20) or (1.21). These formulas are clearly meaningful for any σ, b that satisfy the conditions of the introduction to this chapter. We will show in the present section that Euler's method is suitable for proving the solvability of stochastic equations not only under the assumptions of Theorem 1.1. Later we will see that solvability by Euler's method implies numerous properties of the solutions. This justifies the desire to enlarge the class of equations that can be solved by the method.

1. DEFINITION. We say that equation (0.2) is *solvable by Euler's method* if (σ and b satisfy the conditions of the introduction to this chapter and) for any $t \geq 0$ and any $\mathcal{F}_t$-measurable d-dimensional vector ξ_t^0 equation (0.2) has a solution $\xi_s = \xi_s(t, \xi_t^0)$ for $s \geq t$, which is unique (up to indistinguishability), and, in addition, for any $\mathcal{F}_{t^n}$-measurable vectors $\xi_{t^n}^0 \xrightarrow{P} \xi_t^0$, where $t^n \geq 0$, $n = 1, 2, \ldots$, $t^n \to t$, and any $T \in (0, \infty)$

$$\sup_{s \leq T} \left|\xi_{s \vee t^n}^n(t^n, \xi_{t^n}^0) - \xi_{s \vee t}(t, \xi_t^0)\right| \xrightarrow{P} 0 \quad \text{as } n \to \infty. \tag{1}$$

We say that equation (0.2) is solvable by Euler's method *in the mean* (of arbitrary order) if it is solvable by Euler's method, and for any $p, T \in (0, \infty)$ there exist constants $q, N \in (0, \infty)$ such that for all $n = 1, 2, \ldots$ and (nonrandom) $x \in E_d$, $t \in [0, T]$

$$\mathbf{E} \sup_{t \leq s \leq T} |\xi_s^n(t, x)|^p \leq N(1 + |x|^q). \tag{2}$$

2. REMARK. It follows from Fatou's Lemma and from (1) and (2) that if equation (0.2) is solvable by Euler's method in the mean, then, with the same constants q and N as in (2),

$$\mathbf{E} \sup_{t \leq s \leq T} |\xi_s(t, x)|^p \leq N(1 + |x|^q).$$

Definition 1 is concerned with equation (0.2), which can be regarded for $d \geq 2$ as a system of one-dimensional equations in each of coordinates ξ_s^i. Let us consider a particular case of such a systems for $(d + d_0)$-dimensional process $\zeta_s = (\xi_s, \eta_s)$, where the first d coordinates form a process ξ_s, and the remaining d_0 coordinates a process η_s. We want to find a process ζ_s such that ξ_s satisfies equation (0.2) and η_s the equation

$$\eta_s = \eta_s^0 + \int_t^s \tilde{\sigma}(r, \xi_r, \eta_r)\,dw_r + \int_t^s \tilde{b}(r, \xi_r, \eta_r)\,dr, \qquad s \geq t. \tag{3}$$

Throughout the rest of this section we will assume that $\tilde{\sigma}(r,x,y)$ and $\tilde{b}(r,x,y)$ are functions defined on $\Omega \times (0,\infty) \times E_d \times E_{d_0}$ with values in the set of $(d_0 \times d_1)$-matrices and in E_{d_0}, respectively, continuous in (x,y), of classes H_2 and H_1 respectively for any x, y, and such that for all $\omega \in \Omega$, $T, R \in (0,\infty)$ we have

$$\int_0^T \sup_{|x|\le R, |y| \le R} \left[\|\tilde{\sigma}(r,x,y)\|^2 + |\tilde{b}(r,x,y)| \right] dr < \infty. \tag{4}$$

Briefly, we assume that system (0.2), (3), written as a single equation in ζ_s, satisfies the conditions of the introduction to the chapter. Thus Definition 1 applies to this system.

Suppose, in addition, that $\tilde{\sigma}$ and $\tilde{b}$ satisfy conditions of monotonicity and growth with respect to y, that is, for any $R \in (0,\infty)$ there is a function $K_r(R)$ of class H_1 such that, for all $\omega \in \Omega$, $x \in E_d$, $y, z \in E_{d_0}$, $R_0 \in (0,\infty)$, $|x| \le R_0$, $|y|, |z| \le R$ and $r \in (0,\infty)$,

$$\begin{gathered} 2(y-z, \tilde{b}(r,x,y) - \tilde{b}(r,x,z)) + \|\tilde{\sigma}(r,x,y) - \tilde{\sigma}(r,x,z)\|^2 \le K_r(R \vee R_0)\, |y-z|^2, \\ 2(y, \tilde{b}(r,x,y)) + \|\tilde{\sigma}(r,x,y)\|^2 \le K_r(R_0)\, (1+|y|^2). \end{gathered} \tag{5}$$

In that case equation (3) is solvable for any $\mathcal{F}_t$-measurable initial value η_t^0 and any continuous $\mathcal{F}_s$-adapted process ξ_s defined for $s \ge t$. Indeed, setting $\tilde{\xi}_r = \xi_r I_{t<r}$ and assuming without loss of generality that $K_r(R)$ is an increasing function of R, we obtain from (5) that for $|y|, |z| \le R$,

$$\begin{aligned} 2\Big(y - z, \tilde{b}(r,\tilde{\xi}_r,y) - \tilde{b}(r,\tilde{\xi}_r,z)\Big) &+ \Big\|\tilde{\sigma}(r,\tilde{\xi}_r,y) - \tilde{\sigma}(r,\tilde{\xi}_r,z)\Big\|^2 \\ &\le K_r\Big((1+[|\tilde{\xi}_r|]) \vee R\Big)\, |y-z|^2, \end{aligned} \tag{6}$$

$$2(y, \tilde{b}(r,\tilde{\xi}_r,y)) + \|\tilde{\sigma}(r,\tilde{\xi}_r,y)\|^2 \le K_r\Big(1+[|\tilde{\xi}_r|]\Big)\, (1+|y|^2), \tag{7}$$

where $[a]$ is, as usual, the integer part of a and, as is easily seen, the function $K_r'(R) := K_r\big((1+[|\tilde{\xi}_r|]) \vee R\big)$ is suitably measurable and belongs to H_1. Consequently, the coefficients of equation (3) satisfy the assumptions of Theorem 1.1.

3. REMARK. If the coefficients of equation (0.2) also satisfy the assumptions of Theorem 1.1, this obviously implies the existence of a solution of the *system* (0.2), (3), because we can solve first equation (0.2) and then equation (3). It turns out that in this case the coefficients of the *system* (0.2), (3) may not satisfy the assumptions of Theorem 1.1. An example of such a situation is the system of two one-dimensional equations $d\xi_s = \xi_s\, ds$, $d\eta_s = \xi_s \eta_s\, ds$. If we regard it as one equation in the two-dimensional process (ξ_s, η_s), the left-hand side of (1.2), for example, will be $2(x^1)^2 + 2x^1(x^2)^2$, which is not bounded above by $K(1+|x|^2)$ for all $x = (x^1, x^2) \in E_2$.

4. THEOREM. *If equation* (0.2) *is solvable by Euler's method, then system* (0.2), (3) *is also solvable by Euler's method. If equation* (0.2) *is solvable by Euler's method in the mean and for some constants $K, m \ge 1$ and all $t \ge 0$, $x \in E_d$, $y \in E_{d_0}$, $\omega \in \Omega$,*

$$\|\tilde{\sigma}(t,x,y)\| + |\tilde{b}(t,x,y)| \le K(1+|x|^m + |y|), \tag{8}$$

then system (0.2), (3) *is also solvable by Euler's method in the mean.*

PROOF. The process $\xi_s(t,\xi_t^0)$, as a solution of (0.2), exists by assumption and by Definition 1. The process $\eta_s(t,\xi_t^0,\eta_t^0)$, as a solution of equation (3) with $\xi_s = \xi_s(t,\xi_t^0)$, exists as explained above. Moreover, the solution of system (0.2), (3), that is, $(\xi_s,\eta_s) := (\xi_s(t,\xi_t^0),\eta_s(t,\xi_t^0,\eta_t^0))$, is unique up to indistinguishability (by Definition 1 and Theorem 1.1).

Now let $t^n \geq 0$, $t^n \to t$. Let $\xi_{t^n}^0$ and $\eta_{t^n}^0$ be $\mathcal{F}_{t^n}$-measurable vectors such that $\xi_{t^n}^0 \xrightarrow{P} \xi_t^0$, $\eta_{t^n}^0 \xrightarrow{P} \eta_t^0$, and let $T \in (t+1,\infty)$. Set $\xi_s^n = \xi_s^n(t^n,\xi_{t^n}^0)$ for $s \geq t^n$. We will construct a process $\eta_s^n = \eta_s^n(t^n,\xi_{t^n}^0,\eta_{t^n}^0)$ using Euler's method for *system* (0.2), (3) that is, so that

$$\eta_s^n = \eta_{t^n}^0 + \int_{t^n}^{s} \tilde{\sigma}\big(r,\xi_{\varkappa(n,r)}^n,\eta_{\varkappa(n,r)}^n\big)\,dw_r + \int_{t^n}^{s} \tilde{b}\big(r,\xi_{\varkappa(n,r)}^n,\eta_{\varkappa(n,r)}^n\big)\,dr, \tag{9}$$

where $\varkappa(n,r) = t^n + kn^{-1}$ for $r \in \big(t^n+kn^{-1}, t^n+(k+1)n^{-1}\big]$. For $s \geq t^n$ and $R>0$ denote

$$p_s^n = \xi_{\varkappa(n,s)}^n - \xi_s I_{t<s}, \qquad q_s^n = \eta_{\varkappa(n,s)}^n - \eta_s^n,$$
$$\gamma^n(R) = \inf\{s \geq t^n : |\eta_s^n| \geq R\}.$$

We are given that (1) is true and we want to obtain a similar relationship for η by Theorem 1.5. By this theorem we will prove the first assertion of the present theorem if we can show that, for any $R \in (0,\infty)$,

$$\begin{aligned}&\int_{t^n}^{T\wedge\gamma^n(R)} \sup_{|y|\leq R}\Big\|\tilde{\sigma}\big(r,\xi_r I_{t<r}+p_r^n, y+q_r^n\big) - \tilde{\sigma}(r,\xi_r I_{t<r},y)\Big\|^2 dr \\ &= \int_{t^n}^{T\wedge\gamma^n(R)} \sup_{|y|\leq R}\Big\|\tilde{\sigma}\big(r,\xi_r I_{t<r}+p_r^n I_{t^n<r}, y+q_r^n I_{t^n<r}\big) \\ &\qquad\qquad - \tilde{\sigma}(r,\xi_r I_{t<r},y)\Big\|^2 dr \xrightarrow{P} 0,\end{aligned} \tag{10}$$

and that a similar relationship holds for $\tilde{b}$.

For $R_1 > 0$, define

$$\tau^n(R_1) = \inf\{s \geq t : |\xi_s| \geq R_1\} \wedge \inf\{s \geq t^n : |\xi_s^n| \geq R_1+1\}.$$

It is clear that if $t^n < r \leq \tau^n(R_1)\wedge\gamma^n(R)$, then $|\xi_r^n| \leq R_1+1$, $|\eta_r^n| \leq R$. Using this fact and (4) and arguing as in the case of (1.22) and (1.23), we see that for any $s > t$ the first of the following relationships is true as $n \to \infty$:

$$|q_s^n| I_{t^n<s\leq\tau^n(R_1)\wedge\gamma^n(R)} \xrightarrow{P} 0, \qquad |p_s^n| I_{t^n<s\leq\tau^n(R_1)\wedge\gamma^n(R)} \xrightarrow{P} 0. \tag{11}$$

The second one follows from (1) for $s > t$ by the arguments following (1.24). Formulas (11) are obviously true for $s < t$ as well because $t_n \to t$. In addition, for all s the first expression in (11) is bounded above by $2R$ and the second one by $2R_1+1$.

Consequently, applying Fubini's Theorem to verify condition (III.6.23) of Lemma III.6.19 and applying that lemma to the process $(p_r^n I_{t^n<r}, q_r^n I_{t^n<r})$, the function

$$f(r,p,q) = \sup_{|y|\le R} \left\| \tilde{\sigma}\left(r, \xi_r I_{t<r} + p, y+q\right) - \tilde{\sigma}\left(r, \xi_r I_{t<r}, y\right) \right\|^2,$$

and the random variables $\tau^n(R_1) \wedge \gamma^n(R) \wedge T$, we see that the second relation in (10) is true even when the lower limit of integration is zero, provided that $T \wedge \gamma^n(R) \wedge \tau^n(R_1)$ is substituted for $T \wedge \gamma^n(R)$. This substitution changes the left-hand side of (10) only on the set $\{\tau^n(R_1) \le T\}$, and since

$$\begin{aligned}
\varlimsup_{R_1\to\infty} \varlimsup_{n\to\infty} P\{\tau^n(R_1) \le T\} &\le \lim_{R_1\to\infty} P\{\sup_{[t,T]} |\xi_s| \ge R_1\} \\
&\quad + \lim_{R_1\to\infty} \varlimsup_{n\to\infty} P\Big\{\sup_{s\le T} |\xi_{s\vee t}| < R_1, \sup_{s\le T} |\xi^n_{s\vee t^n}| \ge R_1 + 1\Big\} \\
&\le \lim_{n\to\infty} P\Big\{\sup_{s\le T} |\xi_{s\vee t} - \xi^n_{s\vee t^n}| \ge 1\Big\} = 0,
\end{aligned}$$

it follows that (10) is also true without replacing $T \wedge \gamma^n(R)$. The function $\tilde{b}$ can be treated in the same way. Thus the first assertion of the theorem is proved.

We proceed to the second assertion. It will obviously suffice to prove an inequality similar to (2) for $\eta_s^n(t,x,y)$. Fix $T \in (0,\infty)$, $p \ge 2$, $x \in E_d$, $y \in E_{d_0}$ and for $t \in [0,T]$, $s \in [t,T]$ set

$$\alpha^n = \Big(1 + \sup_{[t,T]} |\xi_r^n(t,x)|^m\Big)^p, \qquad \beta_s^n = \sup_{[t,s]} |\eta_r^n(t,x,y)|^p.$$

To estimate β_s^n, it is natural to take $t^n = t$, $\eta^0_{t^n} = y$, $\xi_s^n = \xi_s^n(t,x)$ in (9), raise the absolute values of both sides of (9) to the pth power and use the inequalities $|a+b+c|^p \le 3^p|a|^p + 3^p|b|^p + 3^p|c|^p$. Now evaluate suprema with respect to s, calculate the expectations, and use inequality (8) and the Burkholder-Davis-Gundy inequalities. We obtain:

$$\mathbf{E}\beta_s^n \le N|y|^p + N\mathbf{E}\Big(\int_t^s \big(\alpha^n + \beta_r^n\big)^{2/p} dr\Big)^{p/2} + N\mathbf{E}\Big(\int_t^s \big(\alpha^n + \beta_r^n\big)^{1/p} dr\Big)^p,$$

where N are constants independent of n, x, y, s, t. In our case $p \ge 2$, and we can apply the Hölder inequality to the inner integrals. Using inequality (2), which holds by assumption, we obtain

$$\mathbf{E}\beta_s^n \le N(1 + |y|^p + |x|^q) + N\int_t^s \mathbf{E}\beta_r^n\, dr, \tag{12}$$

where N, q are constants independent of n, x, y, s, t.

The increasing function $\mathbf{E}\beta_s^n$ is finite on $[t,T]$ for every n. This is readily verified by induction on k, if we rewrite (9) in the form of (1.20). Owing to this fact and the Dominated Convergence Theorem, we can state that $\mathbf{E}\beta_s^n$ is continuous in s. Hence we can apply the Gronwall-Bellman inequality to (12), whence, using also the inequality

$$|y|^p + |x|^q \le N(p,q)\big(1 + (|x| + |y|)^{p+q}\big),$$

we obtain the desired assertion about η^n. It remains to note that for $p \in (0,2]$ inequalities of type (2) can be derived from the case $p = 2$ by using, say, the Hölder inequality or the inequality $|\eta|^p \le N(p) + |\eta|^2$. □

5. REMARK. As stated previously, system (0.2), (3) is a particular case of system (0.2). On the other hand, the opposite obviously is also true. If, for example, $\sigma \equiv 0$, $b \equiv 0$ and $\tilde{\sigma}$ and $\tilde{b}$ are independent of ξ, then Theorem 4 implies a somewhat stronger assertion than Theorem 1.1. Namely, under the assumptions of Theorem 1.1, equation (0.2) is solvable by Euler's method in the sense of Definition 1. Of course, if we are interested in this assertion only, we can obtain it far more easily by following the proof of Theorem 1.1 rather than first proving Theorem 4.

Note also that according to Theorem 4, solvability by Euler's method is "inherited" if we add equation (3) to (0.2). It allows us to add to system (0.2), (3) yet another equation whose coefficients depend on ξ, η as parameters, possess suitable properties of measurability and continuity, and satisfy conditions similar to (5).

4. Some properties of Euler's approximations

In the rest of this book we will study stochastic equations (0.2) with σ and b *independent* of ω. In that case Euler's approximation possess certain properties that will be established in this section and needed later. Fix an integer n and recall the notation $\xi_s^n(t, \xi_t^0)$ for Euler's approximations introduced at the beginning of Sec. V.3. Let $a = \sigma\sigma^*$.

1. LEMMA. (a) *If $t + 1/n \ge s \ge t \ge 0$ and $x \in E_d$, then*

$$\xi_s^n(t,x) \sim \mathcal{N}\Big(x + \int_t^s b(r,x)\,dr, \int_t^s a(r,x)\,dr\Big).$$

(b) *If $G \subset \mathcal{F}_t$ is a σ-algebra and ξ_t^0 a random vector such that ξ_t^0 and G are independent, then the random vector $\xi_s^n(t, \xi_t^0)$ and G are independent for any $s \ge t$. In particular (for $G = \mathcal{F}_t$, $\xi_t^0 \equiv x$), $\xi_s^n(t,x)$ and $\mathcal{F}_t$ are independent for any (nonrandom) $x \in E_d$.*

(c) *If $t^i \ge 0$, $i = 1, 2, \ldots$, $t^i \to t$ and $\xi_{t^i}^0 \xrightarrow{P} \xi_t^0$ as $i \to \infty$, where $\xi_{t^i}^0$ are $\mathcal{F}_{t^i}$-measurable vectors, then for any $T \in (0,\infty)$*

$$\sup_{s \le T} \left|\xi_{s\vee t^i}^n(t^i, \xi_{t^i}^0) - \xi_{s\vee t}^n(t, \xi_t^0)\right| \xrightarrow{P} 0 \quad \text{as } i \to \infty. \tag{1}$$

(d) *Let g be a continuous bounded function defined on E_d. For $s \ge t \ge 0$, $x \in E_d$, set*

$$v^n(t,s,x) = \mathbf{E}\, g\big(\xi_s^n(t,x)\big). \tag{2}$$

Then v^n is bounded and continuous in (t,s,x) and for any $k = 0, 1, 2, \ldots$, $u = t + k/n \le s$, and $\mathcal{F}_u$-measurable bounded variables η we have

$$\mathbf{E}\,\eta g\big(\xi_s^n(t,x)\big) = \mathbf{E}\,\eta v^n\big(u, s, \xi_u^n(t,x)\big). \tag{3}$$

In particular, for $\eta \equiv 1$

$$v^n(t,s,x) = \mathbf{E}\, v^n\big(u, s, \xi_u^n(t,x)\big). \tag{4}$$

PROOF. (a) Fix $\lambda \in E_d$, $t \in [0, \infty)$ and for $s \geq t$ set

$$\varphi_s(x) = \exp\left\{i\left[(\lambda, x) + \int_t^s (\lambda, b(r, x))\,dr\right] - \frac{1}{2}\int_t^s \big(a(r, x)\lambda, \lambda\big)\,dr\right\},$$

$$\rho_s(x) = \varphi_s^{-1} \exp i\big(\lambda, \xi_s^n(t, x)\big).$$

By Itô's formula the bounded process $\rho_{(s\vee t)\wedge t(1,n)}(x)$ is a martingale. Hence, by Lemma II.8.5 (a) for any $A \in \mathcal{F}_t$ and for $t + 1/n \geq s \geq t$, $x \in E_d$

$$\begin{gathered}\mathbf{E}\, I_A \rho_s(x) = \mathbf{E}\, I_A \rho_t(x) = P(A), \qquad \mathbf{E}\,\rho_s(x) = 1, \\ \mathbf{E}\, I_A \exp i\big(\lambda, \xi_s^n(t, x)\big) = P(A)\varphi_s(x) = P(A)\mathbf{E} \exp i\big(\lambda, \xi_s^n(t, x)\big).\end{gathered} \tag{5}$$

The first equality in (5) with $A = \Omega$ proves (a).

(b) Note that if $s \geq t(k, n)$ then, by (1.20),

$$\xi_s^n(t, \xi_t^0) = \xi_s^n\big(t(k, n), \xi_{t(k,n)}^n(t, \xi_t^0)\big) \quad \text{(a.s.)}. \tag{6}$$

Keeping in mind the obvious possibility of induction on k, we conclude that it suffices to consider the case in which $t + 1/n \geq s \geq t$.

Denote $\varkappa_j(x) = \big(\varkappa_j(x^1), \dots, \varkappa_j(x^d)\big)$, where $j = 1, 2, \dots$, $\varkappa_j(p) = 2^{-j}[2^j p]$ for $p \in E_1$. Let Γ_j be the (countable) set of all values of $\varkappa_j(x)$ for $x \in E_d$. Using condition (0.1), from Theorem III.6.6 and the Dominated Convergence Theorem we easily deduce that $\xi_s^n(t, \xi_t^{0j}) \to \xi_s^n(t, \xi_t^0)$ in probability as $j \to \infty$, where $\xi_t^{0j} = \varkappa_j(\xi_t^0)$. In addition, repeating almost literally the arguments following (IV.1.5) we obtain

$$\xi_s^n(t, \xi_t^{0j}) = \sum_{x \in \Gamma_j} I_{\xi_t^{0j} = x}\, \xi_s^n(t, x) \quad \text{(a.s.)}. \tag{7}$$

Finally, by assumption, for $A \in G$,

$$P(A, \xi_t^{0j} = x) = P(A)P(\xi_t^{0j} = x).$$

Reasoning from this and from (5) by Lemma III.6.13 (f) (also using Corollary I.2.8 and the formulas $z = a + ib$, $a = a_+ - a_-$), we obtain

$$\begin{aligned}\mathbf{E}\, I_A \exp i\big(\lambda, \xi_s^n(t, \xi_t^0)\big) &= \lim_{j\to\infty} \mathbf{E}\, I_A \exp i\big(\lambda, \xi_s^n(t, \xi_t^{0j})\big) \\ &= \lim_{j\to\infty} \sum_{x \in \Gamma_j} \mathbf{E}\, I_{A, \xi_t^{0j} = x} \exp i\big(\lambda, \xi_s^n(t, x)\big) \\ &= P(A) \lim_{j\to\infty} \sum_{x \in \Gamma_j} P(\xi_t^{0j} = x)\mathbf{E} \exp i\big(\lambda, \xi_s^n(t, x)\big) \\ &= P(A)\mathbf{E} \exp i\big(\lambda, \xi_s^n(t, \xi_t^0)\big).\end{aligned} \tag{8}$$

Part (b) follows from the equality of the first and last terms, just like the analogous assertion in the proof of Lévy's Theorem III.9.6.

(c) For $k = 0, 1, 2, \dots$, $t \geq 0$ (t, k are parameters), define

$$u(t, k, s) = \begin{cases} t + k/n & \text{for } \; s \leq t + k/n, \\ s & \text{for } \; t + k/n \leq s \leq t + (k+1)/n, \\ t + (k+1)/n & \text{for } \; t + (k+1)/n \leq s. \end{cases}$$

By induction on k, we will first prove that as $i \to \infty$

(9) $$\sup_s \left|\xi^n_{u(t^i,k,s)}(t^i, \xi^0_{t^i}) - \xi^n_{u(t,k,s)}(t, \xi^0_t)\right| \xrightarrow{P} 0.$$

Observe that $u(t,k,s) = u(t+k/n, 0, s)$, and by (6)

$$\xi^n_{u(t^i,k,s)}(t^i, \xi^0_{t^i}) = \xi_{u(t^i+k/n,0,s)}(t^i + k/n, \xi^n_{t^i+k/n}(t^i, \xi^0_{t^i})) \quad \text{(a.s.)}.$$

Moreover, having (9) for $k-1$ instead of k, we see that (for $s \geq t + 1 + k/n$)

$$\xi^n_{t^i+k/n}(t^i, \xi^0_{t^i}) \xrightarrow{P} \xi^n_{t+k/n}(t, \xi^0_t).$$

Consequently, upon passing from $k-1$ to k, we have the same situation as for $k = 0$, except that t^i, t, $\xi^0_{t^i}$, ξ^0_t must be suitably replaced. Thus, to prove (9) it will suffice to investigate only one case: $k = 0$.

It is not difficult to prove using the Dominated Convergence Theorem and part (a) of Lemma III.6.13, that

$$\int_0^\infty \left\| I_{t^i<r\leq t^i+1/n}\sigma(r, \xi^0_{t^i}) - I_{t<r\leq t+1/n}\sigma(r, \xi^0_t)\right\|^2 dr \xrightarrow{P} 0$$

as $i \to \infty$. Proving a similar formula for b and noting that

$$\xi^n_{u(t^i,0,s)}(t^i, \xi^0_{t^i}) = \xi^0_{t^i} + \int_0^s I_{t^i<r\leq t^i+1/n}\sigma(r, \xi^0_{t^i})\, dw_r$$
$$+ \int_0^s I_{t^i<r\leq t^i+1/n} b(r, \xi^0_{t^i})\, dr \quad \text{(a.s.)},$$

we obtain (9) for $k = 0$ by Theorem III.6.6 (and Lemma III.6.13).

The arguments following (1.24) now show that (9) remains valid upon substitution of $t \vee u(t^i,k,s)$ for $u(t,k,s)$. To prove (1) it now remains only to observe that, if $t^i \leq t$, then the left-hand side of (1) is obviously bounded above by the sum over $k = 0, 1, \dots, [(T-t+1)n] + 1$ of the expressions on the left of (9), modified as indicated. If $t^i > t$, then we must add to this sum the expression

$$\sup_{t\leq s\leq t^i} |\xi^0_{t^i} - \xi^n_s(t, \xi^0_t)| \leq |\xi^0_{t^i} - \xi^0_t| + \sup_{t\leq s\leq t^i} |\xi^n_s(t, \xi^0_t) - \xi^0_t|$$

which tends to zero in probability as $i \to \infty$.

(d) That $v^n(t,s,x)$ is bounded is obvious; its continuity follows immediately from (c) and Lemma III.6.13. If

$$\zeta^j = \varkappa_j(\xi^n_u(t,x)),$$

then $\zeta^j \to \xi^n_u(t,x)$ as $j \to \infty$, and by (c) and (6),

$$\xi^n_s(u, \zeta^j) \xrightarrow{P} \xi^n_s(u, \xi^n_u(t,x)) = \xi^n_s(t,x) \quad \text{(a.s.)}.$$

We will use also the fact that, by analogy with (7),

$$\xi^n_s(u, \zeta^j) = \sum_{y\in\Gamma_j} I_{\zeta^j=y}\xi^n_s(u,y) \quad \text{(a.s.)}.$$

Finally, remembering that v is continuous, ζ^j is $\mathcal{F}_u$-measurable and $\xi_s^n(u,y)$ and $\mathcal{F}_u$ are independent, we obtain for $g \geq 0$, $\eta \geq 0$,

$$\mathbf{E}\eta g\left(\xi_s^n(t,x)\right) = \lim_{j\to\infty} \mathbf{E}\eta g\left(\xi_s^n(u,\zeta^j)\right) = \lim_{j\to\infty} \sum_{\Gamma_j} \mathbf{E}\eta I_{\zeta^j=y} g\left(\xi_s^n(u,y)\right)$$

$$= \lim_{j\to\infty} \sum_{\Gamma_j} \mathbf{E}\eta I_{\zeta^j=y} v^n(u,s,y) = \lim_{j\to\infty} \mathbf{E}\eta v^n(u,s,\zeta^j) = \mathbf{E}\eta v^n\left(u,s,\xi_u^n(t,x)\right).$$

Formula (3) can be generalized to η, g of arbitrary sign in an obvious way. □

In the proof of (8) we assumed that $t + 1/n \geq s \geq t$. Nevertheless, by *part* (b) (and Exercise I.4.22) the first and last terms in (8) are also equal for all $s \geq t$. Letting $n \to \infty$ there and using Definition 3.1, we obtain

2. COROLLARY. *If equation* (0.2) *is solvable by Euler's method* (*and* σ, b *are independent of* ω), *then for any* $s \geq t \geq 0$ (*independent of* ω), $x \in E_d$ *the random vector* $\xi_s(t,x)$ *and* σ*-algebra* $\mathcal{F}_t$ *are independent.*

3. LEMMA. (a) *For every* $n = 1,2,\ldots$, $s \geq t \geq 0$, $x \in E_d$, *the distribution of* $\xi_s^n(t,x)$ *is uniquely determined by* n, $s-t$, x *and the functions* $a(t+\cdot,\cdot)$ *and* $b(t+\cdot,\cdot)$, *where* $a = \sigma\sigma^*$.

(b) *Moreover, for every* $n = 1,2,\ldots$, $t \geq 0$, $x \in E_d$, *the distribution of* $\xi_{t+\cdot}^n(t,x)$ *in* $(C,\mathfrak{A}(C))$ *is uniquely determined by* n, x, *and the functions* $a(t+\cdot,\cdot)$, $b(t+\cdot,\cdot)$. *It is unaffected if we change the probability space, the Wiener process* (*together with its dimension*) *or the function* σ, *provided that the latter satisfies the conditions of the introduction to this chapter and yields the same matrix* a. *Finally, if* a *and* b *are independent of time, then the distribution of* $\xi_{t+\cdot}^n(t,x)$ *in* $(C,\mathfrak{A}(C))$ *is also independent of* t.

PROOF. It obviously follows from Lemma 1(a) that for $t+1/n \geq s \geq t$ the function (2) is uniquely determined by g, x, $s-t$, and the functions $a_t(r,y) := a(t+r,y)$, $b_t(r,y) := b(t+r,y)$ as functions in (r,y). This may be written as follows:

$$v^n(t,s,x) = F^n\left(g(\cdot), s-t, x, a_t, b_t\right), \tag{10}$$

where, as above, $t+1/n \geq s \geq t$ and F^n is in fact independent of n. Hence, by (4) with $u = t+1/n$, $t+2/n \geq s \geq t+1/n$,

$$v^n(t,s,x) = F\Big(F\left(g, s-t-1/n, \cdot, a_{t+1/n}, b_{t+1/n}\right), 1/n, x, a_t, b_t\Big).$$

This can be also written as (10), but with F depending on n and defined for $t+2/n \geq s \geq t$. Continuing in the obvious way, we obtain a representation (10) for all $s \geq t$. Applying Theorem I.5.4 and considering different stochastic equations (0.2) with coefficients independent of ω and the same functions a_t, b_t, we see that the distributions of $\xi_s^n(t,x)$ coincide. This proves (a).

To prove (b), let $\lambda(r)$ be a bounded Borel function with values in E_d defined on $(0,\infty)$, and for $s \geq t \geq 0$, $x \in E_d$, $y \in E_1$, define

$$\eta_s^n(t,x,y) = y + \int_t^s \lambda^*(r)\sigma\left(r, \xi_{\varkappa(n,r)}^n(t,x)\right) dw_r + \int_t^s \lambda^*(r) b\left(r, \xi_{\varkappa(n,r)}^n(t,x)\right) dr,$$

where $\varkappa(n,r)$ is defined as in 1.4. It is clear that the pair $\left(\xi_s^n(t,x), \eta_s^n(t,x)\right)$ is the Euler approximation process for the suitably extended system (0.2), where the functions a, b for the new system can be expressed (by elementary means) in terms of a, b, λ. By (10),

$$\mathrm{E} \exp i\eta_s^n(t,x,0) = \Phi^n\left(s-t, x, a_t, b_t, \lambda_t\right),$$

where $\lambda_t(r) := \lambda(t+r)$. Taking arbitrary $\lambda_1, \dots, \lambda_{j-1} \in E_d$ and numbers $s(1), \dots, s(j)$ such that $0 = s(1) < s(2) < \dots < s(j)$, putting $\lambda(r) = \lambda_k$ for $r \in \left(t+s(k), t+s(k+1)\right]$, $k = 1, \dots, j-1$, $\lambda(r) = 0$ for $r > s(j)$, and $s = t + s(j)$, we have

$$\eta_s^n(t,x,0) = \sum_{k=1}^{j-1} \lambda_k^* \left[\xi_{t+s(k+1)}^n(t,x) - \xi_{t+s(k)}^n(t,x)\right] \quad \text{(a.s.)}, \tag{11}$$

$$\mathrm{E} \exp i\eta_s^n(t,x,0) = \Psi^n\left(x, a_t, b_t, \lambda_1, \dots, \lambda_{j-1}, s(1), \dots, s(j)\right). \tag{12}$$

It follows by Theorem I.4.12 that the distribution of the random (column) vector

$$\left(\xi_{t+s(2)}^{n*}(t,x) - \xi_{t+s(1)}^{n*}(t,x), \dots, \xi_{t+s(j)}^{n*}(t,x) - \xi_{t+s(j-1)}^{n*}(t,x)\right)^*$$

is uniquely determined by $x, a_t, b_t, s(1), \dots, s(j), n$. Taking into account that $s(1) = 0$, $\xi_t^n(t,x) = x$ (a.s.), we can find an obvious affine transformation that will map this vector (a.s.) into the vector

$$\left(\xi_{t+s(1)}^{n*}(t,x), \dots, \xi_{t+s(j)}^{n*}(t,x)\right)^*,$$

whose distribution, therefore, also depends only on $x, a_t, b_t, s(1), \dots, s(j), n$.

We have thus proved the following assertion. Let $B_1, \dots, B_j \in \mathfrak{B}(E_d)$ be arbitrary sets, and define $A = \left\{x_\cdot \in C : x_{s(1)} \in B_1, \dots, x_{s(j)} \in B_j\right\}$. (The set A thus defined uniquely determines $B_1, \dots, B_j$ and $s(1), \dots, s(j)$ since $s(1) < \dots < s(j)$.) Then

$$P\left(\xi_{t+\cdot}^n(t,x)\right)^{-1}(A) = P\left\{\xi_{t+\cdot}^n(t,x) \in A\right\}$$

depends only on A and on x, a_t, b_t, n. Applying the lemma on π- and λ-systems, we can generalize this result to all $A \in \mathfrak{A}(C)$, proving (b). □

Recalling Definition 3.1, letting $n \to \infty$ in (11) and (12) and repeating the final part of the proof of the lemma, we obtain

4. COROLLARY. *If equation* (0.2) *is solvable by Euler's method* (*and* σ, b *are independent of* ω), *then for any* $t \geq 0$, $x \in E_d$ *the distribution of* $\xi_{t+\cdot}(t,x)$ *in* $\left(C, \mathfrak{A}(C)\right)$ *is uniquely determined by* x *and the functions* $a(t+\cdot,\cdot)$, $b(t+\cdot,\cdot)$. *If, in addition,* a, b *depend only on the spatial variable, this distribution is independent of* t.

In the next lemma we use the notation $u_{(y)}(x)$ introduced in (I.4.15).

5. LEMMA. *Let* σ, b *be continuously differentiable with respect to* x *for any* $r > 0$. *Assume that*

$$\int_0^T \sup_{|x| \leq R} \left[\left\|\sigma_{x^i}(r,x)\right\|^2 + \left|b_{x^i}(r,x)\right|\right] dr < \infty \tag{13}$$

for all $T, R \in (0,\infty)$, $i = 1,\dots,d$. Let g be a given continuously differentiable function on E_d such that

$$|g(x)| + |g_x(x)| \le K(1+|x|^m) \tag{14}$$

on E_d, where $m, K \in (0,\infty)$ are constants. Then for any n, the function v^n defined by (2) is differentiable with respect to x, continuously in (t,s,x) on the set $0 \le t \le s \le t+1/n$, $x \in E_d$, and for any $x, y \in E_d$, $0 \le t \le s \le t+1/n$,

$$v^n_{(y)}(t,s,x) = \mathbf{E}\, g_{(\eta_s)}(\xi^n_s(t,x)), \tag{15}$$

where

$$\eta_s = y + \int_t^s \sigma_{(y)}(r,x)\,dw_r + \int_t^s b_{(y)}(r,x)\,dr. \tag{16}$$

PROOF. First note that definition (2) is meaningful by (14), Lemma 1(a), and Exercise I.4.13. For the same reasons, the right-hand side of (15) is also meaningful and finite.

Now, if $u \in (0,1]$, then by Lagrange's Theorem

$$\frac{1}{u}\Big[v^n(t,s,x+uy) - v^n(t,s,x)\Big] = \mathbf{E}\, g_{(\eta_s(u))}(\tilde{\xi}_s(u)),$$

where $\tilde{\xi}_s(u)$ is a point depending on ω (possibly in a nonmeasurable way) on the interval between $\xi^n_s(t,x)$ and $\xi^n_s(t,x+uy)$,

$$\eta_s(u) = \frac{1}{u}\Big[\xi^n_s(t,x+uy) - \xi^n_s(t,x)\Big].$$

By condition (13), Theorem III.6.6 and the Dominated Convergence Theorem $\eta_s(u) \xrightarrow{P} \eta_s$ as $u \downarrow 0$. Of course, $\xi^n_s(t,x+uy) \xrightarrow{P} \xi^n_s(t,x)$ as well. Hence, by Lemma III.6.13(a),

$$g_{(\eta_s(u))}(\tilde{\xi}_s(u)) \xrightarrow{P} g_{(\eta_s)}(\xi^n_s(t,x)). \tag{17}$$

In addition, the left-hand side of (17) is bounded above by

$$K|\eta_s(u)|\Big(1+|\tilde{\xi}_s(u)|^m\Big) \le 2^m K|\eta_s(u)|\Big(1+|\xi^n_s(t,x+uy)|^m + |\xi^n_s(t,x)|^m\Big),$$

and, as follows from the Cauchy-Bunyakovsky inequality and Exercise I.4.13, the expectation of the last expression is bounded above for $u \leq 1$ by

(18)

$$N\Big(\mathbf{E}\left|\eta_s(u)\right|^2\Big)^{1/2}\Big(\mathbf{E}\Big(1+\left|\xi_s^n(t,x+uy)\right|^{2m}+\left|\xi_s^n(t,x)\right|^{2m}\Big)\Big)^{1/2}$$

$$\leq N\Bigg[|y|^2+\int_t^s \sup_{u\leq 1}\left\|\sigma_{(y)}(r,x+uy)\right\|^2 dr+\Bigg(\int_t^s \sup_{u\leq 1}\left|b_{(y)}(r,x+uy)\right| dr\Bigg)^2\Bigg]^{1/2}$$

$$\times\Bigg[1+|x|^m+|y|^m+\sup_{u\leq 1}\Bigg(\int_t^s \left\|\sigma(r,x+uy)\right\|^2 dr\Bigg)^{m/2}$$

$$+\sup_{u\leq 1}\Bigg(\int_t^s \left|b(r,x+uy)\right| dr\Bigg)^m\Bigg],$$

where N are constants independent of u.

The right-hand side of (18) is finite, and by Lemma III.6.13(f), the convergence in (17) holds in $\mathcal{L}_1(\mathcal{F},P)$. Hence

$$\frac{1}{u}\Big[v^n(t,s,x+uy)-v^n(t,s,x)\Big]\longrightarrow \mathbf{E}\,g_{(\eta_s)}(\xi_s^n(t,x)).$$

If we put $y=\pm e_i$ in this relation, where e_i are basis vectors in E_d, and note that η_s is (a.s.) a linear function of y, we see that the partial derivatives of v with respect to x exist and formula (15) is true. The continuity of its right-hand side with respect to t, s, x is easily proved by arguments similar to those used for (17) and (18). □

6. REMARK. It is sometimes important that the representation of $v_{(y)}^n(t,s,x)$ as (15), where η_s is given by (16), is by no means unique. For example, under the assumptions of Lemma 5,

$$v_{(y)}^n(t,s,x)=\mathbf{E}\left[g_{(\eta_s)}(\xi_s^n(t,x))-g(\xi_s^n(t,x))(\pi,w_s-w_t)\right]$$

for any (nonrandom) vector $\pi\in E_{d_1}$, where

$$\eta_s=y+\int_t^s \sigma_{(y)}(r,x)\,dw_r+\int_t^s \left[b_{(y)}(r,x)+\sigma(r,x)\pi\right]dr.$$

To see this, we need only put $\eta=w_s-w_t$, $\xi=\xi_s^n(t,x)-\mathbf{E}\,\xi_s^n(t,x)$, $z=\mathbf{E}\,\xi_s^n(t,x)$, note that

$$\sigma:=\mathbf{E}\,\xi\eta^*=\int_t^s \sigma(r,x)\,dr,\qquad \sigma\pi=\int_t^s \sigma(r,x)\pi\,dr,$$

and use Problem I.4.26, by which

$$\mathbf{E}\,g_{(\sigma\pi)}(z+\xi)=\mathbf{E}\,g(z+\xi)(\pi,\eta).$$

5. Strong Markov property of solutions of stochastic equations

In addition to the general assumptions of the whole chapter, we will now assume that σ, b are *independent of* ω and that equation (0.2) is solvable by Euler's method in the sense of Definition 3.1. As we know from Remark 3.5, that is the case if, say, the assumptions of Theorem 1.1 hold.

Let $\xi_s(t,x) = \xi_s(\omega,t,x)$ denote the solution of equation (0.2) with initial value $\xi_t^0 = x$ for nonrandom $x \in E_d$, $s \geq t \geq 0$. Clearly, if $0 \leq t \leq u \leq s$, then $\xi_s = \xi_s(t,x)$ satisfies the relation

$$\xi_s = \left(x + \int_t^u \sigma(r,\xi_r)\,dw_r + \int_t^u b(r,\xi_r)\,dr\right) + \int_u^s \sigma(r,\xi_r)\,dw_r + \int_u^s b(r,\xi_r)\,dr$$

$$= \xi_u + \int_u^s \sigma(r,\xi_r)\,dw_r + \int_u^s b(r,\xi_r)\,dr \quad \text{(a.s.)}.$$

Together with (0.2) this makes the following equality natural:

$$(1) \qquad \xi_s(\omega,t,x) = \xi_s\big(\omega,u,\xi_u(\omega,t,x)\big).$$

This equality states the *evolutionary property* of solutions of stochastic equations. Without going into a rigorous discussion, we will only explain how this yields the Markov property of $\xi_s(t,x)$.

It is clear from the structure of equation (0.2) that for fixed u, y the random vector $\xi_s(u,y)$ is determined, in a sense, only by the increments of w after time u and up to time s. Therefore, for any function g the equality

$$(2) \qquad g\big(\xi_s(t,x)\big) = g\Big(\xi_s\big(u,\xi_u(t,x)\big)\Big)$$

represents the left-hand side as a composite function $g\big(\xi_s(u,y)\big)$, which is determined by $\theta_u w.$ and where y is replaced by the $\mathcal{F}_u$-measurable function $\xi_u(t,x)$. By the rule of stepwise calculation of expectations (see Theorem II.5.5), we can evaluate the expectation of the left-hand side of (2) by calculating the expectation of $g\big(\xi_s(u,y)\big)$, substituting $y = \xi_u(t,x)$ and then calculating the expectation of the result. To be precise: for any nonnegative Borel function g, for $t \leq u \leq s$,

$$(3) \qquad v(t,s,x) := \mathbf{E}\,g\big(\xi_s(t,x)\big) = \mathbf{E}\,v\big(u,s,\xi_u(t,x)\big).$$

This equality can be rewritten in a more suggestive form by defining an operator $T_{u,s}$ that associates to any function $g(y)$ the function of y equal to $T_{u,s}g(y) = \mathbf{E}\,g\big(\xi_s(u,y)\big)$. Then $T_{t,s} = T_{t,u}T_{u,s}$.

After this somewhat informal discussion, we will proceed to the rigorous treatment. We will use the notation introduced before Theorem II.5.5 and for every $u \geq 0$, we let $C^u = C^u(E_d)$ denote the set of all continuous functions x_t defined at least for $t \in [u,\infty)$ with values in E_d. The elements of C^u will be denoted by $x.$, $y.$ and so on, the value of a function $x. \in C^u$ at $t \geq u$, as usual, by x_t. For every $u \geq 0$ we define a mapping $\hat{\theta}_u\colon C^u \to C$ by $\hat{\theta}_u(x.) =: \hat{\theta}_u x.$, $(\hat{\theta}_u x.)(t) =: \hat{\theta}_u x_t := x_{u+t}$, $t \geq 0$.

In other words, $\hat{\theta}_u$ simply shifts the graph x_t a distance u toward the point $t = 0$, parallel to the t-axis.

1. THEOREM. *As above, let $t \geq 0$, $x \in E_d$. Suppose that τ is a Markov time with respect to $\{\mathcal{F}_s\}$, $\tau \geq t$ and let $\widetilde{F}(\omega, u, x.)$ be a nonnegative $\mathcal{F}_\tau \otimes \mathfrak{B}([0,\infty)) \otimes \mathfrak{A}(C)$-measurable function on $\Omega \times [0,\infty) \times C$. For $\omega \in \Omega$, $u \geq 0$, $x. \in C^u$, define*

$$F(\omega, u, x.) = \widetilde{F}(\omega, u, \hat{\theta}_u x.). \tag{4}$$

Then

$$\mathbf{E}\, I_{\tau<\infty} F\big(\tau, \xi.(t,x)\big) = \mathbf{E}\, I_{\tau<\infty} \Phi\big(\tau, \xi_\tau(t,x)\big), \tag{5}$$

where the $\mathcal{F}_\tau \otimes \mathfrak{B}([0,\infty)) \otimes \mathfrak{B}(E_d)$-measurable function $\Phi(\omega, u, y)$ is defined by

$$\Phi(\omega, u, y) = \int_\Omega F\big(\omega, u, \xi.(\omega', u, y)\big)\, P(d\omega'). \tag{6}$$

2. REMARK. It is not difficult to see that for $d = d_1$, $\sigma^{ij} = \delta^{ij}$, $b \equiv 0$ this theorem, which establishes the strong Markov property, contains the main assertion of Theorem II.5.5. Note also that formulas (5) and (6) involve only F, and $\widetilde{F}$ was needed solely to describe the measurability requirements. Finally, it is useful to keep in mind that formula (5) is consistent with our previous discussion of the intuitively legitimate formula (1). Indeed, by (1),

$$F\big(\tau, \xi.(t,x)\big) = F\Big(\tau, \xi.\big(\tau, \xi_\tau(t,x)\big)\Big)$$

for $\tau < \infty$.

To prove the theorem we will need two lemmas.

3. LEMMA. *If $T \in (0,\infty)$, $t^n \in [0,T]$, $x^n \in E_d$, $n = 1,2,\dots$, $t^n \to t$, $x^n \to x$, then*

$$\sup_{s\leq T} \big|\xi_{s\vee t^n}(t^n, x^n) - \xi_{s\vee t}(t,x)\big| \xrightarrow{P} 0. \tag{7}$$

In particular, $\xi_s(t,x)$ possesses the Feller property, *that is, for any continuous function g, bounded on E_d, the function $T_{t,s}g(x) := \mathbf{E}\, g\big(\xi_s(t,x)\big)$ is bounded and continuous in (t,s,x) for $s \geq t \geq 0$, $x \in E_d$. In addition, for the same function g*

$$\mathbf{E}\, g\big(\xi^n_s(t,x)\big) \longrightarrow \mathbf{E}\, g\big(\xi_s(t,x)\big)$$

as $n \to \infty$ uniformly in (t,s,x) on any bounded set of arguments $s \geq t \geq 0$, $x \in E_d$, where $\xi^n_s(t,x)$ are the Euler approximations, defined by (1.21) for $\xi^0_t \equiv x$.

PROOF. Take $s^n \geq t^n \geq 0$, $x^n \in E_d$, $n = 1,2,\dots$, so that $s^n \to s$, $t^n \to t$, $x^n \to x$ as $n \to \infty$. Then by Definition 3.1 and Lemma III.6.13

$$\xi^n_{s^n}(t^n, x^n) \xrightarrow{P} \xi_s(t,x), \qquad \mathbf{E}\, g\big(\xi^n_{s^n}(t^n,x^n)\big) \longrightarrow \mathbf{E}\, g\big(\xi_s(t,x)\big).$$

This obviously implies the last assertion of the lemma. The second assertion follows from the first one and from Lemma III.6.13; alternatively, it follows from the third assertion by Lemma 4.1(d). To prove the first assertion, take $k(n)$ such that $k(n) \geq n$ and

$$P\Big\{\sup_{s\leq T} \big|\xi^{k(n)}_{s\vee t^n}(t^n,x^n) - \xi_{s\vee t^n}(t^n,x^n)\big| \geq \frac{1}{n}\Big\} \leq \frac{1}{n}. \tag{8}$$

This is possible by our assumptions and by Definition 3.1. Since $k(n) \geq n$, it follows from Definition 3.1 that

(9) $$\sup_{s \leq T} \left| \xi^{k(n)}_{s \vee t^n}(t^n, x^n) - \xi_{s \vee t}(t, x) \right| \xrightarrow{P} 0.$$

It remains to note that, by (8), the supremum appearing on the left of (8) tends to zero in probability and the expression obtained by adding it to the left-hand side of (9) is greater than the left-hand side of (7). □

4. Lemma. *Let g be a nonnegative Borel function on E_d. Then $v(t, s, x)$, as defined in (3), is a Borel function of (t, s, x) for $s \geq t \geq 0$, $x \in E_d$. Let $\eta \geq 0$ be a $\mathcal{F}_u$-measurable function for $s \geq u \geq t$. Then*

(10) $$\mathbf{E} \eta g(\xi_s(t, x)) = \mathbf{E} \eta v(u, s, \xi_u(t, x)).$$

In particular (take $\eta \equiv 1$), equality (3) is true.

Proof. Standard arguments, as described, for example, in the proof of Theorem I.5.4 and of Fubini's Theorem, show that it suffices to consider bounded continuous g and bounded η. Actually, we will need this lemma only for such g and η. In that case v is even continuous in (t, s, x). Moreover, since (10) becomes obvious for $u = s$, we will assume that $u < s$. In addition, by Lemma 4.1(d), we have equality (4.3). Since $v^n \to v$ uniformly on every bounded set of arguments, and $\xi^n_s(t, x) \xrightarrow{P} \xi_s(t, x)$, $\xi^n_{u(n)}(t, x) \xrightarrow{P} \xi_u(t, x)$, where $u(n) = t + n^{-1}[n(u - t)] + n^{-1} \geq u$, we obtain (10), by letting $n \to \infty$ in (4.3) and using the Dominated Convergence Theorem and Lemma III.6.13(a). □

5. Proof of Theorem 1. As in the proof of Theorem II.5.5, in order to prove (5) and, in particular, to prove that it is meaningful, we need only consider the case in which

$$\widetilde{F}(\omega, u, x.) = I_A(\omega) h(u) g_1(x_{t^1}) \cdot \ldots \cdot g_m(x_{t^m}),$$

where $A \in \mathcal{F}_\tau$, the functions h, g_i are bounded nonnegative and continuous, and $0 \leq t^1 \leq \cdots \leq t^m$. Define τ_n as in that proof.

Omitting the arguments (t, x) of $\xi.(t, x)$, we see that

$$\begin{aligned} I_{\tau<\infty} \widetilde{F}(\tau, \hat{\theta}_\tau \xi.) &= I_{A,\tau<\infty} h(\tau) g_1(\xi_{\tau+t^1}) \cdot \ldots \cdot g_m(\xi_{\tau+t^m}) \\ &= \lim_{n\to\infty} I_{A,\tau<\infty} h(\tau_n) g_1(\xi_{\tau_n+t^1}) \cdot \ldots \cdot g_m(\xi_{\tau_n+t^m}) \\ &= \lim_{n\to\infty} \sum_{u \geq t} I_{A,\tau_n=u} h(u) g_1(\xi_{u+t^1}) \cdot \ldots \cdot g_m(\xi_{u+t^m}). \end{aligned}$$

Moreover, putting $t^0 = 0$, $\varphi_m \equiv 1$,

$$\varphi_{i-1}(u, y) = \mathbf{E} g_i(\xi_{u+t^i}(u + t^{i-1}, y)) \varphi_i(u, \xi_{u+t^i}(u + t^{i-1}, y))$$

for $i = m, \ldots, 1$ and noting that $g_j(\xi_{u+t^j})$ are $\mathcal{F}_{u+t^{i-1}}$-measurable functions for $j \leq i - 1$, we use the induction to obtain from Lemma 4 that

$$\begin{aligned} &\mathbf{E} I_{A,\tau_n=u} h(u) g_1(\xi_{u+t^1}) \cdot \ldots \cdot g_m(\xi_{u+t^m}) \\ &\quad = \mathbf{E} I_{A,\tau_n=u} h(u) g_1(\xi_{u+t^1}) \cdot \ldots \cdot g_i(\xi_{u+t^i}) \varphi_i(u, \xi_{u+t^i}) \\ &\quad = \mathbf{E} I_{A,\tau_n=u} h(u) \varphi_0(u, \xi_u) = \mathbf{E} I_{A,\tau_n=u} h(\tau_n) \varphi_0(\tau_n, \xi_{\tau_n}). \end{aligned}$$

The first assertion of Lemma 3 shows that the functions $\varphi_i(u,y)$ are continuous in (u,y) for $i=m,\dots,0$. Hence, as in the proof of Theorem II.5.5 we conclude that

$$\mathbf{E}\, I_{\tau<\infty} F(\tau,\xi.) = \mathbf{E}\, I_{\tau<\infty} I_A h(\tau)\varphi_0(\tau,\xi_\tau).$$

On the other hand, using exactly the same arguments, we can deduce from Lemmas 3 and 4 that the $\mathcal{F}_\tau \otimes \mathfrak{B}([0,\infty)) \otimes \mathfrak{B}(E_d)$-measurable function $I_A h(u)\varphi_0(u,y)$ is precisely the right-hand side of (6). □

6. Exercise. A particular case of Theorem 1 states that (5) is true for any $\mathfrak{B}([0,\infty)) \otimes \mathfrak{A}(C)$-measurable nonnegative function $\widetilde{F}(u,x.)$, where Φ is the Borel function of (u,y) defined by $\Phi(u,y) = \mathbf{E} F\big(u,\xi.(u,y)\big)$. Proceeding as in the proof of Lemma II.8.5(a) and using the lemma on π- and λ-systems, derive the general case from this statement.

In what follows we will essentially need only one particular case of Theorem 1. Consider a domain

$$Q \subset E_{d+1} = \big\{(t,x)\colon t\in E_1, x\in E_d\big\}$$

and for $u\ge 0$, $x\in C^u$, define

$$\gamma = \gamma(u,x.) = \inf\big\{s\ge u\colon (s,x_s)\notin Q\big\}. \tag{11}$$

7. Definition. Let $\widetilde{F}(u,x.)$ be a $\mathfrak{B}([0,\infty)) \otimes \mathfrak{A}(C)$-measurable (and therefore, independent of ω) function. Defining $F(u,x.)$ by (4), we will say that F is (pre-exit) *Q-insensitive* if $F(u,x.) = F(s,x.)$ for any (finite) $s\ge u\ge 0$, $x\in C^u$ such that $\gamma(u,x.)\ge s$.

8. Theorem. *Let F be a nonnegative Q-insensitive function. For $t\ge 0$, $x\in E_d$, define*

$$v(t,x) = \mathbf{E} F\big(t,\xi.(t,x)\big). \tag{12}$$

Then v is a Borel function of (t,x). In addition, if $t\ge 0$, $x\in E_d$, τ is a Markov time and $t\le\tau\le\gamma\big(t,\xi.(t,x)\big)$, then

$$v(t,x) = \mathbf{E}\, I_{\tau<\infty} v\big(\tau,\xi_\tau(t,x)\big) + \mathbf{E}\, I_{\tau=\infty} F\big(t,\xi.(t,x)\big). \tag{13}$$

Finally, if η is a $\mathcal{F}_\tau$-measurable nonnegative random variable, then

$$\mathbf{E}\,\eta F\big(t,\xi.(t,x)\big) = \mathbf{E}\, I_{\tau<\infty}\eta v\big(\tau,\xi_\tau(t,x)\big) + \mathbf{E}\, I_{\tau=\infty}\eta F\big(t,\xi.(t,x)\big). \tag{14}$$

This theorem follows immediately from Theorem 1, since the measurability of v follows from that of Φ, while (13) and (14) can be obtained from (5) by observing that $F\big(\tau,\xi.(t,x)\big) = F\big(t,\xi.(t,x)\big)$ at points where $\tau<\infty$.

Equality (13) obviously implies the following corollary.

9. Corollary (cf. Remark 9.5). *Let $\Gamma = \big\{(t,x)\colon t\ge 0,\ x\in E_d,\ v(t,x)<\infty\big\}$, $(t,x)\in\Gamma$ and let τ be a finite Markov time such that $t\le\tau\le\gamma\big(t,\xi.(t,x)\big)$. Then $\big(\tau,\xi_\tau(t,x)\big)\in\Gamma$ (a.s).*

The next result implies that we can take $\tau = \gamma\big(t,\xi.(t,x)\big)$.

10. LEMMA. *If* $t \geq 0$, $x \in E_d$, *then* $\gamma\big(t, \xi.(t,x)\big)$ *is a Markov time with respect to* $\{\mathcal{F}_s\}$.

This lemma, in turn, follows from the fact that $\gamma\big(t, \xi.(t,x)\big)$ is the first exit time from the domain $\big((-\infty, t) \times E_d\big) \cup Q$ of the continuous $\mathcal{F}_s$-adapted process $\big(s, \tilde{\xi}_s(t,x)\big)$ defined by $\tilde{\xi}_s(t,x) = \xi_s(t,x)$ for $s \geq t$, $\tilde{\xi}_s(t,x) = \xi_t(t,x)$ $(= x$ a.s.$)$ for $s \in [0,t]$.

It is slightly more difficult to prove the following lemma.

11. LEMMA. *The functions defined on* $[0,\infty) \times C$ *by*

$$\tilde{\gamma}(u, x.) = \inf\big\{s \geq u\colon (s, x_{s-u}) \notin Q\big\}, \qquad \widetilde{F}(u, x.) = x_{\tilde{\gamma}(u,x.)-u} I_{\tilde{\gamma}(u,x.)<\infty} \tag{15}$$

are $\mathfrak{B}\big([0,\infty)\big) \otimes \mathfrak{A}(C)$*-measurable.*

PROOF. Let $Q_n \subset E_{d+1}$, $n = 1, 2, \ldots$, be open sets such that $\bar{Q}_n \subset Q_{n+1}$, $Q = \bigcup_n Q_n$. Let ρ_s, $s \geq 0$, denote the set of all rational points in $[0, s]$. It is not difficult to see that

$$\big\{(u, x.)\colon \tilde{\gamma}(u, x.) > s\big\} = \bigcup_n \bigcap_{r \in \rho_s} \big\{(u, x.)\colon \big(r \vee u, x_{(r\vee u)-u}\big) \in Q_n\big\}. \tag{16}$$

In addition, $\widetilde{F}_1(u, x.) := x_u$ is continuous in u and $\mathfrak{A}(C)$-measurable with respect to $x.$ for every u. Consequently, it is measurable with respect to $(u, x.)$. Furthermore, for every r the mapping $u \to (r \vee u) - u$ is Borel (continuous). The mappings $(u, x.) \to (r \vee u) - u$ and $(u, x.) \to x.$ are also measurable, hence so are the functions $\widetilde{F}_2(u, x.) := \big((r \vee u) - u, x.\big)$ and $\widetilde{F}_1(\widetilde{F}_2)(u, x.) = x_{(r\vee u)-u}$. Finally, the mapping $u \to r \vee u$, the function $\widetilde{F}_3(u, x.) = r \vee u$ and the function $\big(\widetilde{F}_3, \widetilde{F}_1(\widetilde{F}_2)\big)(u, x.) = \big(r \vee u, x_{(r\vee u)-u}\big)$ are measurable. This proves that all the sets in (16) are elements of the σ-algebra $\mathfrak{B}\big([0,\infty)\big) \otimes \mathfrak{A}(C)$. The rest of the proof that $\tilde{\gamma}(u, x.)$ is measurable was actually the content of Exercise II.4.4. The proof that the second function in (15) is measurable is similar. □

We will now give some examples of functions satisfying the assumptions of Definition 7.

12. EXAMPLE. Obviously, for the functions defined in (15),

$$\gamma(u, x.) = \tilde{\gamma}(u, \hat{\theta}_u x.) = \inf\big\{s \geq u\colon (s, x_s) \notin Q\big\},$$
$$F(u, x.) = \widetilde{F}(u, \hat{\theta}_u x.) = x_{\gamma(u,x.)} I_{\gamma(u,x.)<\infty}$$

and γ and F are Q-insensitive. This is also true for

$$g\big(\gamma(u, x.), x_{\gamma(u,x.)}\big) I_{\gamma(u,x.)<\infty},$$

where g is a *Borel* function on $[0,\infty) \times E_d$. Assume also that $g \geq 0$ and let $\tau(t,x)$, $t \geq 0$, $x \in E_d$, be *the first exit time after time* t of the process $\big(s, \xi_s(t,x)\big)$ from the domain Q. Recall that it is defined by the formula

$$\tau(t,x) = \gamma\big(t, \xi.(t,x)\big) = \inf\big\{s \geq t\colon \big(s, \xi_s(t,x)\big) \notin Q\big\}. \tag{17}$$

Define (cf. (12))

$$v(t,x) = \mathbf{E}\, g\big(\tau(t,x), \xi_{\tau(t,x)}(t,x)\big) I_{\tau(t,x)<\infty},$$

where τ is a finite Markov time such that $t \le \tau \le \tau(t,x)$. Then, by Theorem 8,

$$v(t,x) = \mathbf{E}\, v\big(\tau, \xi_\tau(t,x)\big).$$

If $Q = (-\infty, s) \times E_d$, $t \le u \le s$, $\tau \equiv u$, then we once again obtain (3), since in this case $\tau(t,x) = s$.

13. EXAMPLE. Let g be a real-valued Borel function on $[0,\infty) \times E_d$ and define

$$F(u, x.) = \varlimsup_{n\to\infty} g\big(n \wedge \gamma(u,x.), x_{n\wedge\gamma(u,x.)}\big), \tag{18}$$

where n runs over the integers. It is not difficult to verify that this function, which may also take infinite values, is Q-insensitive. To do that, it suffices to define $\widetilde{F}(u,x.)$ by replacing $n \wedge \gamma(u,x.)$ in (18) with $n \wedge \tilde{\gamma}(u,x.)$, $\big(n \wedge \tilde{\gamma}(u,x.) - u\big) \vee 0$, and to recall that the upper limit of measurable functions is measurable.

14. PROBLEM. For fixed $t \ge 0$, $x \in E_d$, let $\mathcal{N}_s$, $s \in [0,\infty]$, denote the completion with respect to $\mathcal{F}, P$ of the minimal σ-algebra that contains all the sets $\{\omega\colon \xi_{t+u}(t,x) \in \Gamma\}$, where $\Gamma \in \mathfrak{B}(E_d)$, $u \in [0,s]$, $u < \infty$. Let $\mathcal{N}_{s+} = \bigcap_{\varepsilon>0} \mathcal{N}_{s+\varepsilon}$. We can define the projection of any random variable $\alpha \in \mathcal{L}_2(\mathcal{N}_\infty, P)$ onto $\mathcal{L}_2(\mathcal{N}_{s+}, P)$ as a random variable $\beta \in \mathcal{L}_2(\mathcal{N}_{s+}, P)$ such that $\mathbf{E}\,\alpha\delta = \mathbf{E}\,\beta\delta$ for any $\delta \in \mathcal{L}_2(\mathcal{N}_{s+}, P)$. Deduce from Theorem 1 and Lemma 3 (and the fact that $\mathcal{N}_{s+} \subset \mathcal{F}_{t+s+\varepsilon}$) that if

$$\alpha = g_1\big(\xi_{t+u^1}(t,x)\big) \cdot \ldots \cdot g_m\big(\xi_{t+u^m}(t,x)\big),$$

then β is $\mathcal{N}_s$-measurable. Generalize this result to all $\alpha \in \mathcal{L}_2(\mathcal{N}_\infty, P)$ and deduce that $\mathcal{L}_2(\mathcal{N}_{s+}, P) = \mathcal{L}_2(\mathcal{N}_s, P)$ and $\mathcal{N}_{s+} = \mathcal{N}_s$. For $s = 0$ this obviously implies Blumenthal's zero-one law for $\xi_{t+u}(t,x)$, since the completion $\mathcal{N}_0$ of the σ-algebra generated by a nonrandom vector x contains only events, whose probability is either zero or one.

15. PROBLEM. Let c, g and f be Borel functions on $[0,\infty) \times E_d$, define $\tau(t,x)$ as in (17) and put

$$\varphi_s(t,x) = \int_t^s c\big(r, \xi_r(t,x)\big)\, dr, \tag{19}$$

$$\begin{aligned} v(t,x) = \mathbf{E}\Bigg[&\int_t^{\tau(t,x)} f\big(s, \xi_s(t,x)\big) \exp\big(-\varphi_s(t,x)\big)\, ds \\ &+ I_{\tau(t,x)<\infty}\, g\big(\tau(t,x), \xi_{\tau(t,x)}(t,x)\big) \exp\big(-\varphi_{\tau(t,x)}(t,x)\big)\Bigg]. \end{aligned} \tag{20}$$

In the sequel we will frequently encounter such expressions, subject to various conditions that make them meaningful. For simplicity, we will assume here that $c, g, f \ge 0$. Reasoning as in the proof of Lemma II.8.2, prove that the last definition is legitimate. Deduce from Theorem 1, defining $\widetilde{F}$ in a suitable way, that v is a Borel function and that for any t, x and Markov times τ such that $t \le \tau \le \tau(t,x)$,

$$(21) \qquad v(t,x) = \mathbf{E}\left[\int_t^\tau f\big(s, \xi_s(t,x)\big)\exp\big(-\varphi_s(t,x)\big)\,ds + I_{\tau<\infty} v\big(\tau, \xi_\tau(t,x)\big)\exp\big(-\varphi_\tau(t,x)\big)\right].$$

6. The Kolmogorov equations

Among the most effective methods for studying functions are those that take derivatives into consideration. In particular, important information may be derived from differential equations satisfied by the functions under investigation. We will show in this section that if $\xi_\cdot(t,x)$ is a solution of a stochastic equation (0.2) with $\xi_t^0 = x$ then, under fairly general assumptions, the expectations of certain functionals of $\xi_\cdot(t,x)$ satisfy partial differential equations with respect to (t,x). The derivation of these equations is based on Theorem 5.8. We will therefore assume throughout the section that σ, b are independent of ω and equation (0.2) is solvable by Euler's method. Fix a domain $Q \subset (-\infty,\infty) \times E_d$ such that

$$Q_+ := Q \cap \{t \ge 0\} \ne \emptyset$$

and a real-valued function $F(u, x.)$ which is Q-insensitive. In addition, we will assume that definition (5.12) makes sense for any $(t,x) \in Q_+$ and defines in Q_+ a function $v(t,x)$ which is finite and continuous in Q_+ together with its derivatives v_{x^i}, $v_{x^i x^j}$. Finally, we set $a = \sigma\sigma^*$,

$$(1) \qquad L(t,x)u(x) = \frac{1}{2}\sum_{i,j=1}^d a^{ij}(t,x)u_{x^i x^j}(x) + \sum_{i=1}^d b^i(t,x)u_{x^i}(x)$$

and assume that a, b are *defined* and *continuous* in (t,x) in Q_+.

We will need the following lemma.

1. LEMMA. *For $\varepsilon > 0$, denote*

$$(2) \qquad Q(\varepsilon,t,x) = \{(s,y)\colon t-\varepsilon^3 < s < t+\varepsilon^3,\ |y-x| < \varepsilon\},$$

$$(3) \qquad \tau_\varepsilon(t,x) = \inf\{s \ge t\colon (s, \xi_s(t,x)) \notin Q(\varepsilon,t,x)\}.$$

Then for any compact set $\Gamma \subset Q_+$

$$(4) \qquad \varepsilon^{-3}P\{\tau_\varepsilon(t,x) - t < \varepsilon^3\} \longrightarrow 0, \qquad \varepsilon^{-3}\mathbf{E}\left[\tau_\varepsilon(t,x) - t\right] \longrightarrow 1$$

as $\varepsilon \downarrow 0$ uniformly in $(t,x) \in \Gamma$.

PROOF. For the sake of simplicity, we write $\tau_\varepsilon = \tau_\varepsilon(t,x)$, $\xi_r = \xi_r(t,x)$ and $s = t + \varepsilon^3$. We first note that the second assertion in (4) follows from the first one, since

$$\mathbf{E}(\tau_\varepsilon - t) = \mathbf{E}I_{\tau_\varepsilon<s}(\tau_\varepsilon - t) + \varepsilon^3 P\{\tau_\varepsilon = s\},$$
$$P\{\tau_\varepsilon = s\} = 1 - P\{\tau_\varepsilon - t < \varepsilon^3\},$$
$$\mathbf{E}I_{\tau_\varepsilon<s}(\tau_\varepsilon - t) \le \varepsilon^3 P\{\tau_\varepsilon - t < \varepsilon^3\}.$$

To prove the first assertion in (4) we will use the relation

$$P\{\tau_\varepsilon - t < \varepsilon^3\} \le P\Big\{\sup_{[t,\tau_\varepsilon]} |\xi_r - x| \ge \varepsilon\Big\}$$

$$\le P\Big\{\sup_{[0,s]}\Big|\int_0^r \sigma(u,\xi_u) I_{t<u\le\tau_\varepsilon}\, dw_u\Big| \ge \frac{\varepsilon}{2}\Big\}$$

$$+ P\Big\{\sup_{[t,s]}\Big|\int_t^r b(u,\xi_u) I_{u\le\tau_\varepsilon}\, du\Big| \ge \frac{\varepsilon}{2}\Big\}.$$

The last probability equals zero for small ε since $s - t = \varepsilon^3$ and b is bounded in the intersection of some neighborhood of the compact set Γ with the set $\{t \ge 0\}$, and this intersection contains (u, ξ_u) for $t \le u \le \tau_\varepsilon$. To estimate the first probability on the right we use Chebyshev's inequality and the Burkholder-Davis-Gundy inequalities. This gives an upper bound

$$2^8\varepsilon^{-8}\mathbf{E}\sup_{[0,s]}\Big|\int_0^r \sigma(u,\xi_u) I_{t<u\le\tau_\varepsilon}\, dw_u\Big|^8 \le N\varepsilon^{-8}\mathbf{E}\Big(\int_t^s \|\sigma(u,\xi_u)\|^2 I_{t<u\le\tau_\varepsilon}\, du\Big)^4$$

$$\le N\varepsilon^{-8}\varepsilon^{12} = N\varepsilon^4,$$

where N are constants independent of $(t,x) \in \Gamma$ and ε for small ε. This gives the first assertion in (4). □

2. THEOREM. *Under the conditions formulated above, the derivative* $\frac{\partial}{\partial t}v(t,x)$ *exists in* Q_+ *and is continuous there, and* v *satisfies the* backward Kolmogorov equation *in* Q_+:

$$\frac{\partial v}{\partial t}(t,x) + L(t,x)v(t,x) = 0. \tag{5}$$

PROOF. Fixing $(t,x) \in Q^0_+ := Q \cap \{t > 0\}$, we will use notation (2), (3) and omit arguments (t,x) of τ_ε, ξ_r and $Q(\varepsilon)$ as in the proof of Lemma 1. Let ε be so that $\overline{Q(\varepsilon)} \subset Q^0_+$. By Theorem 5.8 (applied first to F_+ and then to F_-),

$$v(t,x) = \mathbf{E}\, v(\tau_\varepsilon, \xi_{\tau_\varepsilon}). \tag{6}$$

We want to apply Itô's formula (Theorem IV.1.1) to transform the right-hand side. To do this, we need v to be smooth for all (t,x). Since (6) involves the values of v in $\overline{Q(\varepsilon)}$ only, we will assume *for a moment* that $\partial v/\partial t$ is continuous in Q^0_+, and modify v outside $\overline{Q(\varepsilon)}$, multiplying it, say, by a smooth truncating function to ensure its necessary smoothness everywhere. Formula (6) will not be affected thereby. Now observing that τ_ε is bounded ($\tau_\varepsilon \le t + \varepsilon^3$), we can repeat the arguments used to derive (2.4) (with obvious modifications). We obtain

$$\mathbf{E}\int_t^{\tau_\varepsilon} \Big(\frac{\partial v}{\partial t} + Lv\Big)(s,\xi_s)\, ds = 0. \tag{7}$$

If we suppose, say, that the left-hand side of (5) is strictly positive at the point (t,x), this immediately contradicts (7), by virtue of the continuity of the left-hand side of (5) and the obvious inequalities $\tau_\varepsilon > t$, $\mathbf{E}(\tau_\varepsilon - t) > 0$. We have thus proved (5) in $Q_+^0 \subset Q_+$, under the additional assumption that $\partial v/\partial t$ exists and is continuous in Q_+^0. But this result and the continuity of Lv in Q_+ quickly imply (for example, by l'Hôpital's rule) that $\partial v/\partial t$ exists and is continuous in Q_+, and equality (5) holds in Q_+.

If we only know that v, v_x, v_{xx} are continuous in Q_+, we will use a somewhat different transformation of (6). Let $s = t + \varepsilon^3$ and observe that by Itô's formula

$$\begin{aligned} v(t,x) - v(s,x) &= \mathbf{E}\left[v(\tau_\varepsilon, \xi_{\tau_\varepsilon}) - v(s, \xi_{\tau_\varepsilon})\right] + \mathbf{E}\left[v(s, \xi_{\tau_\varepsilon}) - v(s, \xi_t)\right] \\ &= \mathbf{E}\, I_{\tau_\varepsilon < s}\left[v(\tau_\varepsilon, \xi_{\tau_\varepsilon}) - v(s, \xi_{\tau_\varepsilon})\right] + \mathbf{E}\int_t^{\tau_\varepsilon} L(r, \xi_r) v(s, \xi_r)\, dr. \end{aligned} \tag{8}$$

Let us vary t, s, ε so that $t, s \to u$, where $(u, x) \in Q_+^0$, and so that always $s > t$, $\varepsilon^3 = s - t$. In addition, divide the outer terms of (8) by $s - t$. Since v is bounded in the neighborhood of (u, x), the first term on the right is bounded from above by a constant independent of t, s, multiplied by $P\{\tau_\varepsilon < s\}$. The second term on the right of (8) equals

$$L(\tilde{t}, \tilde{x}) v(s, \tilde{x})\, \mathbf{E}(\tau_\varepsilon - t),$$

where $(\tilde{t}, \tilde{x})$ is a point in $\overline{Q(\varepsilon)}$. This follows from the Mean Value Theorem, which carries over to our case together with its standard proof, since $L(r, y)v(s, y)$ is continuous in (r, y).

Hence, Lemma 1 shows that these manipulations with (8) yield the existence of $\partial v/\partial t$ and equality (5) at the point (u, x). This point can be chosen arbitrarily in Q_+^0 and the continuity of $\partial v/\partial t$ in Q_+^0 follows from (5). We have already shown that this implies the assertion of the theorem. □

Equation (5) is an equation of parabolic type. In some cases v is independent of t and satisfies an elliptic equation.

3. THEOREM. *Let a, b be independent of time, $Q = (-\infty, \infty) \times D$, where $D \subset E_d$ is an open set, $F(u, x.) = F(0, \hat{\theta}_u x.)$ for $u \geq 0$, $x. \in C^u$. Then v does not depend on t, $v = v(x)$, $L(t, x) = L(x)$, and $L(x)v(x) = 0$ in D.*

PROOF. By the definition and Theorem I.4.3,

$$v(t, x) = \mathbf{E}\, F(0, \hat{\theta}_t \xi.(t, x)) = \int_C F(0, x.) P(\hat{\theta}_t \xi.(t, x))^{-1}(dx.).$$

By Corollary 4.4, the measure in the last integral is independent of t. Hence v is independent of t, and the rest follows from Theorem 1. □

Now we deduce from Theorem 1 several assertions concerning more general equations than (5). Throughout the remainder of this section we assume that $g(t, x)$, $f(t, x)$, $c(t, x)$ are given real-valued Borel functions on $[0, \infty) \times E_d$ such that $c \geq 0$ and f, c are continuous in (t, x).

4. THEOREM. (a) *Assume that $G \subset Q$ is a domain, $G_+ := G \cap \{t \geq 0\} \neq \emptyset$, and for any $(t,x) \in G_+$ we have*

$$\int_t^{\tau(t,x)} \left| f\big(s, \xi_s(t,x)\big) \right| e^{-\varphi_s(t,x)}\, ds < \infty \quad (a.s.), \tag{9}$$

where $\tau(t,x)$ is defined by (5.17) and $\varphi_s(t,x)$ by (5.19). Assume also that the right-hand side of (5.20) is well defined for $(t,x) \in G_+$ and defines in G_+ a finite function $v(t,x)$ that is continuous together with its derivatives v_x, v_{xx}. Then $\partial v/\partial t$ exists and is continuous in G_+, and in G_+

$$\frac{\partial v}{\partial t}(t,x) + L(t,x)v(t,x) - c(t,x)v(t,x) + f(t,x) = 0. \tag{10}$$

(b) *Let a, b, g, f, c be independent of time, B and D domains, $B \subset D \subset E_d$. Suppose that for any $x \in B$*

$$\int_0^{\tau(x)} \left| f\big(\xi_s(x)\big) \right| e^{-\varphi_s(x)}\, ds < \infty \quad (a.s.), \tag{11}$$

where

$$\xi_s(x) = \xi_s(0,x), \quad \varphi_s(x) = \int_0^s c\big(\xi_r(x)\big)\, dr, \quad \tau(x) = \inf\big\{s \geq 0 \colon \xi_s(x) \notin D\big\}. \tag{12}$$

Assume also that the right-hand side of the formula

$$\begin{aligned} v(x) := \mathbf{E}\Bigg[&\int_0^{\tau(x)} f\big(\xi_s(x)\big) \exp\big(-\varphi_s(x)\big)\, ds \\ &+ I_{\tau(x)<\infty} g\big(\xi_{\tau(x)}(x)\big) \exp\big(-\varphi_{\tau(x)}(x)\big)\Bigg] \end{aligned} \tag{13}$$

is meaningful for $x \in B$. Assume that $v(x)$ is finite and continuous in B together with its derivatives v_x, v_{xx}. Then $L(t,x) = L(x)$ and in B

$$L(x)v(x) - c(x)v(x) + f(x) = 0. \tag{14}$$

PROOF. (a) It would be possible to repeat the proof of Theorem 2 with slight modifications, using (5.21) instead of (6). But we prefer to demonstrate a device that will be used repeatedly later on.

Together with equation (0.2), let us consider "equations"

$$\xi_s^{d+1} = \xi_t^{d+1,0} - \int_t^s c(r, \xi_r)\, dr, \qquad s \geq t, \tag{15}$$

$$\xi_s^{d+2} = \xi_t^{d+2,0} + \int_t^s f(r, \xi_r) e(\xi_r^{d+1})\, dr, \qquad s \geq t, \tag{16}$$

where $e(u)$ is an infinitely differentiable function on $(-\infty, \infty)$ such that $e(u) = e^u$ for $u \le 1$ and $e(u) = 0$ for $u \ge 2$.

Theorem 3.4 implies, first, that system (0.2), (15) is solvable by Euler's method, and then (see Remark 3.5) that system (0.2), (15), (16) is also solvable by Euler's method. Denote the solution of the latter system by

$$\hat{\xi}_s(t, \hat{\xi}_t^0) = \left(\xi_s(t, \xi_t^0), \xi_s^{d+1}(t, \hat{\xi}_t^0), \xi_s^{d+2}(t, \hat{\xi}_t^0)\right),$$

where

$$\hat{\xi}_t^0 = (\xi_t^0, \xi_t^{d+1,0}, \xi_t^{d+2,0}).$$

Let $\gamma(u, x.)$ be as defined in (5.11) for $u \ge 0$, $x. \in C^u(E_d)$; for $\hat{x}. = (x., x.^{d+1}, x.^{d+2}) \in C^u(E_d) \times C^u(E_1) \times C^u(E_1)$, define

$$F(u, \hat{x}.) = \varlimsup_{n\to\infty} x^{d+2}_{n\wedge\gamma(u,x.)} + g\left(\gamma(u, x.), x_{\gamma(u,x.)}\right) e\left(x^{d+1}_{\gamma(u,x.)}\right) I_{\gamma(u,x.)<\infty}.$$

Taking $\widehat{Q} = Q \times E_2$ for Q in Examples 5.12, 5.13, we see that $F(u, \hat{x}.)$ is $\widehat{Q}$-insensitive, hence also $\widehat{G} = G \times (-\infty, 1) \times E_1$-insensitive. Moreover, it is easily deduced from (15), (16) and condition (9) that if $(t, \hat{x}) \in \widehat{G}_+ = \widehat{G} \cap \{t \ge 0\}$, $\hat{x} = (x, x^{d+1}, x^{d+2})$, then $F(t, \hat{\xi}.(t, \hat{x}))$ is almost surely equal to the sum of x^{d+2} and the expression in square brackets in (5.20) multiplied by $\exp x^{d+1}$. Thus, in $\widehat{G}_+$,

$$\hat{v}(t, \hat{x}) := \mathbf{E} F\left(t, \hat{\xi}.(t, x)\right) = x^{d+2} + v(t, x) \exp x^{d+1}. \tag{17}$$

Finally, the coefficients of system (0.2), (15), (16) do not depend on ω and by Theorem 2 the function $\hat{v}$ is continuously differentiable with respect to t in $\widehat{G}_+$ and satisfies in $\widehat{G}_+$ an appropriate Kolmogorov equation, that is (see, for example, the arguments following (IV.1.2)),

$$\frac{\partial \hat{v}}{\partial t}(t, \hat{x}) + L(t, x)\hat{v}(t, \hat{x}) - c(t, x)\frac{\partial \hat{v}(t, \hat{x})}{\partial x^{d+1}} + f(t, x)e(x^{d+1})\frac{\partial \hat{v}(t, \hat{x})}{\partial x^{d+2}} = 0.$$

Together with (17), this immediately yields (10) in G_+ and proves (a).

To prove (b), it suffices to take $Q = E_1 \times D$, $G = E_1 \times B$, and observe that $F(u, \hat{x}.) = F(0, \hat{\theta}_u \hat{x}.)$. By Corollary 4.4, the distribution of $F(t, \hat{\xi}.(t, x))$ is independent of t and hence

$$\mathbf{E}\left|F\left(t, \hat{\xi}.(t, \hat{x})\right)\right| = \mathbf{E}\left|F\left(0, \hat{\xi}.(0, \hat{x})\right)\right| < \infty$$

if $x \in B$, $x^{d+1} < 1$. It now remains only to apply Theorem 3 or part (a). □

5. Problem. There are other situations in which it may be useful to introduce additional coordinates. For example, let $p(t, x)$ be a continuous function on $[0, \infty) \times E_d$ with values in E_{d_1}. For $s \ge t$, define

$$\rho(t, s, x) = \exp\left\{ \int_t^s p^*\left(r, \xi_r(t, x)\right) dw_r - \frac{1}{2}\int_t^s \left|p\left(r, \xi_r(t, x)\right)\right|^2 dr \right\}. \tag{18}$$

Multiply f in (5.20), (9) by $\rho(t, s, x)$ and g in (5.20) by $\rho(t, \tau(t, x), x)$. Assume that the new condition (9) holds in G_+, and that the new right-hand side of (5.20) defines a function $\tilde{v}$ in G_+ which is twice differentiable with respect to x continuously with respect to (t, x). Also let σ be continuous in G_+. Treating the exponent in (18) as a new coordinate of ξ_s, prove that $\partial\tilde{v}/\partial t$ is continuous in G_+ and in G_+

$$\frac{\partial \tilde{v}}{\partial t} + L\tilde{v} + \sum_{i=1}^{d} (\sigma p)^i \tilde{v}_{x^i} - c\tilde{v} + f = 0.$$

Compare this result with Girsanov's Theorem.

7. Derivation of the Kolmogorov equation in the inhomogeneous case

Theorems 6.2–6.4 are rather conditional, since they involve assumptions about the existence and continuity of the second-order derivatives of v with respect to x. In this section we will impose conditions on σ, b, g only, which will guarantee both that such functions as $\mathbf{E}\, g(\xi_T(t, x))$ are smooth and that they satisfy appropriate Kolmogorov equations. We will assume that σ, b are independent of ω and fix constants $m, K, T \in (0, \infty)$. We will frequently use notation (I.4.15) and

$$v_{(y)(z)}(x) = \sum_{i,j=1}^{d} v_{x^i x^j}(x) y^i z^j. \tag{1}$$

A natural rule for differentiating $\mathbf{E}\, g(\xi_T(t, x))$ with respect to x is formulated in the following lemma.

1. Lemma. *Let the functions $g(x)$, $\sigma(r, x)$, $b(r, x)$ be continuously differentiable with respect to x, the latter two for every $r > 0$. Suppose that for all $x, y \in E_d$ such that $|y| = 1$*

$$|g(x)| + |g_{(y)}(x)| \le K(1 + |x|^m). \tag{2}$$

Suppose that the system comprising equation (0.2) and the equation

$$\eta_s = \eta_t^0 + \int_t^s \sigma_{(\eta_r)}(r, \xi_r)\, dw_r + \int_t^s b_{(\eta_r)}(r, \xi_r)\, dr \tag{3}$$

is solvable by Euler's method in the mean (and in particular, that its coefficients satisfy the conditions of the introduction to this chapter). Then the function $v(t, x) := \mathbf{E}\, g(\xi_T(t, x))$, considered in $[0, T] \times E_d$, is differentiable with respect to x continuously in (t, x), and for all $x, y \in E_d$, $t \in [0, T]$,

$$v_{(y)}(t, x) = \mathbf{E}\, g_{(\eta_T)}(\xi_T(t, x)), \tag{4}$$

where η_T is the value at the point $s = T$ of the solution η_s of the equation

$$\eta_s = y + \int_t^s \sigma_{(\eta_r)}(r, \xi_r(t, x))\, dw_r + \int_t^s b_{(\eta_r)}(r, \xi_r(t, x))\, dr. \tag{5}$$

PROOF. Let $(\xi_s^n(t,x), \eta_s^n(t,x,y))$ be Euler approximations for system (0.2), (3) with $\xi_t^0 = x, \eta_t^0 = y$. Define

$$v^n(t,x) = \mathbf{E}\, g(\xi_T^n(t,x)), \qquad u^n(t,x,y) = \mathbf{E}\, g_{(\eta_T^n(t,x,y))}(\xi_T^n(t,x)). \tag{6}$$

By (2),

$$|g_{(y)}(x)| \le K|y|(1+|x|^m) \le K(1+|y|+|y|^2+|x|^{2m}). \tag{7}$$

Hence, by Definition 3.1, the expectations in (6) are finite. Using the same arguments together with Lemma 4.1(c) and Lemma III.6.13(g), we see that v^n, u^n are continuous in (t,x,y) for any n and (cf. the beginning of the proof of Lemma 5.3) tend to v and to the right-hand side of (4), respectively, which in particular, are finite and continuous in (t,x,y) uniformly on any compact set in $[0,T] \times E_d$, $[0,T] \times E_d \times E_d$. Thus, it remains to prove (4). Moreover, since the uniform limit of the derivatives of a sequence of functions is the derivative of the limit, it suffices to prove that v^n is continuously differentiable with respect to x for any $t \in [0,T]$, $n = 1, 2, \ldots$, and that $v^n_{(y)}(t,x) = u^n(t,x,y)$.

Fix t, n and let $t_k = t + k/n$, $p = [n(T-t)]$. By Lemma 4.5, $v^n(t_p, x)$ is continuously differentiable with respect to x and $v^n_{(y)}(t_p,x) = u^n(t_p,x,y)$. Hence, by our assumption about system (0.2), (3) and by Definition 3.1,

$$|v^n(t_p,x)| + |v^n_x(t_p,x)| \le N(1+|x|^q) \tag{8}$$

for suitable constants N, q and all x. Now we claim that

$$v^n(t_{p-1},x) = \mathbf{E}\, v^n(t_p, \xi^n_{t_p}(t_{p-1},x)). \tag{9}$$

If g is continuous and bounded, this equality is just one of the assertions of Lemma 4.1. Standard arguments uaing the lemma on π- and λ-systems extend it first to all nonnegative Borel functions g and then, via the equality $g = g_+ - g_-$, to all Borel functions g for which v^n is a finite function. But we have already seen that v^n is finite when condition (2) is satisfied, and so equality (9) is indeed true. Similarly, applying Lemma 4.1 to $(\xi_s^n(t,x), \eta_s^n(t,x,y))$, one can prove that even if $g_{(y)}(x)$ in the second definition in (6) is replaced by an arbitrary Borel function $g(x,y)$ which increases as $|x|+|y| \to \infty$ not faster than a polynomial, then again

$$u^n(t_{p-1},x,y) = \mathbf{E}\, u^n(t_p, \xi^n_{t_p}(t_{p-1},x), \eta^n_{t_p}(t_{p-1},x,y)). \tag{10}$$

By Lemma 4.5, it follows from (8), (9) that $v^n(t_{p-1},x)$ is continuously differentiable with respect to x and

$$v^n_{(y)}(t_{p-1},x) = \mathbf{E}\, v^n_{(\eta^n)}(t_p, \xi^n_{t_p}(t_{p-1},x)),$$

where $\eta^n = \eta^n_{t_p}(t_{p-1},x,y)$. Hence, by (8) and Definition 3.1, it follows that (8) remains true if t_{p-1} is substituted for t_p, possibly with different constants N, q.

Together with the equality $v^n_{(y)}(t_p,x) = u^n(t_p,x,y)$ and (10), this implies that $v^n_{(y)}(t_{p-1},x) = u^n(t_{p-1},x,y)$. Similarly, we can proceed from t_{p-1} to t_{p-2}, and so on. After finitely many steps we obtain $v^n_{(y)}(t,x) = v^n_{(y)}(t_0,x) = u^n(t_0,x,y) = u^n(t,x,y)$. □

The assertion of Lemma 1 carries over to more complicated functions v.

2. LEMMA. *Under the assumptions of Lemma* 1, *let* $c(r,x)$, $f(r,x)$ *be Borel functions of* (r,x) *defined on* $(0,\infty)\times E_d$ *and continuously differentiable with respect to* x *for any* r. *Let* $c \geq 0$ *be such that for* $r>0$, $x,y\in E_d$, $|y|=1$,

$$|c(r,x)| + |c_{(y)}(r,x)| + |f(r,x)| + |f_{(y)}(r,x)| \leq K(1+|x|^m).$$

Then the function

$$v(t,x) = \mathbf{E}\left[\int_t^T f(s,\xi_s(t,x))e^{-\varphi_s(t,x)}\,ds + g(\xi_T(t,x))e^{-\varphi_T(t,x)}\right], \tag{11}$$

where $\varphi_s(t,x) = \int_t^s c(r,\xi_r(t,x))\,dr$, *considered for* $t\in[0,T]$, $x\in E_d$, *is finite, continuous in* (t,x), *and differentiable with respect to* x *continuously in* (t,x). *Moreover, for* $t\in[0,T]$, $x,y\in E_d$,

$$\begin{aligned} v_{(y)}(t,x) = \mathbf{E}\Big\{&\int_t^T \Big[f_{(\eta_s)}(s,\xi_s(t,x)) + f(s,\xi_s(t,x))\eta_s^{d+1}\Big]e^{-\varphi_s(t,x)}\,ds \\ &+ \Big[g_{(\eta_T)}(\xi_T(t,x)) + g(\xi_T(t,x))\eta_T^{d+1}\Big]e^{-\varphi_T(t,x)}\Big\}, \end{aligned} \tag{12}$$

where η_s *is the solution of equation* (5) *and* $\eta_s^{d+1} = -\int_t^s c_{(\eta_r)}(r,\xi_r(t,x))\,dr$.

PROOF. We will use the same device as in the proof of Theorem 6.4. Successive applications of Theorem 3.4 (and inequalities like (7)) show that Euler's method solves both system (0.2), (3), (6.15), (6.16) and the system obtained from it by adding the following two "equations":

$$\eta_s^{d+1} = \eta_t^{d+1,0} - \int_t^s c_{(\eta_r)}(r,\xi_r)\,dr, \tag{13}$$

$$\eta_s^{d+2} = \eta_t^{d+2,0} + \int_t^s \Big[f_{(\eta_r)}(r,\xi_r)e(\xi_r^{d+1}) + f(r,\xi_r)e'(\xi_r^{d+1})\eta_r^{d+1}\Big]dr. \tag{14}$$

Next, using the same notation as in the proof of Theorem 6.4, with $G=Q=(-\infty,T)\times E_d$, $t\in[0,T]$, $\hat g(\hat x) = x^{d+2} + e(x^{d+1})g(x)$, we obtain

$$F(t,\hat x.) = x_T^{d+2} + g(x_T)e(x_T^{d+1}) = \hat g(\hat x_T),$$
$$\hat v(t,\hat x) = \mathbf{E}\hat g(\hat\xi_T(t,\hat x)).$$

In addition, equality (6.17) is true for $x^{d+1}\leq 1$. Together with Lemma 1, this proves the first assertion of the lemma, as well as the equality $v_{(y)}(t,x) = \hat v_{(\hat y)}(t,\hat x)$ for $x^{d+1}=0$, $y\in E_d$, $\hat y = (y,0,0)$. Finally, $\hat v_{(\hat y)}(t,\hat x)$ is known from Lemma 1, and it is easily verified that it equals the right-hand side of (12) for $x^{d+1}=0$, $y\in E_d$, $\hat y=(y,0,0)$. □

3. REMARK. Let σ, b be continuously differentiable with respect to x for any r and such that for all $r, x, i = 1, \dots, d$,

$$\|\sigma_{x^i}\| + |b_{x^i}| \le K, \qquad \|\sigma\| + |b| \le K(1 + |x|). \tag{15}$$

Then it follows from Theorem 3.4 (see also Remark 3.5) that equation (0.2) and system (0.2), (3) are solvable by Euler's method in the mean. Thus we can formulate "explicit" conditions on σ, b, c, f, g, under which the function (11) is continuously differentiable with respect to x.

We will now formulate conditions under which v is twice continuously differentiable with respect to x.

4. THEOREM. *Let σ, b be twice continuously differentiable with respect to x for any $r > 0$ and satisfy condition (15). Let $c(r,x) \ge 0$, $f(r,x)$, $g(x)$ be Borel functions of (r,x) that are twice continuously differentiable with respect to x for any $r > 0$. Assume that these functions, all their partial derivatives with respect to x up to the second order, and all second-order partial derivatives with respect to x of the functions σ^{ij}, b^i, $i = 1, \dots, d$, $j = 1, \dots, d_1$, are bounded above in absolute value by $K(1 + |x|^m)$. Then the function $v(t,x)$ defined in $[0,T] \times E_d$ by (11) is meaningful and twice differentiable with respect to x continuously in (t,x). Moreover, if $a = \sigma\sigma^*, b, c, f$ are defined and continuous in (r,x) in $[0,T] \times E_d$, then v is continuously differentiable with respect to t in $[0,T] \times E_d$ and satisfies equation (6.10) there.*

PROOF. If the first assertion of the theorem is proved, then Theorem 6.4 implies that (6.10) holds in $[0,T) \times E_d$. Since Lv, cv, f are continuous in $[0,T] \times E_d$, l'Hôpital's rule yields the existence and continuity of $\partial v/\partial t$ in $[0,T] \times E_d$. Hence, (6.10) is true on that set, too. This gives the last assertion of the theorem.

The first assertion of the theorem we derive from Lemma 2. To that end, let us consider not equation (0.2) but the system comprising (0.2), (3) and equation (13) as an equation for the process $\hat{\xi}_s = (\xi_s, \eta_s, \eta_s^{d+1})$. As we have seen, the latter is solvable by Euler's method in the mean. Equation (3), corresponding to the equation defining $\hat{\xi}_s$, is the following system

$$\bar{\eta}_s = \bar{\eta}_t^0 + \int_t^s \sigma_{(\bar{\eta}_r)}(r, \xi_r)\, dw_r + \int_t^s b_{(\bar{\eta}_r)}(r, \xi_r)\, dr, \tag{16}$$

$$\begin{aligned} \zeta_s = \zeta_t^0 &+ \int_t^s \left[\sigma_{(\bar{\eta}_r)(\eta_r)} + \sigma_{(\zeta_r)}\right](r, \xi_r)\, dw_r \\ &+ \int_t^s \left[b_{(\bar{\eta}_r)(\eta_r)} + b_{(\zeta_r)}\right](r, \xi_r)\, dr, \end{aligned} \tag{17}$$

$$\zeta_s^{d+1} = \zeta_t^{d+1,0} - \int_t^s \left[c_{(\bar{\eta}_r)(\eta_r)} + c_{(\zeta_r)}\right](r, \xi_r)\, dr. \tag{18}$$

Successive application of Theorem 3.4 proves that system (0.2), (3), (13), (16)–(18) is solvable by Euler's method in the mean. Finally, defining $\hat{\xi}_s(t, x, y, y^{d+1})$ in the natural way for $x, y \in E_d$, $y^{d+1} \in (-\infty, \infty)$, $s \geq t \geq 0$, and setting

$$\hat{f}(s, x, y, y^{d+1}) = f_{(y)}(s, x) + f(s, x)y^{d+1}, \quad \hat{g}(x, y, y^{d+1}) = g_{(y)}(x) + g(x)y^{d+1},$$

$$\hat{v}(t, x, y, y^{d+1}) = \mathbf{E}\left[\int_t^T \hat{f}(s, \hat{\xi}_s(t, x, y, y^{d+1}))e^{-\varphi_s(t,x)}\, ds + \hat{g}(\hat{\xi}_T(t, x, y, y^{d+1}))e^{-\varphi_T(t,x)}\right],$$

we conclude, using Lemma 2, that $\hat{v}$ is differentiable with respect to (x, y, y^{d+1}), continuously in (t, x, y, y^{d+1}). It remains only to compare the definition of $\hat{v}$ with the right-hand side of (12) and to verify the obvious equality

$$\hat{v}(t, x, y, y^{d+1}) = v(t, x)y^{d+1} + v_{(y)}(t, x). \quad \square \tag{19}$$

5. REMARK. By Lemma 2, we can find the derivatives of $\hat{v}$ with respect to x, y, y^{d+1} and derive an expression for $v_{(y)(\bar{y})}$ from (19). Simple computations show that, under the conditions of Theorem 4, for $x, y, \bar{y} \in E_d$, $t \in [0, T]$

$$\begin{aligned} v_{(y)(\bar{y})}(t, x) = \mathbf{E}\Big\{ & \Big[g_{(\zeta_T)}(\xi_T) + g_{(\eta_T)(\bar{\eta}_T)}(\xi_T) + g_{(\bar{\eta}_T)}(\xi_T)\eta_T^{d+1} \\ & + g_{(\eta_T)}(\xi_T)\bar{\eta}_T^{d+1} + g(\xi_T)\zeta_T^{d+1} + g(\xi_T)\eta_T^{d+1}\bar{\eta}_T^{d+1}\Big]e^{-\varphi_T} \\ & + \int_t^T \Big[f_{(\zeta_r)}(r, \xi_r) + f_{(\eta_r)(\bar{\eta}_r)}(r, \xi_r) + f_{(\bar{\eta}_r)}(r, \xi_r)\eta_r^{d+1} \\ & + f_{(\eta_r)}(r, \xi_r)\bar{\eta}_r^{d+1} + f(r, \xi_r)\zeta_r^{d+1} + f(r, \xi_r)\eta_r^{d+1}\bar{\eta}_r^{d+1}\Big]e^{-\varphi_r}\, dr\Big\}, \end{aligned} \tag{20}$$

where the arguments (t, x) of $\xi_\cdot(t, x), \varphi_\cdot(t, x)$ are omitted for simplicity; the processes η_s, η_s^{d+1} are as in Lemma 2; $\bar{\eta}_s^{d+1}$ is defined by substituting $\bar{\eta}_r$ for η_r in the definition of η_s^{d+1}; the process $\bar{\eta}_s$ is defined as the solution of equation (5) with y replaced by $\bar{y}$, that is, $\bar{\eta}_s$ is the solution of equation (16) for $\bar{\eta}_t^0 \equiv \bar{y}$; finally, ζ_s, ζ_s^{d+1} are defined by (17), (18) for $\zeta_t^0 \equiv 0$, $\zeta_t^{d+1,0} \equiv 0$, $\xi_r = \xi_r(t, x)$.

Note also that $v_{(y)(y)}(t, x)$ can be represented by a slightly shorter expression than (20), if one notices that $\bar{\eta}_s$, $\bar{\eta}_s^{d+1}$ are indistinguishable from η_s, η_s^{d+1} respectively, for $\bar{y} = y$, by the uniqueness theorem for solutions of stochastic equations.

6. PROBLEM. Lemma 2 and Remark 5 state that the first and the second-order differentiation of the right-hand side of (11) with respect to x can be performed inside the expectation and the integral signs, and one can apply the usual rules for differentiating composite functions treating the derivatives of $\xi_s(t, x)$ with respect to x just as if $\xi_s(t, x)$ would satisfy an ordinary rather than stochastic equation. In connection with this prove that under the assumptions of Remark 3 we have $u^{-1}[\xi_s(t, x + uy) - \xi_s(t, x)] \to \eta_s$ in probability as $u \to 0$, $s \geq t$, that is, η_s is the *derivative of $\xi_s(t, x)$ in probability with respect to x along the vector y*. State and prove an analogous assertion for ζ_s under the assumptions of Theorem 4.

7. REMARK. Formula (20) and Lemma 2 enable us to analyze the problems pertaining to the existence of derivatives of v with respect to x of order higher than two, provided that suitable conditions are imposed on σ, b, c, f, g.

In Sec. VI.1 we will need representations of $v_{(y)(\bar{y})}$ similar to (12), (20), but with processes η, ζ, defined by *different* equations. This motivates the following remark.

8. REMARK. Let us assume that the functions g, σ, b in Lemma 1 depend also on a parameter q with values in E_{d_0}, and that $g(x,q)$, $\sigma(r,x,q)$, $b(r,x,q)$ are continuous in (x,q) and differentiable with respect to x continuously in (x,q) for any $r>0$, and that $g(x,q)$ satisfies condition (2) for $x, y \in E_d$, $|y|=1$, for every q. Consider the following stochastic equation in E_{d_0}:

$$q_s = q_t^0 + \int_t^s \tilde{\sigma}(r, q_r)\, dw_r + \int_t^s \tilde{b}(r, q_r)\, dr, \qquad s \geq t \tag{21}$$

assuming that the coefficients $\tilde{\sigma}$, $\tilde{b}$ satisfy the condition of the introduction to this chapter and do not depend on ω. Suppose that the system comprising equation (21) and the two equations

$$\xi_s = \xi_t^0 + \int_t^s \sigma(r, \xi_r, q_r)\, dw_r + \int_t^s b(r, \xi_r, q_r)\, dr, \qquad s \geq t, \tag{22}$$

$$\eta_s = \eta_t^0 + \int_t^s \sigma_{(\eta_r)}(r, \xi_r, q_r)\, dw_r + \int_t^s b_{(\eta_r)}(r, \xi_r, q_r)\, dr, \qquad s \geq t, \tag{23}$$

is solvable by Euler's method in the mean. Incidentally, let us point out that (23) involves the derivatives of σ, b with respect to x only and $\xi_s, \eta_s \in E_d$. Let $q_s(t,q)$, $\xi_s(t,x,q)$, $\eta_s(t,x,y,q)$ denote a solution of system (21)–(23) with $q_t^0 \equiv q$, $\xi_t^0 \equiv x$, $\eta_t^0 \equiv y$. We assert that under these conditions the function

$$v(t,x,q) := \mathbf{E}\, g(\xi_T(t,x,q), q_T(t,q))$$

is differentiable with respect to x, continuously in (t,x,q), in $[0,T] \times E_d \times E_{d_0}$, and for any $x, y \in E_d$, $t \in [0,T]$,

$$v_{(y)}(t,x,q) := \sum_{i=1}^{d} y^i v_{x^i}(t,x,q) = \mathbf{E}\, g_{(\eta_T)}(\xi_T, q_T),$$

where $\xi_T = \xi_T(t,x,q)$, $\eta_T = \eta_T(t,x,y,q)$, $q_T = q_T(t,q)$.

In view of the differentiation rule formulated in Problem 6, this assertion is quite natural, since the derivative of $q_s(t,q)$ in probability with respect to x is zero. A formal proof can be obtained by almost literally repeating the proof of Lemma 1, using an obvious modification of Lemma 4.5.

9. EXERCISE. As we saw in the proofs of Lemma 2 and Theorem 4, both ultimately follow from Lemma 1. Following Remark 8, state and prove similar assertions in the case when c, f depend on q as well.

8. Derivation of the Kolmogorov equations in the homogeneous case

We will now study the smoothness of the functions like (6.13) when $D = E_d$, $g = 0$, and the functions σ, b are independent not only of ω but also of time. Thus, we will assume that σ, b depend only on x. Clearly, for such functions the condition that their first derivatives be bounded implies the validity of inequalities (1.3), (1.4) and, therefore, of the assumptions of Theorem 1.1. Moreover, the solutions of equation (0.2) will possess the homogeneity property formulated in Corollary 4.4.

Fix constants $\delta, K \in (0, \infty)$ and recall that the notations $v_{(y)}, v_{(y)(z)}$ were introduced in (I.4.15), (7.1).

1. THEOREM. *Let the first derivatives of $\sigma(x), b(x), c(x), f(x)$ be continuous and bounded on E_d. Assume also that for all $x, y \in E_d$ with $|y| = 1$,*

$$|f(x)| + |c_{(y)}(x)| + |f_{(y)}(x)| \le K, \tag{1}$$

$$c(x) \ge \delta, \tag{2}$$

$$\|\sigma_{(y)}(x)\|^2 - |\sigma^*_{(y)}(x)y|^2 + 2(y, b_{(y)}(x)) \le 2(c(x) - \delta). \tag{3}$$

Then the function

$$v(x) = \mathbf{E} \int_0^\infty f(\xi_s(x)) e^{-\varphi_s(x)}\, ds, \tag{4}$$

where $\xi_s(x), \varphi_s(x)$ are as in (6.12), *is defined and continuously differentiable on E_d, and $|v(x)| + |v_{(y)}(x)| \le N(\delta, K)$ for all $x, y \in E_d$, $|y| = 1$.*

PROOF. By Remark 7.3 and Lemma 7.2, for any $T \in (0, \infty)$ the function

$$v_T(x) = \mathbf{E} \int_0^T f(\xi_s(x)) e^{-\varphi_s(x)}\, ds \tag{5}$$

is continuously differentiable with respect to x and its derivative $v_{T(y)}(x)$ is equal to the right-hand side of (7.12) with $t = 0, g = 0$, which we denote by $u_T(x, y)$ for $T \in [0, \infty]$. Suppose that for all $x, y \in E_d$, $|y| = 1$, $s \ge 0$, we have

$$\mathbf{E} \int_s^\infty (|\eta_r| + |\eta_r^{d+1}|) e^{-\varphi_r(x)}\, dr \le N e^{-\varepsilon s}, \tag{6}$$

where $N = N(\delta, K)$, $\varepsilon = \varepsilon(\delta, K) > 0$. Then $u(x, y) := u_\infty(x, y)$ is finite by assumption (1) and, moreover, $|u(x, y)| \le N(\delta, K)$ and $u(x, y) - u_T(x, y) \to 0$ uniformly in $x, y \in E_d$, $|y| = 1$, as $T \to \infty$. In addition, by (1) and (2),

$$\int_s^\infty |f(\xi_r(x))| e^{-\varphi_r(x)}\, dr \le K \int_s^\infty e^{-\delta r}\, dr = K\delta^{-1} e^{-\delta s},$$

$$|v| \le K\delta^{-1}, \qquad |v - v_T| \le K\delta^{-1} e^{-\delta T},$$

and $v_T \to v$ uniformly on E_d. It follows by a well-known theorem of calculus from the uniform convergence of v_T, u_T to v, u that v is continuously differentiable and

$v_{(y)}(x) = u(x,y)$. Since estimates of $|v|$, $|u|$ have already been obtained, it remains only to prove (6).

Note that

$$
\begin{aligned}
\mathbf{E}\int_s^\infty |\eta_r^{d+1}|e^{-\varphi_r(x)}\,dr &\le K\delta^{-1}\mathbf{E}\int_s^\infty\int_0^r |\eta_u|\,du\,c(\xi_r(x))e^{-\varphi_r(x)}\,dr \\
&= K\delta^{-1}\mathbf{E}|\eta_s|e^{-\varphi_s(x)} + K\delta^{-1}\mathbf{E}\int_s^\infty |\eta_r|e^{-\varphi_r(x)}\,dr,
\end{aligned}
\tag{7}
$$

and we need to consider only the first term on the right.

For convenience, we will omit the argument x. It follows from (7.5) by Itô's formula (it is convenient first to compute $d|\eta_s|^2$ by Lemma III.8.5, then to compute $d\sqrt{\alpha_s+\varepsilon}$ for the scalar process $\alpha_s = |\eta_s|^2$ by using Theorem III.8.7, and finally to compute $d\left((|\eta_s|^2+\varepsilon)^{1/2}e^{-\varphi_s}\right)$ again by Lemma III.8.5) that for any $\varepsilon > 0$ (a.s.)

$$
\begin{aligned}
(|\eta_s|^2+\varepsilon)^{1/2}e^{-\varphi_s} &= (|y|^2+\varepsilon)^{1/2} + \frac{1}{2}\int_0^s \Big[(|\eta_r|^2+\varepsilon)^{-1/2}\|\sigma_{(\eta_r)}(\xi_r)\|^2 \\
&\quad - (|\eta_r|^2+\varepsilon)^{-3/2}|\sigma^*_{(\eta_r)}(\xi_r)\eta_r|^2 + 2(|\eta_r|^2+\varepsilon)^{-1/2}(\eta_r, b_{(\eta_r)}(\xi_r)) \\
&\quad - 2c(\xi_r)(|\eta_r|^2+\varepsilon)^{1/2}\Big]e^{-\varphi_r}\,dr + \int_0^s (|\eta_r|^2+\varepsilon)^{-1/2}\eta_r^*\sigma_{(\eta_r)}(\xi_r)e^{-\varphi_r}\,dw_r.
\end{aligned}
\tag{8}
$$

Letting $\varepsilon \downarrow 0$ and using Theorem III.6.6 and the Dominated Convergence Theorem $(0^{-1}\cdot 0 := 0)$, we obtain

$$
\begin{aligned}
|\eta_s|e^{-\varphi_s} &= |y| + \frac{1}{2}\int_0^s \Big[|\eta_r|^{-1}\|\sigma_{(\eta_r)}(\xi_r)\|^2 - |\eta_r|^{-3}|\sigma^*_{(\eta_r)}(\xi_r)\eta_r|^2 \\
&\quad + 2|\eta_r|^{-1}(\eta_r, b_{(\eta_r)}(\xi_r)) - 2c(\xi_r)|\eta_r|\Big]e^{-\varphi_r}\,dr + \int_0^s |\eta_r|^{-1}\eta_r^*\sigma_{(\eta_r)}(\xi_r)e^{-\varphi_r}\,dw_r
\end{aligned}
\tag{9}
$$

(a.s). Since system (0.2), (7.3) is solvable by Euler's method in the mean, it follows by Remark 3.2 that the last stochastic integral is a martingale. Calculating expectations in (9) and using condition (3), we conclude that

$$
\mathbf{E}|\eta_s|e^{-\varphi_s} + \delta\mathbf{E}\int_0^s |\eta_r|e^{-\varphi_r}\,dr \le |y|. \tag{10}
$$

Finally, under assumptions (3), (2) the pair c,δ can be replaced by $c-\delta/2, \delta/2$. Hence, by (10),

$$
\mathbf{E}|\eta_s|e^{-\varphi_s+\delta s/2} \le |y|, \qquad \mathbf{E}|\eta_s|e^{-\varphi_s} \le |y|e^{-\delta s/2},
$$

and together with (7) this immediately implies (6). □

2. REMARK. As stated before (7), $v_{(y)}(x)$ is exactly the right-hand side of (7.12) with $t=0$, $T=\infty$, $g\equiv 0$.

In Sec. 7, which was concerned with diffusion processes on a finite time interval, conditions like (3) did not arise. It turns out that when $T=\infty$, condition (3) is, generally speaking, necessary for the differentiability of v.

3. EXAMPLE. Take $d=d_1=1$, and let $\sigma(x)=x, b(x)=\beta x$ for $|x|\le 2$ and $c(x)=\nu$, $f(x)=0$ for $|x|\le 1$, where $\nu>0$, $\beta\in(-\infty,\infty)$ are constants. Define $c(x)$, $f(x)$ outside the interval $[-1,1]$ in such a way that $c(x)\ge\nu$, $f(x)>0$, and c, f be smooth on $(-\infty,\infty)$, bounded and have bounded derivatives. Similarly, define σ, b outside the interval $[-2,2]$ so that they have bounded derivatives on $(-\infty,\infty)$ and are, say, also bounded. Define

$$\tau_\rho(x)=\inf\{s\ge 0\colon |\xi_s(x)|\ge\rho\},\qquad \tau(x)=\tau_1(x).$$

For $|x|\le 2$ we have $|\xi_s(x)|\le 2$ if $s\le\tau_2(x)$. Hence

$$\begin{aligned} d\xi_{s\wedge\tau_2(x)}(x) &= \sigma(\xi_{s\wedge\tau_2(x)}(x))I_{s\le\tau_2(x)}\,dw_s+b(\xi_{s\wedge\tau_2(x)}(x))I_{s\le\tau_2(x)}\,ds\\ &=\xi_{s\wedge\tau_2(x)}(x)\big(I_{s\le\tau_2(x)}\,dw_s+\beta I_{s\le\tau_2(x)}\,ds\big).\end{aligned}$$

Together with Lemma IV.3.1, this enables us to find $\xi_{s\wedge\tau_2(x)}(x)$, and assert that for all $s\in[0,\tau_2(x)]$ (a.s.)

$$\xi_s(x)=xe^{w_s+(\beta-\frac12)s}. \tag{11}$$

Hence it is easily deduced (for example, by contradiction) that

$$v(1)\ge \mathbf{E}\int_0^{\tau_2(1)} f(\xi_s(1))e^{-\varphi_s(1)}\,ds>0,\qquad v(-1)>0.$$

In addition, it follows from (11) that for any $x\in(0,1]$ the process $\xi_s(x)$ takes (a.s.) the value 1 at time $\tau(x)$ (if the latter is finite), while the time $\tau(x)$ itself is almost surely the first exit time of the process $w_s+(\beta-\frac12)s+\ln x$ from $(-\infty,0)$. In particular, by Exercise II.6.7 (or Exercise II.10.3), for $x\in(0,1]$

$$\mathbf{E}\,e^{-\nu\tau(x)}=x^\lambda,\qquad \lambda=\Big[\Big(\beta-\frac12\Big)^2+2\nu\Big]^{1/2}-\beta+\frac12.$$

Furthermore, obviously,

$$\begin{aligned}\int_0^\infty f(\xi_s(x))e^{-\varphi_s(x)}\,ds &= e^{-\varphi_{\tau(x)}(x)}\int_{\tau(x)}^\infty f(\xi_s(x))\exp\Big(-\int_{\tau(x)}^s c(\xi_r(x))\,dr\Big)ds\\ &= e^{-\nu\tau(x)}\int_0^\infty f(\xi_{\tau(x)+s}(x))\exp\Big(-\int_0^s c(\xi_{\tau(x)+r}(x))\,dr\Big)ds.\end{aligned}$$

It now follows by Theorem 5.1 (see also Lemma II.8.2) and Corollary 4.4 that

$$v(x) = \mathbf{E}\left[e^{-\nu\tau(x)}\mathbf{E}\int_0^\infty f\left(\xi_{u+s}(u,y)\right)\right.$$

$$\left.\times\exp\left(-\int_0^s c\left(\xi_{u+r}(u,y)\right)dr\right)ds\Big|_{y=\xi_u(x),u=\tau(x)}\right]$$

$$= \mathbf{E}\,e^{-\nu\tau(x)}v\left(\xi_{\tau(x)}(x)\right).$$

If $x \in (0,1]$, this equals $v(1)x^\lambda$ by the aforesaid. Similarly, $v(x) = v(-1)|x|^\lambda$ for $x \in [-1,0)$, and obviously $\xi_s(0) = 0$ (a.s.), $v(0) = 0$. Thus a necessary condition for v to be differentiable at zero is that $\lambda > 1$. It is easily verified that this is equivalent to the inequality $\beta < \nu$, which, in turn, is true if and only if condition (3) holds for some $\delta > 0$. In this case, therefore, condition (3) cannot be weakened or, a fortiori, eliminated.

4. EXERCISE. The functions σ, b, c, f of the previous example can be extended in such a way that, besides the original conditions, they will satisfy the relations $\sigma(-x) = -\sigma(x)$, $b(-x) = -b(x)$, $c(-x) = c(x)$, $f(-x) = 2f(x)$ for $x \geq 0$. Prove that in that case $v(-x) = 2v(x)$ for $x \geq 0$.

5. THEOREM. *Under the assumptions of Theorem 1, let σ, b, c, f be twice continuously differentiable with respect to x, and suppose that for all $x, y \in E_d$, $y = 1$, we have*

$$\|\sigma_{(y)}(x)\| + \|\sigma_{(y)(y)}(x)\| + |b_{(y)(y)}(x)| + |c_{(y)(y)}(x)| + |f_{(y)(y)}(x)| \leq K, \tag{12}$$

$$\|\sigma_{(y)}(x)\|^2 + 2(y, b_{(y)}(x)) \leq c(x) - \delta. \tag{13}$$

Then v as defined in (4) is twice continuously differentiable with respect to x, its second derivatives are bounded from above in absolute value by a constant $N(\delta, K)$, and v satisfies equation (6.14) on E_d.

PROOF. The last assertion follows from the first two and Theorem 6.4(b). As for the remaining assertions, we will again use notation (5), observing that now, by Theorem 7.4, the functions $v_T(x)$ are twice continuously differentiable with respect to x. We saw in the proof of Theorem 1 that v_{Tx^i} tends to v_{x^i} uniformly in x. Thus it suffices to prove that the functions $v_{Tx^ix^j}$ have limits as $T \to \infty$, uniformly in x, and to show that $|v_{Tx^ix^j}| \leq N(\delta, K)$. It is convenient to consider $v_{T(y)(y)}$ instead of $v_{Tx^ix^j}$. Since

$$4v_{Tx^ix^j} = v_{T(e_i+e_j)(e_i+e_j)} - v_{T(e_i-e_j)(e_i-e_j)},$$

where e_k are the basis vectors, it is actually sufficient to consider only $v_{T(y)(y)}$. Using Remark 7.5, we express $v_{T(y)(y)}$ as in (7.20), taking $t = 0$, $g \equiv 0$, $\bar{y} = y$, $\bar{\eta} = \eta$, $\bar{\eta}^{d+1} = \eta^{d+1}$. Starting from this formula and using arguments similar to those based on (6) and the inequalities

$$|f_{(\eta)(\eta)}| \leq K|\eta|^2, \quad |f_{(\eta)}\eta^{d+1}| \leq K|\eta|^2 + K|\eta^{d+1}|^2, \quad |\eta_s^{d+1}| \leq Ks\sup_{r\leq s}|\eta_r|, \tag{14}$$

we see that to prove the theorem it suffices to show that

$$\text{(15)}\qquad \mathbf{E}\, e^{-\varphi_s} \sup_{r\le s} |\eta_r|^2 \le N e^{-\delta s/2}, \qquad \mathbf{E}\, e^{-\varphi_s} \sup_{r\le s} |\zeta_r| \le N e^{-\delta s/2} |y|^2,$$

where $N = N(\delta, K)$, η_s is, as before, the solution of equation (7.5) with $t = 0$ and ζ_s the solution of equation (7.17) with $t = 0$, $\zeta_t^0 = 0$, $\xi_r = \xi_r(x)$, $\bar\eta_r = \eta_r$.

Using Itô's formula (we omit the argument x), we see that (a.s.)

$$\text{(16)}\qquad \begin{aligned} |\eta_s|^2 e^{-\varphi_s} = |y|^2 &+ \int_0^s \Big[\big\|\sigma_{(\eta_r)}(\xi_r)\big\|^2 + 2(\eta_r, b_{(\eta_r)}(\xi_r)) - c(\xi_r)|\eta_r|^2\Big] e^{-\varphi_r}\, dr \\ &+ 2\int_0^s \eta_r^* \sigma_{(\eta_r)}(\xi_r) e^{-\varphi_r}\, dw_r. \end{aligned}$$

After calculating the expectation of both sides of this equality, and taking into account condition (13), we obtain

$$\text{(17)}\qquad \mathbf{E}\,|\eta_s|^2 e^{-\varphi_s} + \delta \mathbf{E} \int_0^s |\eta_r|^2 e^{-\varphi_r}\, dr \le |y|^2.$$

Hence, again using (16), dropping the second term on the right and using the Davis inequality, we obtain

$$\text{(18)}\qquad \begin{aligned} \mathbf{E} \sup_{r\le s} |\eta_r|^2 e^{-\varphi_r} &\le |y|^2 + 6K\,\mathbf{E}\Big(\int_0^s |\eta_r|^4 e^{-2\varphi_r}\, dr\Big)^{1/2} \\ &\le |y|^2 + 6K\,\mathbf{E} \sup_{r\le s} |\eta_r| e^{-\frac{1}{2}\varphi_r} \Big(\int_0^s |\eta_r|^2 e^{-\varphi_r}\, dr\Big)^{1/2} \\ &\le |y|^2 + \frac{1}{2}\mathbf{E} \sup_{r\le s} |\eta_r|^2 e^{-\varphi_r} + 18K^2\,\mathbf{E}\int_0^s |\eta_r|^2 e^{-\varphi_r}\, dr \\ &\le |y|^2 + \frac{1}{2}\mathbf{E} \sup_{r\le s} |\eta_r|^2 e^{-\varphi_r} + 18K^2\delta^{-1}|y|^2. \end{aligned}$$

Combining like terms, we obtain

$$\text{(19)}\qquad \mathbf{E} \sup_{r\le s} |\eta_r|^2 e^{-\varphi_r} \le N(\delta, K)|y|^2.$$

By what was said about (7.17), we see by Itô's formula that for any $\tau \in \mathfrak{M}$ (a.s.)

$$\text{(20)}\qquad \begin{aligned} |\zeta_\tau|^2 e^{-2\varphi_\tau} = &\int_0^\tau \Big[\big\|\sigma_{(\eta_r)(\eta_r)} + \sigma_{(\zeta_r)}\big\|^2 + 2(\zeta_r, b_{(\eta_r)(\eta_r)} + b_{(\zeta_r)}) - 2c|\zeta_r|^2\Big] e^{-2\varphi_r}\, dr \\ &+ 2\int_0^\tau \zeta_r^* \big(\sigma_{(\eta_r)(\eta_r)} + \sigma_{(\zeta_r)}\big) e^{-2\varphi_r}\, dw_r, \end{aligned}$$

where the arguments ξ_r in c and the derivatives of σ, b are also omitted for brevity. Evaluate the expectation of both sides of this equality and use conditions (2), (12), (13), and the inequalities

$$\|\sigma_{(\eta_r)(\eta_r)} + \sigma_{(\zeta_r)}\|^2 \le (1+\varepsilon)\|\sigma_{(\zeta_r)}\|^2 + \Big(1+\frac{1}{\varepsilon}\Big)\|\sigma_{(\eta_r)(\eta_r)}\|^2$$
$$\le \|\sigma_{(\zeta_r)}\|^2 + \varepsilon K^2|\zeta_r|^2 + \Big(1+\frac{1}{\varepsilon}\Big)K^2|\eta_r|^4,$$

$$2\big|(\zeta_r, b_{(\eta_r)(\eta_r)})\big| \le \varepsilon|\zeta_r|^2 + \frac{1}{\varepsilon}\big|b_{(\eta_r)(\eta_r)}\big|^2 \le \varepsilon|\zeta_r|^2 + \frac{1}{\varepsilon}K|\eta_r|^4,$$

where ε is an arbitrary positive number. Suitably selecting ε, depending only on δ, K and proceeding as described, we deduce from (20) and (13) that

$$\mathbf{E}\,|\zeta_\tau|^2 e^{-2\varphi_\tau} \le N(\delta, K)\,\mathbf{E}\int_0^\tau |\eta_r|^4 e^{-2\varphi_r}\,dr.$$

It follows by Theorem III.6.8 and inequalities (18), (19) that

$$\mathbf{E}\sup_{r\le s}|\zeta_r|e^{-\varphi_r} \le N(\delta, K)\,\mathbf{E}\Big(\int_0^s |\eta_r|^4 e^{-2\varphi_r}\,dr\Big)^{1/2} \le N(\delta, K)|y|^2. \tag{21}$$

To derive (15) it suffices, as at the end of the proof of Theorem 1, to replace c, δ in inequalities (19), (21) by $c - \delta/2, \delta/2$ and to observe that, for example,

$$e^{\delta s/2}\,\mathbf{E}\,e^{-\varphi_s}\sup_{r\le s}|\zeta_r| \le \mathbf{E}\sup_{r\le s}|\zeta_r|e^{-\varphi_r+\delta r/2}. \quad \square$$

6. REMARK. Exercise 4 shows that a necessary condition for the second derivatives of v in Example 3 to exist and to be continuous is that $\lambda > 2$. It is easy to prove that the inequality $\lambda > 2$ is equivalent to $1 + 2\beta < \nu$, which in our case is equivalent, in turn, to condition (13) for some δ $(= \nu - 1 - 2\beta)$. Therefore, we cannot, generally speaking, weaken condition (13) if we want the second derivatives of v to be bounded and continuous.

Thus, the situation with regard to conditions (3) and (13) seems to be quite satisfactory. Nevertheless, we stress that they fail to hold in a great many cases when the *conclusions* of Theorems 1, 5 remain, however, valid. For instance, if we replace b in Example 3 by $b + \gamma$, where γ is a constant of sufficiently large absolute value, then v will be continuously differentiable for any ν. This is a particular case of the results of Sec. VI.1, where we will modify conditions (3) and (13) by introducing in the new conditions the functions σ, b themselves, besides their derivatives.

7. REMARK. Inequalities (14), (15) imply not only that $v_{T(y)(y)} \to v_{(y)(y)}$, but also that $v_{(y)(y)}$ can be determined from the formula for $v_{T(y)(y)}$ by formally letting $T = \infty$.

9. Probabilistic solutions of partial differential equations

As we did starting from Sec. 4, we assume that σ, b are independent of ω and that equation (0.2) is solvable by Euler's method.

Using the notation (6.1), let us consider a domain $Q \subset (-\infty, \infty) \times E_d$ and discuss the first part of Theorem 6.4. If there exists a sufficiently smooth function u such that

$$\frac{\partial u}{\partial t} + Lu - cu + f = 0 \tag{1}$$

in $Q_+ = Q \cap \{t \geq 0\}$ and $u = g$ on $\{t \geq 0\} \cap \partial Q$, then under broad assumptions, we can deduce from Itô's formula that u coincides with v in Q_+ and *therefore* v satisfies (1). Similar results are contained in Theorems II.8.6 and II.10.1, and we in fact discussed them in Sec. IV.1 as well. In a sense, Theorem 6.4 contains the converse assertion: the existence of a solution of differential equation is not required but stated (albeit under the assumption that v is a sufficiently smooth function). It thus provides us with means to study the solvability of equations like (6.10), (6.14) by probabilistic methods. In Secs. 7, 8 we saw that in some situations probabilistic methods also enable us to prove the necessary smoothness of v. Thus, if the right-hand side of (5.20) is meaningful and, in particular, if condition (6.9) is satisfied for all $(t, x) \in Q_+$, then v is called the *probabilistic solution* of equation (1), regardless of whether it is smooth or not. Similarly, if condition (6.11) is satisfied and the right-hand side of (6.13) is meaningful, it is called the probabilistic solution of the equation

$$Lu - cu + f = 0. \tag{2}$$

In comparison with methods of the theory of partial differential equations, as far as the existence of solutions is concerned, the probabilistic approach boasts the frequently important advantage that there is always a prospective solution, and one needs only prove that it is sufficiently smooth. As frequently occurs, once one has guessed the solution of a problem, it is far easier to show that it is indeed a solution than simply prove the existence of a solution.

It must be mentioned that in some cases, even for smooth domains and smooth functions g, f, c, the functions (5.20), (6.13) are not even continuous, and equations (1), (2) with the appropriate boundary conditions have no smooth or, for that matter, even continuous solutions. In such cases the probabilistic approach enables us to understand how to define the generalized solution.

1. EXAMPLE. For $d = 2$, $d_1 = 1$ consider the "equations" $d\xi_s^1 = dw_s$, $d\xi_s^2 = dw_s$. Let $D = \{|x| < 2\} \backslash \{|x| \leq 1\}$, $g(x) = |x| - 1$ and define $v(x)$ by formula (6.13) with $f = c = 0$. Noting that $D \subset \{|x^1| < 2\}$ we deduce, say, from Theorem II.10.1 that $\tau(x) < \infty$ (a.s.). Hence it obviously follows that for $x \in D$

$$v(x) = P\Big\{|\xi_{\tau(x)}(x)| = 2\Big\}.$$

In addition, $d(\xi_s^1 - \xi_s^2) = 0$, so that with probability one $\xi_s(x)$ will always remain on the straight line through x parallel to the bisector of the first quadrant. It follows that

$$v(x) = 1 \quad \text{if} \quad |x^1 - x^2| > \sqrt{2}, \quad x \in D. \tag{3}$$

If $|x^1-x^2| \le \sqrt{2}, x^1+x^2 \ge 0, x \in D$, then $v(x)$ is the probability that the process $\xi_s^1(x) + \xi_s^2(x) = x^1 + x^2 + 2w_s$ will reach the first of the points

$$(8-(x^1-x^2)^2)^{1/2}, \qquad (2-(x^1-x^2)^2)^{1/2}$$

earlier than the second one. Together with Example II.6.11 and the obvious fact that $v(x)$ is even, this shows that

$$v(x) = \frac{|x^1+x^2| - (2-(x^1-x^2)^2)^{1/2}}{(8-(x^1-x^2)^2)^{1/2} - (2-(x^1-x^2)^2)^{1/2}}, \tag{4}$$
$$\text{if } |x^1-x^2| \le \sqrt{2},\ x \in D.$$

It follows from (3) and (4) that v is discontinuous on the parts of $\{|x^1-x^2| = \sqrt{2}\}$ contained in D. Note also that equation (2) takes the form

$$\frac{1}{2}u_{x^1x^1} + u_{x^1x^2} + \frac{1}{2}u_{x^2x^2} = 0, \tag{5}$$

and, by the discussion at the beginning of this section, this equation does not have twice continuously differentiable solution in D that is continuous in $\bar{D}$, equal to unity if $|x| = 2$ and to zero if $|x| = 1$. The function v defined by (3) and (4) should be considered as a right generalized solution of equation (5) in D with the specified boundary conditions. Although it indeed does not assume the prescribed boundary value at the point $x_0 = (-\sqrt{2}/2, \sqrt{2}/2)$ since $v(x) \to 1 \ne 0$ as $x \in D$ goes to x_0 and satisfies $x^2 - x^1 > \sqrt{2}$, it is easy to see that it nonetheless satisfies equation (5) if, after introducing new coordinates $y^1 = x^1 + x^2$ and $y^2 = x^1 - x^2$, we rewrite it as $u_{y^1y^1} = 0$.

It turns out that even in more general situations we can describe the probabilistic solutions in terms of derivatives and the operator L (see Crandall, Lions [7] (1983), Lions [35] (1983)). We will not show how to do this, confining ourselves to the following discussion which aims to explain that the function v from (5.20) is in a sense continuous and *assumes* boundary value g.

From now on, whenever we say that a function $\psi(s)$ is continuous on $[a,b]$, $[a,b)$ or $(a,b]$, we will be referring solely to its values on those intervals, even if it is defined on a larger set. Thus, in the case $[a,b]$, $\psi(s)$ will be continuous on (a,b), right continuous at a and left continuous at b.

2. THEOREM. *Let $g(t,x)$ be a Borel function, $f(t,x)$ and $c(t,x)$ functions continuous in x and defined for $t > 0$, $x \in E_d$ (independent of ω). Assume that $c \ge 0$ and f, c belong to H_1 for any x, and for all $T, R \in (0,\infty)$*

$$\int_0^T \sup_{|x|\le R} \big[|f(t,x)| + c(t,x)\big]\,dt < \infty.$$

(a) *Let Γ denote the set of all (t,x) for which condition (6.9) is satisfied and the expression $v(t,x)$ as defined in (5.20) makes sense and is finite. Let $(t,x) \in \Gamma$. Then there exists a set $\Omega' \subset \Omega$ such that $P(\Omega') = 1$ and for any $\omega \in \Omega'$ the function*

$$\alpha(s) := v\big(s, \xi_s(t,x)\big)$$

is defined and continuous in s *on* $[t,\tau(t,x)]\cap[t,\infty)$. *If, in addition,* $\tau(t,x)=\infty$ *at some* $\omega\in\Omega'$, *then at the same* ω

$$\lim_{s\to\infty}\alpha(s)e^{-\varphi_s(t,x)}=0. \tag{6}$$

(b) *Let* g, f, c *be independent of* t *and* $D\subset E_d$ *be a domain. Let* Γ *denote the set of all* x *for which condition* (6.11) *is satisfied and* $v(x)$ *from* (6.13) *is defined and finite. Let* $x\in\Gamma$. *Then there exists a set* $\Omega'\subset\Omega$ *such that* $P(\Omega')=1$ *and for any* $\omega\in\Omega'$ *the function* $\alpha(s):=v(\xi_s(x))$ *is defined and continuous in* s *on* $[0,\tau(x)]\cap[0,\infty)$. *If, in addition,* $\tau(x)=\infty$ *at some* $\omega\in\Omega'$, *then equality* (6), *with* $\varphi_s(x)$ *substituted for* $\varphi_s(t,x)$, *holds at the same* ω.

3. REMARK. Under the assumptions in (a), if $(t,x)\in\Gamma$, $\omega\in\Omega'$, and $\tau(t,x)<\infty$ at ω, then

$$\lim_{s\uparrow\tau(t,x)} v(s,\xi_s(t,x))=v(\tau(t,x),\xi_{\tau(t,x)}(t,x))=g(\tau(t,x),\xi_{\tau(t,x)}(t,x)),$$

since, obviously, $v=g$ outside Q_+ and $(\tau(t,x),\xi_{\tau(t,x)}(t,x))\notin Q_+$. In this sense v takes the boundary value g. A similar situation holds under the assumptions in (b).

To illustrate Theorem 2, we advise the reader to return to Example 1. Theorem 2 is derived immediately from Theorem 4 below by using the device described in the proof of Theorem 6.4 and a simple fact that our assumptions on g, f, c are sufficient to carry this device out.

4. THEOREM. *Let* $F(u,x.)$ *be a real-valued* Q*-insensitive function. Denote*

$$\Gamma=\Big\{(t,x)\colon t\geq 0, x\in E_d,\ \mathbf{E}\big|F(t,\xi.(t,x))\big|<\infty\Big\}.$$

Assume that $\Gamma\neq\emptyset$ *and set* $v(t,x)=\mathbf{E}F(t,\xi.(t,x))$ *for* $(t,x)\in\Gamma$. *Fix* $(t,x)\in\Gamma$ *and define* $\tau(t,x)$, *as usual, as in* (5.17). *Then there exists a set* $\Omega'\subset\Omega$ *such that* $P(\Omega')=1$ *and for* $\omega\in\Omega'$

(a) *the function* $\alpha(s):=v(s,\xi_s(t,x))$ *is defined and continuous in* s *on* $[t,\tau(t,x)]\cap[t,\infty)$,

(b) *if, in addition,* $\tau(t,x)=\infty$, *then*

$$\lim_{s\to\infty}\alpha(s)=F(t,\xi.(t,x)), \tag{7}$$

and the expression on the right is finite.

Postponing the proof of this theorem until the next section, we confine ourselves here to few comments.

5. REMARK. It is stated in (a), in particular, that $(s,\xi_s(t,x))\in\Gamma$ for all $\omega\in\Omega'$, $s\in[t,\tau(t,x)]\cap[t,\infty)$ (cf. Corollary 5.9). In addition, for all $s\in[0,\infty]$ define

$$\beta(s)=I_{\Omega'}\alpha((t\vee s)\wedge\tau(t,x)),$$

where for $\omega\in\Omega'$, $s=\tau(t,x)=\infty$ the right-hand side is understood as the right-hand side of (7). Since $P(\Omega')=1$, it follows that $\Omega'\in\mathcal{F}_s$ for any s. In addition, $v_{(\pm)}(t,x):=\mathbf{E}F_\pm(t,\xi.(t,x))$ are Borel functions by Theorem 5.8, and hence $\Gamma=\{(t,x)\colon v_{(+)}(t,x)+v_{(-)}(t,x)<\infty\}$ is a Borel set and $I_\Gamma v$ a Borel function. Hence, by Exercises II.4.16 and II.4.17, $\beta(s)$ is $\mathcal{F}_s$-measurable for $s\geq t$. If $s\leq t$, then obviously $\beta(s)=v(t,\xi_t(t,x))=v(t,x)$ (a.s.). Hence $\beta(s)$ is $\mathcal{F}_s$-measurable for $s\leq t$ as well.

Finally, the $\mathcal{F}_s$-adapted process $\beta(s)$ is continuous by Theorem 4, and if we apply Theorem 5.8 first to F_+ and then to F_-, we see that it is a *martingale* since for any Markov time τ (even if it assumes infinite values)

$$\mathbf{E}\,\beta(\tau) = \mathbf{E}\,\alpha\big((t \vee \tau) \wedge \tau(t,x)\big) = v(t,x).$$

6. REMARK. Assertions like (7) are much stronger than (6) and sometimes lead to somewhat unexpected conclusions. For example, let f be a bounded, say, continuous function on $[0,\infty) \times E_d$. For $u \geq 0$, $x. \in C^u$ define

$$F(u,x.) = \overline{\lim_{T\to\infty}} \frac{1}{T} \int_0^T f(t,x_t)\,dt.$$

Obviously, F is $(-\infty,\infty) \times E_d$-insensitive. Consequently, by Theorem 4, there exists a Borel function v on $[0,\infty) \times E_d$ such that for any $t \geq 0$ and $x \in E_d$ we have

$$\lim_{s\to\infty} v\big(s, \xi_s(t,x)\big) = F\big(t, \xi.(t,x)\big) \tag{8}$$

with probability one. This equality is not trivial, since, for example, $F(t,x.)$ cannot generally be expressed for all $x. \in C^t$ as the limit of $v(s,x_s)$ if we want to have the same v for all $x.$. The function v in (8) *depends* on f, a, b, and if we change a or b, then, generally speaking, v will also change.

10. Proof of Theorem 9.4

In this section we assume that the assumptions of Theorem 9.4 are satisfied. Note that the assertion of the theorem is trivial unless $(t,x) \in Q$ when $\tau(t,x) \neq t$. Therefore, we assume that $(t,x) \in Q$. In addition, bearing in mind that $F = F_+ - F_-$, we easily see that Theorem 9.4 needs to be proved only for nonnegative F. Therefore, below we consider the case $F \geq 0$. The following lemma enables us to restrict further to the case of bounded F.

1. LEMMA. *Let $\beta^n(s)$ be martingales, and assume that $\beta^n(s) \leq \beta^{n+1}(s)$ for all $s \geq 0$, $n = 1, 2, \dots$, $\omega \in \Omega$ and $\sup_n \mathbf{E}\,\beta^n(0) < \infty$. Then there exist a martingale $\beta(s)$ and an event $\Omega' \subset \Omega$ such that $P(\Omega') = 1$ and for any $\omega \in \Omega'$*

$$\lim_{n\to\infty} \sup_{s\geq 0} \big|\beta^n(s) - \beta(s)\big| = 0. \tag{1}$$

PROOF. By Theorem III.6.11(a), for $n \geq m$, $\varepsilon > 0$ we have

$$P\Big\{\sup_s \big|\beta^n(s) - \beta^m(s)\big| \geq \varepsilon\Big\} \leq \frac{1}{\varepsilon}\Big[\mathbf{E}\,\big(\beta^n(0) - \beta^1(0)\big) - \mathbf{E}\,\big(\beta^m(0) - \beta^1(0)\big)\Big].$$

By the Monotone Convergence Theorem, the right-hand side vanishes as $n, m \to \infty$. As in Lemma III.6.18, this implies the existence of a continuous $\mathcal{F}_s$-adapted process $\beta(s)$ for which (1) is valid in the sense of convergence in probability. Then by Exercise III.6.20 and Fatou's Lemma equality (1) is true a.s., provided we take the lower limit instead of the ordinary limit. Let Ω' denote the set on which this modification of (1) is valid.

Clearly, $P(\Omega') = 1$. Moreover, since $\beta^n(s)$ increases with n, it follows that $\beta^n(s) \leq \beta(s)$. Hence the expression after the limit sign in (1) decreases with n on Ω'. In view of the aforesaid, this proves (1) on Ω'. Finally, for any $\tau \in \mathfrak{M}$, we have $\mathbf{E}\,\beta^n(\tau) = \mathbf{E}\,\beta^n(0)$.

Letting $n \to \infty$ here and keeping in mind that $\beta^n(\tau) \uparrow \beta(\tau)$, $\beta^n(0) \uparrow \beta(0)$ (a.s), we conclude by the Monotone Convergence Theorem that $\mathbf{E}\,\beta(\tau) = \mathbf{E}\,\beta(0)$, and moreover $\mathbf{E}\,\beta^1(0) \le \mathbf{E}\,\beta(0) \le \sup_n \mathbf{E}\,\beta^n(0) < \infty$. □

If Theorem 9.4 is true for bounded F, then for any nonnegative F we can define $F^n = F \wedge n$, $n = 1, 2, \dots$, and let v^n, Ω'_n, β^n denote the function v, set Ω', and martingale β from Theorem 9.4 and Remark 9.5 for the function F^n. Let $\Omega_0 = \cap_n \Omega'_n$, $\tilde{\beta}^n = \beta^n$ on Ω_0, $\tilde{\beta}^n = 0$ outside Ω_0. Since $v^n \uparrow v$ by the Monotone Convergence Theorem, it follows that

$$(2) \qquad \tilde{\beta}^n(s) = v^n\big((s \vee t) \wedge \tau(t,x), \xi_{(s\vee t)\wedge\tau(t,x)}(t,x)\big) \uparrow \alpha\big((s \vee t) \wedge \tau(t,x)\big)$$

for $s \ge 0$, $\omega \in \Omega_0$. In addition, $\tilde{\beta}^n$ are still martingales since $P(\Omega_0) = 1$ and $\mathbf{E}\,\tilde{\beta}^n(0) = v^n(t,x) \le v(t,x) < \infty$, and the convergence in (2) is uniform in $s \in [0, \infty)$ by Lemma 1 if $\omega \in \Omega_0 \cap \Omega'_0 =: \Omega'$ with some Ω'_0 such that $P(\Omega'_0) = 1$. Observe that also $P(\Omega') = 1$. Hence all the assertions of the theorem about F obviously follow. It remains, therefore, to consider bounded F.

We now consider a particular pair Q, F.

2. Lemma. *Let g be a bounded Borel function on E_d. Let $T \in (0, \infty)$, $v(t,x) := \mathbf{E}\,g\big(\xi_T(t,x)\big)$ for $t \in [0,T]$. Then for any $t \in [0,T)$, $x \in E_d$, for almost all ω the function $v\big(s, \xi_s(t,x)\big)$ is continuous in s on $[t,T]$.*

Proof. If g is continuous, our assertion follows from Lemma 5.3, by which v is continuous in $[0,T] \times E_d$. Let $g = I_\Gamma$, where Γ is a closed set. As in the proof of Theorem I.5.4, construct continuous functions g^n such that $0 \le g^{n+1} \le g^n \le 1$, $g^n \to g$, and define

$$v^n(t,x) = \mathbf{E}\,g^n\big(\xi_T(t,x)\big)$$

for $t \in [0,T]$. By the Dominated Convergence Theorem, $v^n \downarrow v$. Hence, for any $t \in [0,T)$, $x \in E_d$, $s \in [0,T]$ we have

$$(3) \qquad -\alpha^n(s) := -v^n\big(s \vee t, \xi_{s\vee t}(t,x)\big) \uparrow -v\big(s \vee t, \xi_{s\vee t}(t,x)\big).$$

It follows from the continuity of $\alpha^n(s)$ on $[0,T]$ and from Example 5.12 that $\alpha^n(s \wedge T)$ is a martingale on $[0,\infty)$ for any n. By Lemma 1 the convergence in (3) is uniform in $s \in [0,T]$ on a set of probability 1. This proves the lemma in the particular case.

Now let Λ denote the set of all $A \in \mathfrak{B}(E_d)$ such that the assertion of the lemma holds for $g = I_A$. It follows immediately from Lemma 1 and Remark 9.5 that Λ is a λ-system. Since Λ contains the π-system of all closed sets, $\Lambda = \mathfrak{B}(E_d)$. Having proved the lemma for indicators of all Borel sets, we obviously obtain that it is true for $g \in S\big(\mathfrak{B}(E_d)\big)$ as well. It remains to observe that any bounded Borel function g can be approximated *uniformly* by functions from $S\big(\mathfrak{B}(E_d)\big)$. □

Using this lemma, we can prove one part of Theorem 9.4.

3. Lemma. *If F is bounded, then $\alpha(s)$ is continuous in s on $[t, \tau(t,x))$ with probability one.*

PROOF. We define functions $v_n(s,y)$ for $s \geq 0$, $y \in E_d$, letting $t(k,n) = k/n$, $\varkappa_n(s) = t(k+1,n)$ for $s \in \big[t(k,n), t(k+1,n)\big)$, $k = 0, 1, \ldots$, $n = 1, 2, \ldots$,

$$v_n(s,y) = \mathbf{E}\, v\big(\varkappa_n(s), \xi_{\varkappa_n(s)}(s,y)\big). \tag{4}$$

Let Ω' be the set of all ω such that for any n the function $\alpha_n(s) := v_n\big(s, \xi_s(t,x)\big)$ is right-continuous at all points $t(k,n)$, continuous on any interval $\big(t(k,n), t(k+1,n)\big)$, and such that for all k

$$\alpha_n\big(t(k+1,n)-\big) = v\big(t(k+1,n), \xi_{t(k+1,n)}(t,x)\big). \tag{5}$$

By Lemma 2, $P(\Omega') = 1$. We will see below that for any $\omega \in \Omega'$ and any $T \in \big[t, \tau(t,x)\big)$ we have

$$\sup_{s\in[t,T]} \Big(\big|\alpha_n(s) - \alpha(s)\big| + \big|\alpha_n(s-) - \alpha(s)\big|\Big) \longrightarrow 0, \tag{6}$$

where $\alpha_n(t-) := \alpha_n(t)$. Consequently, for $\omega \in \Omega'$, $T \in \big[t, \tau(t,x)\big)$, the function $\alpha(s)$ is right-continuous on $[t,T]$, as a uniform limit of the right-continuous functions $\alpha_n(s)$, and $\alpha(s)$ is left-continuous on $[t,T]$ as a uniform limit of the left-continuous functions $\alpha_n(s-)$. Hence $\alpha(s)$ is continuous on $[t,T]$, and since T is arbitrary, the assertion of the lemma now follows.

Thus it remains to prove (6). Fix $\omega \in \Omega'$ and $T \in \big[t, \tau(t,x)\big)$. The trajectory of $\xi_s(t,x)$ on $[t,T]$ is a compact set $\Gamma \subset Q$, separated from ∂Q by a strictly positive distance. Therefore there exists $\varepsilon > 0$ such that for any $(s,y) \in \Gamma$, $s \leq u \leq s + \varepsilon$, $|z - y| \leq 2\varepsilon$ we have $(u,z) \in Q$. Denote

$$\delta_n(s,y) = P\Big\{\sup_{[s,s+n^{-1}]} \big|\xi_u(s,y) - y\big| \geq 2\varepsilon\Big\}.$$

It follows easily from Lemma 5.3 that if $(s_0, y_0) \in [0,\infty) \times E_d$ and $s_n \longrightarrow s_0$, $y_n \longrightarrow y_0$, then

$$\varlimsup_{n\to\infty} \delta_n(s_n, y_n) \leq P\Big\{\big|\xi_s(s,y) - y\big| \geq \varepsilon\Big\} = 0.$$

In particular, $\delta_n \longrightarrow 0$ uniformly on any compact set in $[0,\infty) \times E_d$, and

$$\Delta_n(s) := \delta_n\big(s, \xi_s(t,x)\big) \longrightarrow 0 \tag{7}$$

uniformly on $[t,T]$ (for any $\varepsilon > 0$ and any $T \in [t,\infty)$).

Note also that by Theorem 5.8

$$\begin{aligned} v(s,y) &= \mathbf{E}\, v\big(\tau(s,y) \wedge \varkappa_n(s), \xi_{\tau(s,y)\wedge\varkappa_n(s)}(s,y)\big) \\ &= \mathbf{E}\, v\big(\varkappa_n(s), \xi_{\varkappa_n(s)}(s,y)\big) I_{\tau(s,y)>\varkappa_n(s)} \\ &\quad + \mathbf{E}\, v\big(\tau(s,y), \xi_{\tau(s,y)}(s,y)\big) I_{\tau(s,y)\leq\varkappa_n(s)}. \end{aligned}$$

Expressing the expectation in (4) as a similar sum, we subtract one representation from another and use the fact that v is bounded and that by the choice of ε the obvious inequality $P\{\tau(s,y) \leq \varkappa_n(s)\} \leq \delta_n(s,y)$ holds for $n \geq \varepsilon^{-1}$, $(s,y) \in \Gamma$. Then we see that $\big|v(s,y) - v_n(s,y)\big| \leq N\delta_n(s,y)$ for $(s,y) \in \Gamma$, $n \geq \varepsilon^{-1}$, where N is a constant, and consequently

$$\big|\alpha_n(s) - \alpha(s)\big| \leq N\Delta_n(s) \tag{8}$$

on $[t,T]$ for $n \geq \varepsilon^{-1}$.

Finally, since $\omega \in \Omega'$, it follows that (5) is true and

$$\sup_{[t,T]} \left|\alpha_n(s-) - \alpha_n(s)\right| = \sup_{t(k+1,n)\in[t,T]} \left|\alpha\big(t(k+1,n)\big) - \alpha_n\big(t(k+1,n)\big)\right|.$$

Together with (8) and the uniformity of the convergence in (7), this implies (6). □

Our next step will be to study the limit of $\alpha(s)$ as $s \uparrow \tau(t,x)$, under the assumption that it exists. We will need the following lemma.

4. LEMMA. *Let $Q(n)$, $n = 1, 2, \dots$, be domains such that $Q(n) \subset Q(n+1)$ and $Q = \cup_n Q(n)$. Define*

$$\tau^n(t,x) = \inf\left\{s \geq t\colon \big(s, \xi_s(t,x)\big) \notin Q(n)\right\}, \qquad \mathcal{L} = \bigcup_n \mathcal{F}_{\tau^n(t,x)}.$$

Then the random variable

$$\zeta(t,x) := \alpha\big(\tau(t,x)\big) I_{\tau(t,x)<\infty} + F\big(t, \xi_\cdot(t,x)\big) I_{\tau(t,x)=\infty} \tag{9}$$

is measurable with respect to the σ-algebra $\sigma(\mathcal{L})$.

PROOF. It is obvious that $\tau^n(t,x) \to \tau(t,x)$. By Exercise II.4.16, this implies that $\tau(t,x)$ is $\sigma(\mathcal{L})$-measurable. Moreover, for any $s \geq t$,

$$\xi_{s\wedge\tau(t,x)}(t,x) = \lim_{n\to\infty} \xi_{s\wedge\tau^n(t,x)}(t,x). \tag{10}$$

By Exercise II.4.17, the expression after the limit sign is $\mathcal{F}_{s\wedge\tau^n(t,x)}$-measurable. Since $s \wedge \tau^n(t,x) \leq \tau^n(t,x)$, it follows from Exercise II.4.16 that it is $\mathcal{F}_{\tau^n(t,x)}$-measurable and $\sigma(\mathcal{L})$-measurable. The same property holds for the left-hand side of (10), and also for the left-hand side of (10) multiplied by the indicator of the set $\{\tau(t,x) < \infty\} \in \sigma(\mathcal{L})$. Letting $s \to \infty$, we see that $\xi_{\tau(t,x)}(t,x) I_{\tau(t,x)<\infty}$ is also $\sigma(\mathcal{L})$-measurable. This proves that the first term in (9) is $\sigma(\mathcal{L})$-measurable and that

$$g\big(\xi_s(t,x)\big) I_{\tau(t,x)=\infty} = g\big(\xi_{s\wedge\tau(t,x)}(t,x)\big) I_{\tau(t,x)=\infty}$$

is $\sigma(\mathcal{L})$-measurable for any Borel function g. From the latter fact the $\sigma(\mathcal{L})$-measurability of the second term in (9) can be easily derived by using the lemma on π- and λ-systems and by taking into account that by definition $F\big(t, \xi_\cdot(t,x)\big) = \widetilde{F}\big(t, \hat{\theta}_t \xi_\cdot(t,x)\big)$, where $\widetilde{F}(t, x_\cdot)$ is $\mathfrak{A}(C)$-measurable. □

5. LEMMA. *Let F be a bounded function and assume that the limit of $\alpha(s)$ as $s \uparrow \tau(t,x)$ exists (a.s.). Then all the assertions of Theorem 9.4 hold.*

PROOF. Let $\alpha(s) \to \zeta_1$ as $s \uparrow \tau(t,x)$ (a.s.). Let $Q(n)$ be bounded domains such that $\bar{Q}(n) \subset Q(n+1)$, $Q = \cup_n Q(n)$. Defining $\tau^n(t,x)$ as in Lemma 4, we use formula (5.14) with $\tau = \tau^m(t,x), \tau(t,x)$. Then, if $n \leq m$, $A \in \mathcal{F}_{\tau^n(t,x)}$, $\eta = I_A$, we obtain

$$\mathbf{E}\eta\zeta(t,x) = \mathbf{E}\eta\alpha\big(\tau^m(t,x)\big),$$

where ζ is defined by (9). When $m \to \infty$ this gives

$$\mathbf{E}\eta\zeta(t,x) = \mathbf{E}\eta\zeta_1. \tag{11}$$

Since this equality is true for the indicators of all sets in the π-system $\mathcal{L}$ defined as in the previous lemma, it is also true for the indicators of all elements of $\sigma(\mathcal{L})$

and for all $\sigma(\mathcal{L})$-measurable bounded functions η. As ζ_1 is the limit (a.s.) of $\sigma(\mathcal{L})$-measurable quantities $\alpha(\tau^n(t,x))$, it can be chosen $\sigma(\mathcal{L})$-measurable, while $\zeta(t,x)$ is $\sigma(\mathcal{L})$-measurable by the previous lemma. Consequently, we can take $\eta = \zeta(t,x)$ and $\eta = \zeta_1$ in (11). Hence

$$\mathbf{E}\zeta^2(t,x) = \mathbf{E}\zeta(t,x)\zeta_1 = \mathbf{E}\zeta_1^2, \qquad \mathbf{E}\left|\zeta(t,x) - \zeta_1\right|^2 = 0$$

and $\zeta(t,x) = \zeta_1$ (a.s.). It remains only to note that, this fact, together with Lemma 3, proves the continuity of $\alpha(s)$ on $[t,\tau(t,x)] \cap [t,\infty)$ (a.s.) and equality (9.7) on $\{\tau(t,x) = \infty\}$ (a.s.). □

Again, let $\tau^n(t,x)$ be as in the previous proof. If F is bounded, then by Lemma 3 and Remark 9.5 there exists a martingale $\beta^n(s)$ such that $\beta^n(s) = \alpha(s \wedge \tau^n(t,x))$ for all $s \in [t,\infty)$ at once (a.s.). It is easily seen that we can choose β^n so that $\beta^n(s) = \beta^m(s)$ for $s \le \tau^n(t,x)$ and all $m \ge n$. By Lemma 5, to complete the proof of Theorem 9.4, it suffices to apply the following result, which in fact is a very particular case of a general theorem from the theory of martingales.

6. LEMMA. *Let $\tau^n \in \mathfrak{M}$, $n = 1,2,\dots$ be given Markov times such that $\tau^n \le \tau^{n+1}$ and $\beta(s) = \beta(\omega,s)$ a given bounded real-valued function defined on the set $\cup_n\{(\omega,s)\colon 0 \le s \le \tau^n(\omega)\}$. Assume that the process $\beta(s \wedge \tau^n)$ is a martingale for any n. Denote*

$$\tau = \lim_{n\to\infty} \tau^n, \qquad \gamma = \overline{\lim_{s\uparrow\tau}}\, \beta(s).$$

Then there exists a set $\Omega' \subset \Omega$ of probability 1 such that $\beta(s) \to \gamma$ as $s \uparrow \tau$, $\omega \in \Omega'$, and the process $\gamma(s)$ defined as

$$\beta(s)I_{s<\tau} + \gamma I_{s\ge\tau}$$

on Ω' and zero outside Ω', is a martingale.

PROOF. By Lemma II.8.5, $\mathbf{E}\beta^2(\tau^n) \le \mathbf{E}\beta^2(\tau^{n+1})$. In addition, both sides of this inequality are bounded, and therefore have a finite limit as $n \to \infty$. By the same lemma,

$$\begin{aligned}\mathbf{E}\left|\beta(\tau^n) - \beta(\tau^m)\right|^2 &= \mathbf{E}\beta^2(\tau^n) - \mathbf{E}\beta^2(\tau^m) + 2\mathbf{E}\beta(\tau^m)[\beta(\tau^m) - \beta(\tau^n)] \\ &= \mathbf{E}\beta^2(\tau^n) - \mathbf{E}\beta^2(\tau^m),\end{aligned}$$

for $n \ge m$. The last expression vanishes as $n,m \to \infty$. Note also that by Doob's moment inequality (Theorem III.3.4, see also Theorem III.4.9 (d), (j))

$$\begin{aligned}\mathbf{E}\sup_{s\in[\tau^m,\tau^n]} \left|\beta(s) - \beta(\tau^m)\right|^2 &= \mathbf{E}\sup_{s\ge 0}\left|\beta((s\wedge\tau^n)\vee\tau^m) - \beta(\tau^m)\right|^2 \\ &\le 4\lim_{s\to\infty}\mathbf{E}\left|\beta((s\wedge\tau^n)\vee\tau^m) - \beta(\tau^m)\right|^2 = 4\mathbf{E}\left|\beta(\tau^n) - \beta(\tau^m)\right|^2,\end{aligned}$$

for $n \ge m$. Denoting $I(m) = \cup_{n\ge m}[\tau^m,\tau^n]$ and letting first $n \to \infty$ and then $m \to \infty$, we obtain, by Fatou's Lemma,

$$\underline{\lim}_{m\to\infty} \sup_{s\in I(m)} \left|\beta(s) - \beta(\tau^m)\right| = 0 \quad \text{(a.s.)}. \tag{12}$$

Now, letting $\varepsilon(m)$ denote the expression after the limit sign in (12), we easily see that $\varepsilon(n) \le 2\varepsilon(m)$ for $n \ge m$. This shows that we can replace the lower limit in (12) by the usual limit. Let Ω' be the set of all ω for which this modification of equality

(12) is valid. Then $P(\Omega') = 1$. By Cauchy's test, it follows that for any $\omega \in \Omega'$ such that $\tau^m < \tau$ for all m, $\beta(s)$ has a limit as $s \uparrow \tau$. If $\tau^m = \tau$ for some m, the same is true by the continuity of $\beta(s \wedge \tau^m)$. The limit under consideration obviously equals γ on Ω'. In particular, the set $I(m)$ and function $\beta(s)$ can be replaced in the modified equality (12) on Ω' by $[\tau^m, \tau]$ and $\gamma(s)$. It then follows that $\beta(s \wedge \tau^m)$ converges to $\gamma(s)$ for $\omega \in \Omega'$, uniformly on $[0, \infty)$. The assertion that $\gamma(s)$ is a martingale now follows from the Dominated Convergence Theorem (or from Theorem III.4.9). □

CHAPTER VI

Further Methods for Investigating the Smoothness of Probabilistic Solutions of Differential Equations

In this chapter we will continue our investigation, that began in Sec. V.8, of the smoothness of probabilistic solutions of differential equations for temporally homogeneous processes. First of all, we would like to weaken conditions (V.8.3) and (V.8.13), that is, to broaden the applicability of Theorems V.8.1 and V.8.5. The main methods which make this possible are based on change of measure and random time change. It turns out that these operations, which use functions of the solutions of stochastic equations, modify the original stochastic equations and need not commute with formal differentiation of their solutions with respect to the initial data. One thus obtains new equations for the derivatives of the solutions and new conditions to guarantee that the moments of the derivatives increase at an appropriate rate. The results obtained by this method will be presented in Sec. 1, while their proofs will be given partly in Sec. 1 and in Sec. 3. In Sec. 2 we will investigate when the derivatives of probabilistic solutions of differential equations can be represented by formulas like (V.7.12) and (V.7.20), but with processes η and ξ *different* from those used in Sec. V.7.

We will then go on to consider solutions of stochastic equations up to the first exit time from a *domain*. The reader has noticed surely that after the derivation of the Kolmogorov equations in Sec. V.6, under a not very effective assumption that the functions (V.5.20) and (V.6.13) were smooth (see Theorem V.6.4), we proved the necessary smoothness only for very special domains Q and D, namely, $Q = (-\infty, T) \times E_d$, $D = E_d$. Only such domains will be considered in Secs. 1–3. We are forced to do this because of our method of proving the smoothness, which is actually based on differentiation of expressions of type (V.6.13) with respect to the parameter x under the expectation sign. However, in Secs. 1–3 we will slightly modify the expression in question before the differentiation. But in any case, in order to avoid the need to differentiate $\tau(x)$ and $\tau(t,x)$ with respect to x, we are forced to consider only cases when they are *independent* of x. It is worth mentioning here that if $\tau(x)$ and $\tau(t,x)$ vary with x, they are not only nondifferentiable, but even discontinuous in x (for example, the Cauchy process in Problem II.10.8 is discontinuous in t).

It turns out that under broad assumptions the consideration of probabilistic solutions of differential equations in a domain $D \neq E_d$ *can be reduced* to the case $D = E_d$. We will show how to do this in Sec. 5, which may — perhaps it is even instructive — be read before Sec. 4. Indeed, the proof of the main result of Sec. 5 makes no use of the contents of Sec. 4. In my opinion, however, it is impossible to understand the essence of the rather complicated construction from that proof, unless one is familiar with the ideas of the theory of conditional processes introduced in Sec. 4; the latter are undoubtedly interesting irrespective of their application to probabilistic solutions of differential equations.

Throughout this chapter, we will use the notation and assumptions of Chapter V, beginning with Sec. V.4. In particular, $(\Omega, \mathcal{F}, P)$ is a complete probability space and $(w_t, \mathcal{F}_t)$ a d_1-dimensional Wiener process defined on that space for all $t \in [0, \infty)$, where the σ-algebras $\{\mathcal{F}_t\}$ are complete with respect to $\mathcal{F}$ and P. In addition, we will assume that $\sigma(t,x)$, $b(t,x)$, $c(t,x)$ and $f(t,x)$ are *bounded* Borel functions of (t,x) defined for $t \in (0,\infty)$, $x \in E_d$, where c and f are real-valued, b takes values in E_d and σ in the set of $(d \times d_1)$-matrices (σ, b, c and f are independent of ω). We will assume that σ, b, c and f are continuously differentiable with respect to x for any $t \in (0,\infty)$ and *the first derivatives of σ and b with respect to x are bounded functions of* (t,x). By Remark V.1.2, for any $t \in [0,\infty)$, $x \in E_d$ the Itô stochastic equation

$$\xi_s = x + \int_t^s \sigma(r, \xi_r)\, dw_r + \int_t^s b(r, \xi_r)\, dr, \quad s \geq t, \tag{1}$$

has a unique solution (up to indistinguishability), which we will denote by $\xi_s(t,x)$. Let

$$\varphi_s(t,x) = \int_t^s c\big(r, \xi_r(t,x)\big)\, dr.$$

Finally, recall that $\xi_s(x) = \xi_s(0,x)$ and $\varphi_s(x) = \varphi_s(0,x)$.

1. Some generalizations of Theorems V.8.1 and V.8.5

In addition to the conditions listed in the introduction to this chapter, let us assume, as in Sec. V.8, that σ, b, c, and f are independent of t. We will also need a continuous bounded function $\alpha(x)$ on E_d with values in E_d, and a function $\pi(x,y)$ on $E_d \times E_d$, continuous in (x,y), bounded with respect to x for every y, with values in the set $\mathfrak{P}$ of all skew-symmetric $d_1 \times d_1$ matrices and linear in y (so that $\pi(x,0) = 0$). Fix constants $\delta, K \in (0,\infty)$ and, as usual, set $a = \sigma\sigma^*$.

1. THEOREM. *Suppose that for all $x, y \in E_d$ such that $|y| = 1$,*

$$\begin{aligned} &|\sigma(x)| + |\sigma_{(y)}(x)| + |\alpha(x)| + |\pi(x,y)| + |c(x)| \\ &\qquad + |c_{(y)}(x)| + |f(x)| + |f_{(y)}(x)| \leq K, \qquad c(x) \geq \delta, \end{aligned} \tag{1}$$

$$\begin{aligned} &\left\|\sigma_{(y)}(x) + \frac{1}{2}(\alpha(x),y)\sigma(x) + \sigma(x)\pi(x,y)\right\|^2 \\ &\qquad - \left|\Big(\sigma^*_{(y)}(x) + \frac{1}{2}(\alpha(x),y)\sigma^*(x) + \pi^*(x,y)\sigma^*(x)\Big)y\right|^2 \\ &\qquad + 2\Big(y, b_{(y)}(x) + (\alpha(x),y)b(x)\Big) \leq 2\big(c(x) - \delta\big) + K\big(a(x)y,y\big). \end{aligned} \tag{2}$$

Then the function

$$v(x) = \mathbf{E} \int_0^\infty f\big(\xi_s(x)\big) e^{-\varphi_s(x)} ds \tag{3}$$

is defined and continuously differentiable on E_d, *and* $|v(x)| + |v_x(x)| \leq N(\delta, K)$ *on* E_d.

Before we begin the proof, let us compare conditions (1) and (2) with conditions (V.8.1)–(V.8.3) and illustrate the applications of Theorem 1.

Condition (1) is slightly stronger than (V.8.1), because of the presence of $|\sigma|$, $|\sigma_{(y)}|$ and $|c|$. In addition, Theorem 1 requires b to be bounded (see the introduction to the chapter). However, even if $\alpha = 0$ and $\pi = 0$, condition (2) is much weaker than (V.8.3), because of the term $K\big(a(x)y, y\big)$. If, for example, $c(x) \geq \delta$ and the process $\xi_s(x)$ is *uniformly nondegenerate*, that is, $\big(a(x)y, y\big) \geq \delta$ for all $x \in E_d$ and $|y| = 1$, then we can always choose the constant K so that all the inequalities in (1) and (2) are satisfied. Condition (V.8.3) cannot give this effect.

Note also that condition (2), unlike condition (V.8.3), has certain invariance properties. For example, let us replace c, f in (3) by $\beta c, \beta f$ and $\xi_s(x)$ by a solution of the equation

$$d\xi_s = \sqrt{\beta(\xi_s)}\,\sigma(\xi_s)\,dw_s + \beta(\xi_s)b(\xi_s)\,ds,$$

where β is a sufficiently smooth function, $\beta > 0$ and β, β^{-1} are bounded. It follows from the theorem about random time change (see Sec. IV.2) and Corollary V.4.4 that $v(x)$ remains unchanged. It turns out that condition (2) holds for σ, b, c up to a substitution of δ, K, if and only if it holds for $\sqrt{\beta}\sigma$, βb, βc. Moreover, if α, π were appropriate for the initial functions, then for the new ones we can use $\alpha - (\ln\beta)_x, \pi$.

It is slightly more difficult to explain that change of measure (see Sec. IV.3) does not enable us to weaken condition (2). We will not go into details here, noting only that under change of measure the function f in (3) is multiplied by a martingale exponent ρ_s, and we have to estimate $|\eta_s|\rho_s$ rather than $|\eta_s|$.

Neither does change of the Wiener process weaken condition (2). This procedure involves taking $Q(\xi_r)\,d\bar{w}_r$ instead of dw_r in (0.1), where $Q(x)$ is an orthogonal $(d_1 \times d_1)$-matrix which depends on x in a sufficiently regular way; by Lévy's Theorem the $\bar{w}_r$, subject to the condition $\bar{w}_0 = 0$, is also a d_1-dimensional Wiener process. The point is that $Q^*\pi Q$, $Q^*Q_{(y)} \in \mathfrak{P}$ and, as is readily seen, neither side of (2) will be changed if we replace σ and π by σQ and $Q^*\pi Q - Q^*Q_{(y)}$, respectively. The desire that our conditions remain invariant under replacement of σ by σQ is quite natural, because $a = \sigma\sigma^*$ remains unchanged and, by Corollary V.4.4, v is determined by a, b, c, f only .

Now let us consider a particular case of Theorem 1.

2. EXAMPLE. Let $d = d_1 = 1$, $c > 0$. Assume that the set of all values of $\big(a(x), b(x), b'(x), c(x)\big)$, when x runs through the whole real line E_1, is a compact subset of E_4, and at every point where $a(x) = 0$, $b(x) = 0$ we have

$$b'(x) < c(x). \tag{4}$$

We claim that for some $\delta, K \in (0, \infty)$ and a bounded smooth function α and $\pi \equiv 0$, conditions (1) and (2) hold and, consequently, v is continuously differentiable.

Our claim concerning condition (1) is certainly true. To verify (2) we observe that for $\alpha = -nb$, $\delta = 1/n$, $K = 2n$, and $|y| = 1$ condition (2) becomes

$$b'(x) \leq c(x) - \frac{1}{n} + n\Big(a(x) + |b(x)|^2\Big). \tag{5}$$

Suppose that for any $n = 1, 2, \ldots$ we can find a point x_n at which the inequality converse to (5) holds. Then we can extract from the sequence $\big(a(x_n),\ b(x_n),\ b'(x_n),\ c(x_n)\big)$

a subsequence that converges to $(a(x_0), b(x_0), b'(x_0), c(x_0))$ for some x_0. Clearly, $a(x_n)+|b(x_n)|^2 \le Nn^{-1}$, where $N = \sup b' + 1$. Therefore, $a(x_0)+|b(x_0)|^2 = 0$ and $b'(x_n) \ge c(x_n) - 1/n$, $b'(x_0) \ge c(x_0)$. We have obtained a contradiction to (4), so inequality (5) holds in E_1 for some n and condition (2) is indeed satisfied.

Condition (2) on $\{x\colon a = b = 0\}$ is, so to speak, of a local character: it shows that in Example V.8.3 if $\beta \ge \nu$, then the function v is not continuously differentiable, not because condition (V.8.3) fails to hold on $[-2, 2]$, but because if fails to hold at just one point $x = 0$.

We will derive Theorem 1 from the following two lemmas. Before stating them, we fix an infinitely differentiable function $\varkappa(t)$ on E_1 such that $\varkappa(t) = 1$ for $t \le 1$, $\varkappa(t) = 0$ for $t \ge 2$ and $|\varkappa'| \le 2$. In addition, for $T \in (0, \infty)$, $x \in E_d$ we define

$$v^T(x) = \mathbf{E}\int_0^\infty e^{-\varphi_s(x)} f(\xi_s(x))\varkappa\Big(\frac{s}{T}\Big)\,ds. \tag{6}$$

The integration in (6) may obviously be confined to the interval $(0, 2T)$, and by Lemma V.7.2 (see also Remark V.7.3) the function $v^T(x)$ is continuously differentiable with respect to x.

3. Lemma. *Let the assumptions stated before Theorem 1 be satisfied and let c and f be bounded together with their first-order derivatives with respect to x. Then for any $x, y \in E_d$ (and any constant K)*

$$\begin{aligned} v^T_{(y)}(x) = \mathbf{E}\int_0^\infty \Big\{ & f_{(\eta_s)}\varkappa\Big(\frac{s}{T}\Big) + f(\xi_s)\Big[(\alpha(\xi_s), \eta_s)\varkappa\Big(\frac{s}{T}\Big) \\ & + \eta_s^{d+1}\varkappa\Big(\frac{s}{T}\Big) + \frac{1}{T}\varkappa'\Big(\frac{s}{T}\Big)\eta_s^{d+2} + \eta_s^{d+3}\varkappa\Big(\frac{s}{T}\Big)\Big]\Big\} e^{-\varphi_s}\,ds, \end{aligned} \tag{7}$$

where $\xi_s = \xi_s(x)$, $\varphi_s = \varphi_s(x)$, η_s is a solution of the equation

$$\begin{aligned} \eta_s = y + & \int_0^s \Big[\sigma_{(\eta_r)}(\xi_r) + \frac{1}{2}(\alpha(\xi_r), \eta_r)\sigma(\xi_r) + \sigma(\xi_r)\pi(\xi_r, \eta_r)\Big]\,dw_r \\ & + \int_0^s \Big[b_{(\eta_r)}(\xi_r) + (\alpha(\xi_r), \eta_r)b(\xi_r) - \frac{1}{2}Ka(\xi_r)\eta_r\Big]\,dr, \end{aligned} \tag{8}$$

$$\eta_s^{d+1} = \frac{1}{2}K\int_0^s \eta_r^*\sigma(\xi_r)\,dw_r, \qquad \eta_s^{d+2} = \int_0^s (\alpha(\xi_r), \eta_r)\,dr, \tag{9}$$

$$\eta_s^{d+3} = -\int_0^s \Big[c_{(\eta_r)}(\xi_r) + (\alpha(\xi_r), \eta_r)c(\xi_r)\Big]\,dr. \tag{10}$$

We will postpone the proof of this lemma until Sec. 3, noting for the moment that equation (8) is similar to (V.7.5) and that by Theorem V.3.4, the system (0.1), (8) is solvable by Euler's method in the mean. Therefore, by simply replacing $\sigma_{(\eta_r)}(\xi_r)$, $b_{(\eta_r)}(\xi_r)$ in (V.8.8), (V.8.9) by the corresponding integrands from (8) and using condition (2), we again obtain (V.8.10). Returning to the new relationship (V.8.9) and using the Davis inequality, as in the case of (V.8.18) and (V.8.19), we obtain

$$\mathbf{E}\sup_{r\le s}|\eta_r|e^{-\varphi_r}\le |y|+N\mathbf{E}\Big(\int_0^s |\eta_r|^2 e^{-2\varphi_r}\,dr\Big)^{1/2}$$
$$\le |y|+N\mathbf{E}\sup_{r\le s}|\eta_r|^{\frac12}e^{-\frac12\varphi_r}\Big(\int_0^s|\eta_r|e^{-\varphi_r}\,dr\Big)^{1/2},$$
$$\mathbf{E}\sup_{r\le s}|\eta_r|e^{-\varphi_r}\le N|y|,$$

where $N = N(\delta, K)$. Hence, as in the final part of the proof of Theorem V.8.5, it follows that

$$\mathbf{E}\,e^{-\varphi_s}\sup_{r\le s}|\eta_r|\le N(\delta,K)|y|e^{-\delta s/2}. \tag{11}$$

Together with inequalities similar to (V.8.14), this proves inequalities (12) (below) for $\gamma_s = \eta_s, \eta_s^{d+2}, \eta_s^{d+3}$.

4. LEMMA. *Under the assumptions of Theorem* 1, *let* $|y| = 1$ *and define the processes* $\eta_s, \eta_s^{d+1}, \eta_s^{d+2}, \eta_s^{d+3}$ *by* (8)–(10). *Then, for all* $x \in E_d$, $s \ge 0$, *inequality* (V.8.10) *holds and for* $\gamma_s = \eta_s, \eta_s^{d+1}, \eta_s^{d+2}, \eta_s^{d+3}$, $s \ge 0$, *we have*

$$\mathbf{E}\int_s^\infty |\gamma_r|e^{-\varphi_r}\,dr\le Ne^{-\varepsilon s} \tag{12}$$

with some constants $N(\delta, K)$, $\varepsilon(\delta, K) > 0$.

PROOF. By our previous remarks it suffices to prove (12) for $\gamma_s = \eta_s^{d+1}$. To do this we change variables in the integral with respect to r, substituting $r = \psi_u$, where ψ is the function inverse to φ (note that $\varphi_s \ge \delta s$) and then apply Fubini's Theorem and the Davis inequality. The result is that the left-hand side of (12) is bounded by

$$\mathbf{E}\int_{\varphi_s}^\infty c^{-1}(\xi_{\psi_u})|\gamma_{\psi_u}|e^{-u}\,du\le\delta^{-1}\int_{\delta s}^\infty e^{-u}\mathbf{E}|\gamma_{\psi_u}|\,du$$
$$\le\frac32\delta^{-1}K\int_{\delta s}^\infty e^{-u}\mathbf{E}\Big(\int_0^{\psi_u}\big(a(\xi_r)\eta_r,\eta_r\big)dr\Big)^{1/2}du$$
$$\le N\,\mathbf{E}\int_{\delta s}^\infty e^{-u}\sqrt{\psi_u}\sup_{r\le\psi_u}|\eta_r|\,du\le N\,\mathbf{E}\int_{\psi_{\delta s}}^\infty\sqrt{r}e^{-\varphi_r}\sup_{u\le r}|\eta_u|\,dr,$$

where N are constants that depend only on δ, K. It remains once again to use inequality (11) and the fact that $\psi_{\delta s} \ge K^{-1}\delta s$. □

By this lemma, taking $|y| = 1$, omitting $\varkappa$, replacing $\varkappa'$ by zero on the right of (7), we obtain a finite expression, which we denote by $u(x, y)$. Clearly, $|u(x, y)| \le N(\delta, K)$.

It is also obvious that

$$\left|v_{(y)}^T(x) - u(x,y,)\right| \le N\,\mathbf{E}\int_T^\infty \left(|\eta_s| + |\eta_s^{d+1}| + |\eta_s^{d+2}| + |\eta_s^{d+3}|\right)e^{-\varphi_s}\,ds,$$

where $N = N(\delta, K)$, and the right-hand side tends to zero as $T \to \infty$ uniformly in $x, y \in E_d$, $|y| = 1$. A similar assertion is easily proved for $|v^T - v|$. By a theorem of calculus which has been repeatedly used, this implies the statement of Theorem 1 and the following corollary.

5. COROLLARY. *If $x, y \in E_d$, $|y| = 1$, and $u(x,y)$ is the function defined in the previous paragraph, then $v_{(y)}(x) = u(x,y)$.*

Now we continue with the analysis of the second derivatives of v. Naturally, we will strengthen the conditions of Theorem 1.

6. THEOREM. *Let σ, b, c and f be twice continuously differentiable functions, bounded in E_d together with their first and second-order derivatives. Suppose that for all $x, y \in E_d$, $|y| = 1$, inequalities (1) hold and also*

$$\text{(13)}\qquad \begin{aligned} &\|\sigma(x)\| + \|\sigma_{(y)}(x)\| + \|\sigma_{(y)(y)}(x)\| + |b_{(y)}(x)| + |b_{(y)(y)}(x)| \\ &\quad + |c_{(y)(y)}(x)| + |f_{(y)(y)}(x)| + \|\pi(x,y)\| \le K, \end{aligned}$$

$$\text{(14)}\qquad \begin{aligned} &\left\|\sigma_{(y)}(x) + \frac{1}{2}\big(\alpha(x), y\big)\sigma(x) + \sigma(x)\pi(x)\right\|^2 \\ &\quad + 2\Big(y, b_{(y)}(x) + \big(\alpha(x), y\big)b(x)\Big) \le c(x) - \delta + K\big(a(x)y, y\big). \end{aligned}$$

Then the function v, defined by (3) is twice continuously differentiable with respect to x, its second-order derivatives are bounded in absolute value by a constant $N(\delta, K)$, and v satisfies the Kolmogorov equation (V.6.14) on E_d.

As for the advantages of this theorem over Theorem V.8.5, the same remarks hold as regarding Theorem 1. We will therefore restrict ourselves to a single example.

7. EXAMPLE. In addition to the assumptions of Example 2, let the first assumption of Theorem 6 be satisfied and for every x such that $a(x) = b(x) = 0$ we have

$$\text{(15)}\qquad |\sigma'(x)|^2 + 2b'(x) < c(x).$$

It turns out that in this case $v(x)$ is twice continuously differentiable. To verify this, we need only repeat the arguments of Example 2, but taking $K = n^4$ in order to deduce from the inequality inverse to (14) that $n^2a(x_n) \to 0$, $n\sigma(x_n) \to 0$, $b(x_n) \to 0$. Note also that, as stated in Remark V.8.6, condition (15) is, generally speaking, necessary; in the one-dimensional case it is essentially the same as condition (V.8.13), with one important difference: condition (15) is required *only* for x such that $a(x) = b(x) = 0$ (if such x's exist at all).

To prove Theorem 6 we will need two lemmas, the first one of which, together with Lemma 3, will be proved in Sec. 3; we restrict ourselves here to a short comment. Lemma 8 establishes a formula for the second-order derivatives with respect to x of v^T, which is twice continuously differentiable with respect to x by virtue of Theorem V.7.4. As the statement of the lemma is rather tedious, it is certainly useful for the reader to keep in mind the most important features of (16) and (18)–(20): these formulas involve

$\varkappa$, $\varkappa'$, $\varkappa''$ in some appropriate way; the integrands are sums of functions that are linear in $\hat{\zeta}$, ζ^{d+3} and functions that are quadratic in $\hat{\eta}$, η^{d+3}, with bounded coefficients; ζ^{d+i} does not appear on the right of (18)–(20). The only essential feature of (17) is that the terms containing ζ have the same form as in (8), and the "free terms" σ_2, b_2 are quadratic in η with bounded coefficients. These features of (16)–(20) could easily have been derived from Secs. 2, 3 without the detailed calculations whose result is presented in Lemma 8.

8. LEMMA. *Let the first assumption of Theorem 6 be satisfied. Then, for any* $x, y \in E_d$ *(and any* K*)*,

$$\text{(16)}\quad v^T_{(y)(y)}(x) = \mathbf{E}\int_0^\infty \Big[f_1^T\Big(\frac{s}{T},\xi_s,\hat{\zeta}_s\Big) + f_2^T\Big(\frac{s}{T},\xi_s,\hat{\eta}_s\Big) + 2\eta_s^{d+3} f_1^T\Big(\frac{s}{T},\xi_s,\hat{\eta}_s\Big) + f(\xi_s)\varkappa\Big(\frac{s}{T}\Big)\Big(\zeta_s^{d+3} + (\eta_s^{d+3})^2\Big)\Big]e^{-\varphi_s}\,ds,$$

where $\xi_s = \xi_s(x)$, $\varphi_s = \varphi_s(x)$, $\hat{\zeta}_s = (\zeta_s, \zeta_s^{d+1}, \zeta_s^{d+2})$, $\hat{\eta}_s = (\eta_s, \eta_s^{d+1}, \eta_s^{d+2})$, *the process* η_s *is a solution of equation* (8), η_s^{d+i} *are taken from* (9) *and* (10), ζ_s *is a solution of the equation*

$$\text{(17)}\quad \begin{aligned}\zeta_s = &\int_0^s \Big[\sigma_{(\zeta_r)}(\xi_r) + \frac{1}{2}(\alpha(\xi_r),\zeta_r)\sigma(\xi_r) + \sigma(\xi_r)\pi(\xi_r,\zeta_r) + \sigma_2(\xi_r,\eta_r)\Big]dw_r\\ &+\int_0^s \Big[b_{(\zeta_r)}(\xi_r) + (\alpha(\xi_r),\zeta_r)b(\xi_r) - \frac{1}{2}Ka(\xi_r)\zeta_r + b_2(\xi_r,\eta_r)\Big]dr,\end{aligned}$$

$$\text{(18)}\quad \begin{aligned}\zeta_s^{d+1} = K\int_0^s \Big[&\frac{1}{2}\zeta_r^*\sigma(\xi_r) + \eta_r^{d+1}\eta_r^*\sigma(\xi_r)\\ &+ \frac{1}{4}(\alpha(\xi_r),\eta_r)\eta_r^*\sigma(\xi_r) + \eta_r^*\sigma(\xi_r)\pi(\xi_r,\eta_r)\Big]dw_r,\end{aligned}$$

$$\text{(19)}\quad \zeta_s^{d+2} = \int_0^s (\alpha(\xi_r),\zeta_r)\,dr,$$

$$\text{(20)}\quad \begin{aligned}\zeta_s^{d+3} = -\int_0^s \Big[&c_{(\zeta_r)}(\xi_r) + (\alpha(\xi_r),\zeta_r)c(\xi_r)\\ &+ 2(\alpha(\xi_r),\eta_r)c_{(\eta_r)}(\xi_r) + c_{(\eta_r)(\eta_r)}(\xi_r)\Big]dr,\end{aligned}$$

$$\begin{aligned}f_1^T(s,x,\hat{y}) =& f_{(y)}(x)\varkappa(s)\\ &+ f(x)\Big[(\alpha(x),y)\varkappa(s) + y^{d+1}\varkappa(s) + \frac{1}{T}y^{d+2}\varkappa'(s)\Big],\\ \hat{y} =& (y, y^{d+1}, y^{d+2}),\end{aligned}$$

$$
\begin{aligned}
f_2^T(s,x,\hat{y}) = {} & f_{(y)(y)}(x)\varkappa(s) \\
& + 2f_{(y)}(x)\left[y^{d+1}\varkappa(s) + (\alpha(x),y)\varkappa(s) + \frac{1}{T}y^{d+2}\varkappa'(s)\right] \\
& + f(x)\Big[2y^{d+1}\varkappa(s)(\alpha(x),y) \\
& + 2(\alpha(x),y)\frac{1}{T}y^{d+2}\varkappa'(s) + 2\frac{1}{T}y^{d+1}y^{d+2}\varkappa'(s) + \frac{1}{T^2}\left(y^{d+2}\right)^2\varkappa''(s)\Big], \\
\sigma_2(x,y) = {} & \sigma_{(y)(y)}(x) + 2(\alpha(x),y)\sigma_{(y)}(x) + 2\sigma_{(y)}\pi(x,y) \\
& - \frac{1}{4}(\alpha(x),y)^2\sigma(x) + \sigma(x)\pi^2(x,y) + 2(\alpha(x),y)\sigma(x)\pi(x,y), \\
b_2(x,y) = {} & b_{(y)(y)}(x) + 2(\alpha(x),y)b_{(y)}(x) \\
& - K(\alpha(x),y)a(x)y - K\sigma_{(y)}(x)\sigma^*(x)y.
\end{aligned}
$$

9. LEMMA. *Under the assumptions of Theorem 6, there exist constants $N(\delta,K)$, $\varepsilon(\delta,K)>0$ such that inequality (12) holds for $\gamma_s = \zeta_s,\ \zeta_s^{d+1},\ \zeta_s^{d+2},\ \zeta_s^{d+3}$ for all $x,y\in E_d$, $|y|=1$, $s\geq 0$, and it remains valid after substitution of $|\gamma_s|^2$ for $|\gamma_s|$ if $\gamma_s = \eta_s,\ \eta_s^{d+1},\ \eta_s^{d+2},\ \eta_s^{d+3}$.*

PROOF. It was pointed out after Lemma 3 that equation (8) is analogous to (V.7.5). Therefore, inequalities (V.8.18) and (V.8.19) may be derived from conditions (13) and (14) just as in the proof of Theorem V.8.5. Equation (17) for ζ_s is analogous to the equation for ζ_s in Sec. V.7, in the same way as (8) is analogous to (V.7.5). In addition, $\|\sigma_2(x,y)\| + |b_2(x,y)| \leq N(K)|y|^2$ by (13) and (1). Hence, as in the proof of Theorem V.8.5, inequalities (V.8.21) hold. Thus, we again obtain inequalities (V.8.15); from these inequalities and from inequalities like (V.8.14), it easily follows that we need only consider $\gamma_s = \zeta_s^{d+1}$ and $\gamma_s = \eta_s^{d+1}$. If $\gamma_s = \eta_s^{d+1}$, the desired inequality is obtained as in the proof of Lemma 4, using not the Davis inequality but the isometric property of the stochastic integration operator. Finally, the case of $\gamma_s = \zeta_s^{d+1}$ differs from the proof of Lemma 4 only in the use of the inequality $(a+b)^{1/2} \leq a^{1/2} + b^{1/2}$, $a,b\geq 0$. □

The assertion of Theorem 6 follows from this lemma and from (16) by an obvious modification of the arguments given after the proof of Lemma 4. This completes the proof of Theorem 6.

10. REMARK. As in Corollary 5, we obtain a formula for $v_{(y)(y)}$ by replacing $\varkappa, \varkappa', \varkappa''$ in (16) with $1,0,0$ respectively.

2. Quasiderivatives of solutions of stochastic equations

We saw in Sec. V.8 that in order to study the smoothness of probabilistic solutions of differential equations, one needs estimates for the moments of the derivatives in probability of solutions of stochastic equations. Stochastic equations for these derivatives were written down in Sec. V.7 and used later in Sec. V.8. It turns out that in certain cases the processes η_s, $\bar{\eta}_s$, ζ_s in formulas (V.7.12) and (V.7.20) for $v_{(y)}(t,x)$ and $v_{(y)(y)}(t,x)$ may be assumed to satisfy not equations (V.7.5), (V.7.16), (V.7.17) but *other* equations, whose solutions we call *quasiderivatives of $\xi_s(t,x)$ with respect to x*. The efficacy of this approach was demonstrated in Theorems 1.1 and 1.6, Lemmas 1.3 and 1.8 will be proved in the next section using quasiderivatives.

The situation studied in this section is as follows. Let $\hat{\sigma}$ and $\hat{b}$ be some functions defined for $r \in (0,\infty)$, $x \in E_d$, also depending on a parameter $p \in B := \{q \in E_{d_2}: |q| < 1\}$, that is, $\hat{\sigma} = \hat{\sigma}(r,x,p)$, $\hat{b} = \hat{b}(r,x,p)$. Suppose that for any $p \in B$ they satisfy the conditions listed in the introduction to the chapter, and they are also continuously differentiable with respect to (x,p) for any r and their first-order derivatives with respect to (x,p) are bounded as functions of (r,x,p). Then by Theorem V.1.1 there exists a unique solution of the equation

$$\xi_s = x + \int_t^s \hat{\sigma}(r,\xi_r,p_r)\,dw_r + \int_t^s \hat{b}(r,\xi_r,p_r)\,dr \qquad s \geq t \tag{1}$$

for any nonrandom $t \in [0,\infty)$, $x \in E_d$ and any B-valued predictable function p_r.

We will assume that

$$\sigma(r,x) = \hat{\sigma}(r,x,0), \qquad b(r,x) = \hat{b}(r,x,0).$$

Clearly, if $p_r \equiv 0$, the solution of (1) coincides (a.s.) with $\xi_s(t,x)$ for all $s \geq t$ at once.

Next, let $\hat{c}(r,x,p)$, $\hat{f}(r,x,p)$, $g(x)$ be real-valued Borel functions of (r,x,p), defined on $(0,\infty) \times E_d \times B$, continuously differentiable with respect to (x,p) for any $r \geq 0$, and such that $\hat{c} \geq 0$ and for all $r > 0$, $x \in E_d$, $p \in B$, $i = 1,\dots,d$, $j = 1,\dots,d_2$ and certain constants $K, m \in [0,\infty)$ we have

$$\begin{aligned} &|\hat{c}(r,x,p)| + |\hat{c}_{x^i}(r,x,p)| + |\hat{c}_{p^j}(r,x,p)| + |\hat{f}(r,x,p)| \\ &\quad + |\hat{f}_{x^i}(r,x,p)| + |\hat{f}_{p^j}(r,x,p)| + |g(x)| + |g_{x^i}(x)| \leq K(1+|x|^m). \end{aligned} \tag{2}$$

Suppose that

$$c(r,x) = \hat{c}(r,x,0), \qquad f(r,x) = \hat{f}(r,x,0).$$

Fix $T \in (0,\infty)$ and use formula (V.7.11) to define a function $v(t,x)$ for $t \in [0,T]$, $x \in E_d$ on the basis of $\sigma(r,x)$, $b(r,x)$, $c(r,x)$, $f(r,x)$, and $g(x)$. By Lemma V.7.2 and Remark V.7.3, $v(t,x)$ is differentiable with respect to x continuously in (t,x).

Finally, fix a bounded Borel function $P(r,x)$ with values in the set of $d_2 \times d$ matrices, defined on $(0,\infty) \times E_d$ and continuous in x for any $r > 0$. For a sufficiently regular function $u(r,x,p)$ and $x, y \in E_d$, $r > 0$, denote

$$\partial_r(y)u(r,x) = \sum_{i=1}^d \Big(u\big(r,x,P(r,q)(x-q)\big)\Big)_{x^i} y^i \Big|_{q=x}, \tag{3}$$

$$\partial_r^2(y)u(r,x) = \sum_{i,j=1}^d \Big(u\big(r,x,P(r,q)(x-q)\big)\Big)_{x^i x^j} y^i y^j \Big|_{q=x}. \tag{4}$$

1. THEOREM. *Fix $t \in [0,T]$ and a domain $D \subset E_d$ and assume that for any predictable process p_r with values in B and any $x \in D$ we have*

$$\mathbf{E}\left[\int_t^T \hat{f}(s,\xi_s,p_s)\exp\left(-\int_t^s \hat{c}(r,\xi_r,p_r)\,dr\right)ds + g(\xi_T)\exp\left(-\int_t^T \hat{c}(r,\xi_r,p_r)\,dr\right)\right] = v(t,x), \tag{5}$$

where ξ_s is the solution of equation (1). In other words, we are assuming that the left-hand side of (5) does not depend on the process p for $x \in D$. Then, for any $x \in D$, $y \in E_d$ we have

$$v_{(y)}(t,x) = \mathbf{E}\left\{\left[g_{(\eta_T)}(\xi_T(t,x)) + g(\xi_T(t,x))\eta_T^{d+1}\right]e^{-\varphi_T(t,x)} + \int_t^T \left[\partial_s(\eta_s)\hat{f}(s,\xi_s(t,x)) + f(s,\xi_s(t,x))\eta_s^{d+1}\right]e^{-\varphi_s(t,x)}\,ds\right\}, \tag{6}$$

where η_s is a solution of the equation

$$\eta_s = y + \int_t^s \partial_r(\eta_r)\hat{\sigma}(r,\xi_r(t,x))\,dw_r + \int_t^s \partial_r(\eta_r)\hat{b}(r,\xi_r(t,x))\,dr, \tag{7}$$

$$\eta_s^{d+1} = -\int_t^s \partial_r(\eta_r)\hat{c}(r,\xi_r(t,x))\,dr \tag{8}$$

(and, for example, $\partial_s(\eta_s)\hat{f}(s,\xi_s(t,x))$ is the value of $\partial_s(y)\hat{f}(s,z)$ for $y = \eta_s, z = \xi_s(t,x)$).

PROOF. Let $\gamma(p)$ be a bounded infinitely differentiable function on E_{d_2} with values in B such that $\gamma(p) = p$ for $|p| \le \frac{1}{2}$ and $\gamma(p) = 0$ for $|p| \ge 1$. For $r > 0$, $x, q \in E_d$, denote

$$(\tilde{\sigma},\tilde{b},\tilde{c},\tilde{f})(r,x,q) = (\hat{\sigma},\hat{b},\hat{c},\hat{f})\Big(r,x,\gamma\big(P(r,q)(x-q)\big)\Big),$$

$$v(t,x,q) = \mathbf{E}\left[g(\xi_T)\exp\left(-\int_t^T \tilde{c}(r,\xi_r,q_r)\,dr\right) + \int_t^T \tilde{f}(s,\xi_s,q_s)\exp\left(-\int_t^s \tilde{c}(r,\xi_r,q_r)\,dr\right)ds\right],$$

where the pair (ξ_s, q_s) is a solution of the following system

$$q_s = q + \int_t^s \sigma(r, q_r)\, dw_r + \int_t^s b(r, q_r)\, dr, \tag{9}$$

$$\xi_s = x + \int_t^s \tilde{\sigma}(r, \xi_r, q_r)\, dw_r + \int_t^s \tilde{b}(r, \xi_r, q_r)\, dr. \tag{10}$$

By the comments at the end of Sec. V.7 (see Remark V.7.8 and Exercise V.7.9), the function $v(t, x, q)$ is continuously (in (t, x, q)) differentiable with respect to x, and for $x, y \in E_d$ we have

(11)

$$v_{(y)}(t, x, q) = \mathbf{E}\left\{\left[g_{(\eta_T)}(\xi_T) + g(\xi_T)\eta_T^{d+1}\right] \exp\left(-\int_t^T \tilde{c}(r, \xi_r, q_r)\, dr\right)\right.$$
$$\left. + \int_t^T \left[\tilde{f}_{(\eta_s)}(s, \xi_s, q_s) + \tilde{f}(s, \xi_s, q_s)\eta_s^{d+1}\right] \exp\left(-\int_t^s \tilde{c}(r, \xi_r, q_r)\, dr\right) ds\right\},$$

where ξ_s, q_s are, naturally, a solution of system (9), (10), and η_s a solution of the equation

$$\eta_s = y + \int_t^s \tilde{\sigma}_{(\eta_r)}(r, \xi_r, q_r)\, dw_r + \int_t^s \tilde{b}_{(\eta_r)}(r, \xi_r, q_r)\, dr, \tag{12}$$

$$\eta_s^{d+1} = -\int_t^s \tilde{c}_{(\eta_r)}(r, \xi_r, q_r)\, dr, \tag{13}$$

and all the derivatives of $\hat{\sigma}$, $\hat{b}$, $\hat{c}$, $\hat{f}$ involved in these expressions are evaluated only with respect to the second argument, i.e., x.

It remains only to note that if $x \in D$, then $v(t, x, q)$ is independent of q by assumption so that, for example, $v_{(y)}(t, x) = v_{(y)}(t, x, q)$ for $q = x \in D$. But if $q = x$, then by the uniqueness of the solutions of the stochastic equations (9) and (10), we obtain $q_s = \xi_s = \xi_s(t, x)$ (a.s.). It immediately follows that the right-hand sides of (11)–(13) coincide with the right-hand sides of (6)–(8), respectively. □

The proof of the following theorem is based on formula (V.7.20) (also see its discussion in Remark V.7.5) for the function (V.7.11) and is quite analogous.

2. THEOREM. *Let the assumptions stated before and in Theorem* 1 *be satisfied. Also let* $\hat{\sigma}^{ij}$, $\hat{b}^j$, $\hat{c}$, $\hat{f}$, *and* g *be twice continuously differentiable with respect to* (x, p) *for any* $r > 0$ *and assume that their second-order partial derivatives with respect to* (x, p) *are bounded in absolute values by* $K(1 + |x|^m)$. *Then the function* $v(t, x)$ *(which is twice continuously differentiable with respect to* x *by Theorem* V.7.4) *is such that for any*

$x \in D$, $y \in E_d$ we have

$$v_{(y)(y)}(t,x) = \mathbf{E}\Big\{\Big[g_{(\zeta_T)}(\xi_T) + g_{(\eta_T)(\eta_T)}(\xi_T) + 2g_{(\eta_T)}(\xi_T)\eta_T^{d+1}$$

$$+ g(\xi_T)(\zeta_T^{d+1} + (\eta_T^{d+1})^2)\Big]e^{-\varphi_T}$$

$$+ \int_t^T \Big[\partial_r(\zeta_r)\hat{f}(r,\xi_r) + \partial_r^2(\eta_r)\hat{f}(r,\xi_r)$$

$$+ 2\eta_r^{d+1}\partial_r(\eta_r)\hat{f}(r,\xi_r) + f(r,\xi_r)(\zeta_r^{d+1} + (\eta_r^{d+1})^2)\Big]e^{-\varphi_r}\,dr\Big\},$$

where η_s, η_s^{d+1} are defined by (7), (8) *and ζ_s is a solution of the equation*

$$\zeta_s = \int_t^s \Big[\partial_r(\zeta_r)\hat{\sigma}(r,\xi_r) + \partial_r^2(\eta_r)\hat{\sigma}(r,\xi_r)\Big]dw_r$$

$$+ \int_t^s \Big[\partial_r(\zeta_r)\hat{b}(r,\xi_r) + \partial_r^2(\eta_r)\hat{b}(r,\xi_r)\Big]dr,$$

$$\zeta_s^{d+1} = -\int_t^s \Big[\partial_r(\xi_r)\hat{c}(r,\xi_r) + \partial_r^2(\eta_r)\hat{c}(r,\xi_r)\Big]dr,$$

$$\xi_r = \xi_r(t,x), \qquad \varphi_r = \varphi_r(t,x).$$

3. Proofs of Lemmas 1.3 and 1.8

We will use the results of the previous section and introduce auxiliary parameters p and functions $\hat{\sigma}$, $\hat{b}$, $\hat{c}$, $\hat{f}$. Of course, we assume from the very beginning that the first assumption of Lemma 1.3 holds.

As in Sec. 1, let $\mathfrak{P}$ denote the set of all skew-symmetric $(d_1 \times d_1)$-matrices. The set $\mathfrak{P}$ may be treated as a $d_1(d_1-1)/2$-dimensional Euclidean space with inner product $(\pi_1, \pi_2) = (1/2)\,\mathrm{tr}\,\pi_1\pi_2^*$.

The parameter p will takes values in E_{d_2}, where $d_2 = 1 + d_1(d_1-1)/2 + d_1$ and it will be convenient to treat it as a triple $p = (\alpha, \pi, \lambda)$, $\alpha \in E_1$, $\pi \in \mathfrak{P}$, $\lambda \in E_{d_1}$. Putting $\hat{x} = (x, x^{d+1}, x^{d+2}, x^{d+3}) = (x^1, \dots, x^d, x^{d+1}, x^{d+2}, x^{d+3})$, $p = (\alpha, \pi, \lambda)$, we define

$$\hat{c}(\hat{x},p) = \Big(1 + \frac{1}{2}\alpha\Big)c(x), \qquad \hat{f}(\hat{x},p) = \Big(1 + \frac{1}{2}\alpha\Big)f(x)e(x^{d+1})\varkappa\Big(\frac{x^{d+2}}{T}\Big),$$

where e is an infinitely differentiable function on E_1, $e(u) = e^u$ if $u \le 4 + 2T$ and $e(u) = 0$ if $u \ge 8 + 4T$. The functions $\hat{\sigma}$ and $\hat{b}$ will also be independent of r, but rather than write down their entries explicitly, it will be more expressive and instructive to write down the equation corresponding to (2.1), from which the definitions of $\hat{\sigma}$ and $\hat{b}$ will be clear. The role of equation (2.1) for $\hat{\xi}_s = (\xi_s, \xi_s^{d+1}, \xi_s^{d+2}, \xi_s^{d+3})$ will be played by the following system ($p_r = (\alpha_r, \pi_r, \lambda_r)$):

$$\xi_s = x + \int_0^s \left(1 + \frac{1}{2}\alpha_r\right)^{1/2} \sigma(\xi_r) e^{\pi_r}\, dw_r$$

$$+ \int_0^s \left(1 + \frac{1}{2}\alpha_r\right)\left(b(\xi_r) + \sigma(\xi_r)\lambda_r \chi'(\xi_r^{d+3})\right) dr, \tag{1}$$

$$\xi_s^{d+1} = x^{d+1} - \int_0^s \left(1 + \frac{1}{2}\alpha_r\right)^{1/2} \chi'(\xi_r^{d+3})\lambda_r^* e^{\pi_r}\, dw_r$$

$$- \frac{1}{2}\int_0^s \left(1 + \frac{1}{2}\alpha_r\right)\left|\chi'(\xi_r^{d+3})\lambda_r\right|^2 dr, \tag{2}$$

$$\xi_s^{d+2} = x^{d+2} + \int_0^s \left(1 + \frac{1}{2}\alpha_r\right) dr, \tag{3}$$

$$\xi_s^{d+3} = x^{d+3} + \int_0^s \left(1 + \frac{1}{2}\alpha_r\right)^{1/2} \lambda_r^* e^{\pi_r}\, dw_r, \tag{4}$$

where χ is some infinitely differentiable function on E_1 such that $\chi(0) = \chi''(0) = 0$, $\chi'(0) = 1$, $|\chi| \le 1$, $|\chi'| \le 1$ and $|\chi''| \le 1$ (for example, $\chi(t) = \sin t$).

It obviously follows from our assumptions on σ and b that the first derivatives of $\hat{\sigma}$ and $\hat{b}$ with respect to $(\hat{x}, p)$ are continuous and bounded for $p \in B = \{q \in E_{d_2} : |q| < 1\}$. It is easy to see that condition (2.2) is also satisfied for $g \equiv 0$, $\hat{c}(\hat{x}, p)$, $\hat{f}(\hat{x}, p)$ with $m = 0$ for some constant K and $p \in B$. Finally, for $\hat{x} = (x, x^{d+1}, x^{d+2}, x^{d+3})$ we define a function $P(\hat{x}) = P(x)$ with values in the space of $(d_2 \times (d+3))$-matrices such that for any $\hat{y} = (y, y^{d+1}, y^{d+2}, y^{d+3})$

$$P(\hat{x})\hat{y} = \left(2\big(\alpha(x), y\big), \pi(x, y), -\frac{1}{2}K\sigma^*(x)y\right). \tag{5}$$

We have now defined all the objects mentioned before Theorem 2.1. Now take an arbitrary predictable process $p_r = (\alpha_r, \pi_r, \lambda_r)$ with values in B and denote

$$\hat{w}_s = \int_0^s e^{\pi_r}\, dw_r, \qquad \hat{\varphi}(s) = \int_0^s \left(1 + \frac{1}{2}\alpha_r\right) dr,$$

$$\hat{\psi}(s) = \inf\{t \ge 0 : \hat{\varphi}(t) \ge s\}, \qquad \tilde{w}_s = \int_0^{\hat{\psi}(s)} \left(1 + \frac{1}{2}\alpha_r\right)^{1/2} d\hat{w}_r, \qquad \tilde{\mathcal{F}}_s = \mathcal{F}_{\hat{\psi}(s)}.$$

Since $e^{\pi_r}\left(e^{\pi_r}\right)^*$ is the identity matrix, it follows by Lévy's Theorem that $(\hat{w}_s, \mathcal{F}_s)$ is a d_1-dimensional Wiener process. By Theorem IV.2.3 (and Remark IV.2.4), $(\tilde{w}_s, \tilde{\mathcal{F}}_s)$ is a d_1-dimensional Wiener process too. Applying also Lemma IV.2.2 and Remark

III.10.6, we conclude that for any constant $S \in (0,\infty)$

$$\begin{aligned}(6)\qquad &\mathbf{E}\int_0^S \hat f(\hat\xi_s,p_s)\exp\Big(-\int_0^s \hat c(\hat\xi_r,p_r)\,dr\Big)\,ds\\ &=\mathbf{E}\int_0^{\hat\varphi(S)} f(\xi_s)e(\xi_s^{d+1})\varkappa\Big(\frac{\xi_s^{d+2}}{T}\Big)\exp\Big(-\int_0^s c(\xi_r)\,dr\Big)\,ds,\end{aligned}$$

where the process $\hat\xi_s$ on the left is a solution of system (1) – (4) and ξ_s, ξ_s^{d+1}, ξ_s^{d+2}, ξ_s^{d+3} on the right are determined from the following system, in which $\tilde\lambda_r = \lambda_{\hat\psi(r)}$:

$$(7)\qquad \xi_s = x + \int_0^s \sigma(\xi_r)\,d\tilde w_r + \int_0^s \Big[b(\xi_r) + \sigma(\xi_r)\tilde\lambda_r\chi'(\xi_r^{d+3})\Big]\,dr,$$

$$(8)\qquad \xi_s^{d+1} = x^{d+1} - \int_0^s \chi'(\xi_r^{d+3})\tilde\lambda_r^*\,d\tilde w_r - \frac12\int_0^s \Big|\chi'(\xi_r^{d+3})\tilde\lambda_r\Big|^2\,dr,$$

$$(9)\qquad \xi_s^{d+2} = x^{d+2} + s,$$

$$(10)\qquad \xi_s^{d+3} = x^{d+3} + \int_0^s \tilde\lambda_r^*\,d\tilde w_r.$$

Furthermore, in E_{d+3} we consider the domain $D = \big\{|x^{d+1}|+|x^{d+2}|+|x^{d+3}|<1\big\}$ and put $S = 4T+2$. For $\hat x \in D$ it follows from (9), the inequality $|p_r| < 1$, and the properties of $\varkappa$ that $\hat\varphi(S) \geq \frac12 S = 2T+1$ and $\varkappa(\xi_s^{d+2}/T) = 0$ for $s \geq \hat\varphi(S)$ or $s \geq \frac12 S$. Consequently, the second expression in (6) remains unchanged if we replace $\hat\varphi(S)$ by S. Now we reveal the role of the process ξ_s^{d+3}. By Itô's formula it follows from (8) and (10) that

$$\xi_s^{d+1} = x^{d+1} - \chi(\xi_s^{d+3}) + \chi(x^{d+3}) + \frac12\int_0^s (\chi'' - (\chi')^2)(\xi_r^{d+3})|\tilde\lambda_r|^2\,dr.$$

By the definition of $e(u)$ it follows that $e(\xi_s^{d+1}) = \exp\xi_s^{d+1}$ for $\hat x \in D$, $s \leq S$. Thus, for $\hat x \in D$ the right-hand side of (6) is

$$(11)\qquad e^{x^{d+1}}\mathbf{E}\int_0^S f(\xi_s)\rho_s\varkappa\Big(\frac{x^{d+2}+s}{T}\Big)\exp\Big(-\int_0^s c(\xi_r)\,dr\Big)\,ds,$$

where (ξ_s,ξ_s^{d+3}) is the solution of system (7), (10) and

$$\rho_s = \exp\Big[-\int_0^s \chi'(\xi_r^{d+3})\tilde\lambda_r^*\,d\tilde w_r - \frac12\int_0^s \Big|\chi'(\xi_r^{d+3})\tilde\lambda_r\Big|^2\,dr\Big].$$

By Theorem IV.3.5 (a) and Lemma IV.3.2, the process $(\rho_s,\tilde{\mathcal F}_s)$ is a martingale. By Fubini's Theorem we can interchange the expectation and integral signs in (11), then apply Lemma II.8.5 (a) and bring the expectation sign forward again. Then we see that (11) will not change if we replace ρ_s by ρ_S and pull the latter out from the integral with respect to s. Defining a new probability measure $\bar{P}$ by $\bar{P}(d\omega) = \rho_S(\omega)P(d\omega)$ and denoting the integral with respect to this measure by $\bar{\mathbf{E}}$, we can rewrite (11) as

$$(12)\qquad e^{x^{d+1}}\bar{\mathbf{E}}\int_0^S f(\xi_s)\varkappa\Big(\frac{x^{d+2}+s}{T}\Big)\exp\Big(-\int_0^s c(\xi_r)\,dr\Big)ds.$$

By Girsanov's Theorem, the process

$$\bar{w}_s = \tilde{w}_s + \int_0^s \chi'(\xi_r^{d+3})\tilde{\lambda}_r I_{r\le S}\,dr$$

is a d_1-dimensional Wiener process on the probability space $(\Omega, \mathcal{F}, \bar{P})$ with respect to $\{\tilde{\mathcal{F}}_s\}$ and by (7), for all s at once (a.s.),

$$\xi_{s\wedge S} = x + \int_0^s \sigma(\xi_{r\wedge S})I_{r\le S}\,d\bar{w}_r + \int_0^s b(\xi_{r\wedge S})I_{r\le S}\,dr.$$

Finally, by Corollary V.4.4, the processes $\xi_{\cdot\wedge S}$ and $\xi_{\cdot\wedge S}(x)$ are identically distributed on $(C, \mathfrak{A}(C))$, and by Theorem I.4.3 (see also Lemma II.8.2) the expression (12) is not affected if we replace $\bar{\mathbf{E}}$ and $\xi_\cdot$ by $\mathbf{E}$ and $\xi_\cdot(x)$ respectively.

Consequently, the left-hand side of (6) is independent of the process $p_\cdot$ for $\hat{x}\in D$. If we denote it by $\hat{v}(\hat{x})$ for $p_r \equiv 0$, then by Theorem 2.1 with $t = 0$ we can obtain an expression similar to (2.6) for the derivatives of $\hat{v}(\hat{x})$ with respect to $\hat{x}$ in D. To that end, we have to learn how to apply the operator $\partial(\hat{y})$ to functions $u(\hat{x}, p)$ (see (2.3); we are omitting the subscript r, since P in (5) is independent of r). This is easily done, if we observe that

$$(13)\qquad \partial(\hat{y})u(\hat{x}) = \frac{\partial}{\partial t}u\big(\hat{x}+t\hat{y}, tP(\hat{x})\hat{y}\big)\Big|_{t=0}.$$

Hence, for example, $\partial(\hat{y})u(\hat{x}) = u_{(\hat{y})}(\hat{x})$ if $u(\hat{x}, p) = u(\hat{x})$. Using notation which is not entirely rigorous but more expressive, we can also write

$$\partial(\hat{y})e^{\pi}(\hat{x}) = \frac{\partial}{\partial t}e^{t\pi(x,y)}\Big|_{t=0} = \pi(x,y),$$
$$\partial(\hat{y})\alpha(\hat{x}) = 2\big(\alpha(x), y\big), \qquad \partial(\hat{y})\lambda(\hat{x}) = -\frac{1}{2}K\sigma^*(x)y.$$

Formula (13) enables us, in particular, to apply the usual rules for the differentiation of composite functions or products of functions in the calculation of $\partial(\hat{y})u$.

Note also that if $p_r \equiv 0$, then formulas (2)–(4) become the equalities $\xi_s^{d+1} = x^{d+1}$, $\xi_s^{d+2} = x^{d+2}+s$, $\xi_s^{d+3} = x^{d+3}$ and, consequently, $\hat{v}(x, 0, 0, 0) = v^T(x)$. It follows that the expression $\hat{v}_{(y)}(x, 0, 0, 0)$, which we can write down by Theorem 2.1, coincides with $v^T_{(y)}(x)$. To complete the proof of Lemma 1.3 it suffices to perform the calculations already discussed above; we leave that to the reader.

Lemma 1.8 can be proved by applying Theorem 2.2 to calculate the second-order derivatives of $\hat{v}$ and using again the equality $\hat{v}(x, 0, 0, 0) = v^T(x)$. The calculations in this case are somewhat more complicated, since we must evaluate the second-order derivatives of products of several functions (in the case of $\hat{f}$ there are four functions).

These calculations are also left to the reader. To better organize them, we recommend using the following formulas:

$$\partial^2(\hat{y})u(\hat{x}) = \frac{\partial^2}{\partial t^2}u\big(\hat{x} + t\hat{y}, tP(\hat{x})\hat{y}\big)\Big|_{t=0},$$

$$\partial^2(\hat{y})e^{\pi}(\hat{x}) = \pi^2(x,y), \qquad \partial^2(\hat{y})\alpha(\hat{x}) = 0, \qquad \partial^2(\hat{y})\lambda(\hat{x}) = 0.$$

1. REMARK. We see that the results of Sec. 2 concerning processes that satisfy stochastic equations with parameters enable us to improve Theorems V.8.1 and V.8.5, which originally involved no parameters. We have done this by introducing artificial parameters responsible for random time change, change of measure, and change of the Wiener process. We could also have assumed that the original functions σ, b, c, f of Sec. 1 depended on certain parameters in a way that did not affect v^T. This would have led to a natural generalization of Theorems 1.1 and 1.6, at a cost of slightly complicating the presentation. We preferred not to do so for the sake of simplicity, as well as in order to show that even if there were initially no parameters, they can be always *introduced* nontrivially so they satisfy the assumption of Theorem 2.1, however strange this assumption might look at the first glance.

In some situations, exploiting specific features of a, b, c, f one can introduce parameters that are not related to changes of measure, time, or Wiener processes. For example, it may turn out that a, b, c, f are invariant, in a suitable sense, with respect to all rotations in E_d, or with respect to a subgroup of group of rotations. Then the assumptions of Theorems 1.1 and 1.6 may be weakened further after parameterizing the group using an appropriate $\mathfrak{P}$. We leave it to the reader to find an appropriate example.

4. Some ideas from the theory of conditional processes

1. In this section we would like to acquaint the reader with one of the extremely beautiful parts of the theory of stochastic processes: the theory of conditional processes. Since our topic is diffusion processes, we will use them to illustrate the theory of conditional processes. The main reason for singling out conditional processes from many other branches of the theory of diffusion processes, whether mentioned in this book or not, is that the ideas of the theory of conditional processes will help us to explain why Theorem 5.1, to be established in the next section, is natural.

The style of this section differs drastically from the style in the rest of the book— the arguments will often be rather heuristic. The point is that our main purpose is to help the reader to understand the reasoning of Sec. 5 concerning Theorem 5.1. At the same time, there is no suitable rigorous theory of conditional processes, partly because both in Theorem 5.1 and in most other situations where the concept of conditional processes is helpful, it is *much* easier to achieve *technically* what this concept is prompting without justifying the existence of the appropriate conditional process and without investigating its properties. However, if this is done without necessary explanations, the main, often very far from trivial, idea becomes hidden, and all the manipulations that arise from it quite naturally seem to be a mere "bag of tricks", leading to the result in a rather mysterious way.

The presentation style we have chosen will enable us to review fairly quickly a wide range of problems related to conditional processes. Once again, the arguments of the section do not pretend to be rigorous, and the references to other results of the book are only intended to explain that certain conclusions are natural.

2. If $A, B \subset \Omega$ are events and $P(B) > 0$, then the *conditional probability of A given B* is defined as

$$P(A \mid B) := \frac{P(A \cap B)}{P(B)}.$$

We have already discussed the physical interpretation of this ratio as the limit of the frequency of occurrence of event A in an infinite sequence of independent experiments confined to those cases in which B has also occurred. It is obvious that $P(A \mid B)$, as a function of A, is a probability measure on $\mathcal{F}$. We will denote the integral of a random variable ξ with respect to this measure by $\mathbf{E}\{\xi \mid B\}$ and call it *the conditional expectation of ξ given B*. For simple random variables ξ the frequency interpretation of $P(A \mid B)$ shows that $\mathbf{E}\{\xi \mid B\}$ is the limit of the arithmetic means of the values of ξ in the experiments in which B has occurred. This interpretation of $\mathbf{E}\{\xi \mid B\}$ is used in all cases in which this quantity is defined. If $\xi = I_A$, then clearly

$$\mathbf{E}\,\xi I_B = P(B)\,\mathbf{E}\{\xi \mid B\}. \tag{1}$$

This formula can be extended in a straightforward way, first to simple random variables and then to all random variables for which at least one side of the equality is meaningful. It easily follows from (1) that if $B_1, \dots, B_n \in \mathcal{F}$, $B_i \cap B_j = \emptyset$ for $i \neq j$, $P(B_i) > 0$ and $\cup_i B_i = \Omega$, then

$$\mathbf{E}\,\xi = \sum_{i=1}^{n} \mathbf{E}\,\xi I_{B_i} = \sum_{i=1}^{n} P(B_i)\,\mathbf{E}\{\xi \mid B_i\}. \tag{2}$$

The equality between the outer terms is known as the *total expectation formula*.

3. Now consider the stochastic equation (0.1). Let $Q \subset (-\infty, \infty) \times E_d$ be a domain, bounded from the right with respect to the t-axis, and $\Gamma \subset \partial Q$ a Borel set. Finally, define $\tau(t, x)$ by formula (V.5.17) and consider only points (t, x) that belong to Q (assuming that $Q_+ = Q \cap \{t \geq 0\} \neq \emptyset$).

Our first aim is to understand what stochastic equation the process $\xi(t, x)$ must satisfy during the time interval $[t, \tau(t, x)]$ under the condition that its trajectory (that is, the process $(s, \xi_s(t, x))$, $s \geq t$) hits Γ at its first exit time from Q. In other words, we wish to see what stochastic equation must hold for the process $\eta_s(t, x)$, $s \geq t$, up to the first exit time of $(s, \eta_s(t, x))$ from Q in order to meet the following conditions: for any, say, bounded $\mathfrak{A}(C)$-measurable function $F(x.)$ on C such that $F(\hat{\theta}_t x.)$ depends only on the trajectory of x_s in $[t, \gamma(t, x.)]$ (see the definition of $\gamma(t, x.)$ in (V.5.11)), we have

$$\mathbf{E}\Big\{F\big(\hat{\theta}_t \xi.(t, x)\big) \mid \big(\tau(t, x), \xi_{\tau(t,x)}(t, x)\big) \in \Gamma\Big\} = \mathbf{E}\,F\big(\hat{\theta}_t \eta.(t, x)\big). \tag{3}$$

By what was said in Subsec. 2, we must assume that

$$u(t, x) := P\Big\{\big(\tau(t, x), \xi_{\tau(t,x)}(t, x)\big) \in \Gamma\Big\} > 0. \tag{4}$$

1. PROBLEM. Using Theorem IV.1.5 and the frequency interpretation of the left-hand side of (3), explain that if $\eta_s(t, x)$ has a stochastic differential, then the coefficient of the differential of the Wiener process in $d\eta_s(t, x)$ can be taken the same as in $d\xi_s(t, x)$.

Besides the assumption that condition (4) holds for all $(t, x) \in Q_+$, we will also assume that u is a sufficiently smooth function (and that the subsequent manipulations make sense).

Clearly, if $(r, y) \in \partial Q$, $r > 0$, then $\tau(r, y) = 0$ and $u(r, y) = 0$ for $(r, y) \notin \Gamma$ and $u(r, y) = 1$ for $(r, y) \in \Gamma$. Hence,

$$I_{(\tau(t,x),\xi_{\tau(t,x)}(t,x))\in\Gamma} = u\big(\tau(t, x), \xi_{\tau(t,x)}(t, x)\big).$$

By (1) this means that the left-hand side of (3) equals

$$u^{-1}(t, x)\, \mathbf{E} F\big(\hat{\theta}_t \xi.(t, x)\big) u\big(\tau(t, x), \xi_{\tau(t,x)}(t, x)\big). \tag{5}$$

Let us transform (5) fixing $(t, x) \in Q_+$ and omitting the arguments (t, x) in τ and ξ. By Theorem V.6.4 with $c = f = 0$, $g = I_\Gamma$ the function u satisfies equation (V.6.5) in Q_+. Hence, by Itô's formula, for $t \le s < \tau$ we have

$$du(s, \xi_s) = u_x^*(s, \xi_s)\sigma(s, \xi_s)\, dw_s, \tag{6}$$

$$d \ln u(s, \xi_s) = \alpha^*(s, \xi_s)\, dw_s - \frac{1}{2}\big|\alpha(s, \xi_s)\big|^2\, ds, \tag{7}$$

where u_x is the column-vector of the partial derivatives of u with respect to x and

$$\alpha(r, y) = \sigma^*(r, y)\big(\ln u(r, y)\big)_y.$$

These computations yield a *multiplicative representation* of $u(s, \xi_s)$:

$$u(s, \xi_s) = u(t, x) \exp\left[\int_t^s \alpha^*(r, \xi_r)\, dw_r - \frac{1}{2}\int_t^s \big|\alpha(r, \xi_r)\big|^2\, dr\right], \qquad t \le s < \tau. \tag{8}$$

Such multiplicative representations exist for any positive local martingales, of which $u(s, \xi_s)$ is one by (4) and (6) (or by Remark V.9.5). Taking into account that, naturally (or by Remark V.9.3), $u(s, \xi_s) \to u(\tau, \xi_\tau)$ as $s \uparrow \tau$, we obtain from (8)

$$u(\tau, \xi_\tau) = u(t, x) \exp\left[\int_t^\tau \alpha^*(r, \xi_r)\, dw_r - \frac{1}{2}\int_t^\tau \big|\alpha(r, \xi_r)\big|^2\, dr\right]. \tag{9}$$

Hence, in view of our discussion of (3) and (5), we can write the left-hand side of (3) as

$$\mathbf{E} F(\hat{\theta}_t \xi.) \exp\left[\int_t^\infty \alpha^*(r, \xi_r) I_{r<\tau}\, dw_r - \frac{1}{2}\int_t^\infty \big|\alpha(r, \xi_r)\big|^2 I_{r<\tau}\, dr\right]. \tag{10}$$

It is natural now to recall Girsanov's Theorem and perform a change of measure, denoting the last factor by ρ and setting $\bar{P}(d\omega) = \rho(\omega)P(d\omega)$. By the way, computing the expectations in (9) we see that $\bar{P}$ is a probability measure. Consequently,

$$\bar{w}_s := w_s - \int_0^s \alpha(r, \xi_r) I_{t<r<\tau}\, dr$$

is a d_1-dimensional Wiener process on the probability space $(\Omega, \mathcal{F}, \bar{P})$. Then we obtain $dw_s = d\bar{w}_s + \alpha(s, \xi_s)\, ds$ for $t \le s < \tau$, and for the same s values the process ξ_s satisfies the equation

$$\xi_s = x + \int_t^s \sigma(r, \xi_r)\, d\bar{w}_r + \int_t^s \Big[b(r, \xi_r) + \beta(r, \xi_r)\Big]\, dr, \tag{11}$$

where

$$\beta(r, y) := a(r, y)\big(\ln u(r, y)\big)_y, \qquad a = \sigma\sigma^*. \tag{12}$$

Expression (10) can be obviously written as $\bar{\mathbf{E}}\, F(\hat{\theta}_t \xi.)$, where $\bar{\mathbf{E}}$ denotes integration with respect to $\bar{P}$ over Ω. In addition, as is usually the case, equation (11) has a solution up to *the first exit time* of its trajectories from Q on any probability space with any Wiener process, and the distributions of the functionals $F(\hat{\theta}_t \xi.)$, depending only on ξ_s for $s \in [t, \tau]$, are independent of the probability space and the Wiener process (cf. Corollary V.4.4). In this case $\bar{\mathbf{E}}\, F(\hat{\theta}_t \xi.) = \mathbf{E}\, F(\hat{\theta}_t \eta.)$ and we obtain the desired equality (3) for $\eta.$ which is defined for times less than the first exit time of its trajectory from Q as a solution of the equation

$$\eta_s = x + \int_t^s \sigma(r, \eta_r)\, dw_r + \int_t^s \Big[b(r, \eta_r) + \beta(r, \eta_r)\Big]\, dr. \tag{13}$$

Thus, this is the equation that describes the behavior of the process $\xi.(t, x)$ given that its sample paths leave Q via Γ. It is highly significant that the coefficients $\sigma(r, y)$, $b(r, y) + \beta(r, y)$ of this equation are independent of (t, x) and that the sample paths of its solution $\eta_s(t, x)$ do indeed leave Q through Γ, as follows immediately from (3) if $F(\hat{\theta}_t, x.)$ is taken as the indicator of the set

$$\Big\{x.\colon \big(\gamma(t, x), x_{\gamma(t,x)}\big) \in \Gamma\Big\}.$$

4. In applications it is often of most interest to consider solutions of a stochastic equation given that some event of probability zero has occurred. In such situations one must, of course, introduce a suitable limit procedure. For example, let us find a stochastic equation governing the d-dimensional Wiener process w_t, given that $w_T = z$ at some nonrandom fixed moment $T \in (0, \infty)$, where z is a fixed point in E_d. If $z = 0$, the corresponding conditional process is called the *"Brownian bridge."*

In other words, let $t = 0$, $x = 0$, $b = 0$, $d_1 = d$, $\sigma = (\delta^{ij})$, $Q = (-\infty, T) \times E_d$, $\Gamma = \big\{(T, z)\big\}$. Then $u(0, 0) = P\{w_T = z\} = 0$ and condition (4) is not satisfied. In order to overcome this difficulty, take $\varepsilon > 0$ and first write an equation which is satisfied by w_s given that $|z - w_T| < \varepsilon$. Thus, we are replacing Γ by $\Gamma_\varepsilon = \{T\} \times \{y\colon |y - z| < \varepsilon\}$.

The corresponding u and β are obtained from (4) and (12):

$$u_\varepsilon(t,x) := P\{|x + w_{T-t} - z| < \varepsilon\} = (2\pi\varepsilon)^{-d/2} \int\limits_{|p|<\varepsilon} \exp\Big(-\frac{1}{2(T-t)}|x-z-p|^2\Big)\,dp,$$

$$\beta_\varepsilon(r,y) := \big(\ln u_\varepsilon(r,y)\big)_{(y)} = -\Big(\int\limits_{|p|<\varepsilon} \exp\Big(-\frac{|y-z-p|^2}{2(T-r)}\Big)dp\Big)^{-1}$$
$$\times \int\limits_{|p|<\varepsilon} \frac{y-z-p}{T-r}\exp\Big(-\frac{|y-z-p|^2}{2(T-r)}\Big)dp = -\frac{y-z}{T-r}$$
$$+\Big(\int\limits_{|p|<\varepsilon} \exp\Big(-\frac{|y-z-p|^2}{2(T-r)}\Big)dp\Big)^{-1}$$
$$\times \int\limits_{|p|<\varepsilon} \frac{p}{T-r}\exp\Big(-\frac{|y-z-p|^2}{2(T-t)}\Big)dp.$$

The absolute value of the last expression is obviously bounded by $\varepsilon/(T-r)$ and tends to zero as $\varepsilon \downarrow 0$ if $r < T$. Therefore, writing equation (13) for $\beta = \beta_\varepsilon$ and formally letting $\varepsilon \downarrow 0$, we obtain the following equation for η, which yields a representation of the conditional Wiener process w given that $w_T = z$:

$$\eta_s = w_s - \int_0^s \frac{1}{T-r}(\eta_r - z)\,dr, \qquad s < T. \tag{14}$$

2. PROBLEM. Noting that $\eta_s - w_s$ satisfies a first-order linear ordinary differential equation, express η in terms of w and prove that equation (14) is solvable for $s < T$.

3. PROBLEM. Prove that if $s_1, \dots, s_n < T$, then the vector composed of all coordinates of $\eta_{s_1}, \dots, \eta_{s_n}$ is Gaussian. Find $\mathbf{E}\eta^i_s\eta^j_t$ and prove, using Exercise II.5.4, that if we put $\eta_s \equiv z$ for $s \geq T$, then the processes η_s and $\tilde\eta_{s\wedge T}$, where

$$\tilde\eta_s := w_s - \frac{sw_s}{T} + \frac{sz}{T},$$

have the same distribution on C (in particular, prove that η is continuous (a.s.)).

This result gives us one more reason why the solution of (14) can be *called* a d-dimensional Wiener process conditioned upon hitting the point z at $t = T$. Indeed, by Exercise II.5.9, the behavior of the process $\tilde w_s = w_s - sw_T/T$ for $s \leq T$ is independent of the values of w_T. In particular, it follows from the physical interpretation of independence that the distribution of $\tilde w_s + sw_T/T = w_s$, $s \leq T$, given that $w_T = z$, must *by definition* coincide with that of $\tilde w_s + sz/T = \tilde\eta_s$, that is of η_s.

Note that had we started with this definition and then tried to prove that, under the condition $w_T = z$, the behavior of w is the same (in the sense of distributions on C) as that of the solution of (14), it would have been sufficient to prove only the assertions of Problems 2 and 3 saying *nothing* at all about the general ideas of the theory of conditional processes.

4. PROBLEM. Denote the solution of (14) by $\eta_s(z)$ and set $\eta_s(z) = z$ for $s \geq T$. Prove a "continuous" version of formula (2): for any nonnegative $\mathfrak{A}_T(C)$-measurable function $F(x)$,

$$\mathbf{E}F(w.) = \int_{E_d} \mathbf{E}\left\{F(w.)\,\middle|\, w_T = z\right\} P(w_T \in dz), \tag{15}$$

where

$$P(w_T \in dz) := (2\pi T)^{-d/2} e^{-\frac{1}{2T}|z|^2}\, dz,$$
$$\mathbf{E}\left\{F(w.)\,\middle|\, w_T = z\right\} := \mathbf{E}F(\eta.(z)).$$

5. We will present one more example of conditional processes conditioned upon the occurrence of an event of probability zero. Assume that σ and b depend only on x, fix a domain $D \subset E_d$ and, as usual, set $\tau(x) = \inf\{s \geq 0\colon \xi_s(x) \notin D\}$.

We want to see what stochastic equation the process $\xi_s(x)$ must satisfy given that it never leaves D. In contrast to Subsec. 3, we will assume here that $P\{\tau(x) = \infty\} = 0$ for all $x \in D$, that is, our condition has probability zero. Therefore, the search for a process $\eta.(x)$ such that

$$\mathbf{E}\left\{F(\xi.(x))|\tau(x) = \infty\right\} = \mathbf{E}F(\eta.(x)), \tag{16}$$

is complicated by the fact that the left-hand side of (16) is not defined. We will define it to be

$$\lim_{S\to\infty} \mathbf{E}\left\{F(\xi.(x))|\tau(x) > S\right\}, \tag{17}$$

assuming, as in (4), that for all $x \in D$, $t \in (0, \infty)$,

$$p(x,t) := P\{\tau(x) > t\} > 0.$$

Moreover, under no circumstances will (16) hold for all bounded $\mathfrak{A}(C)$-measurable functions F. Indeed, by the very sense of our condition the process $\eta.(x)$ should not leave D at all, while if we take F equal to the indicator of the set

$$\left\{x. \in C\colon \inf\{t \geq 0\colon x_t \notin D\} = \infty\right\}$$

then the right-hand side of (16) equals 1, and the left-hand side, that is, the expression (17), is

$$\lim_{S\to\infty} \frac{P\{\tau(x) = \infty,\ \tau(x) > S\}}{P\{\tau(x) > S\}} = 0.$$

Therefore, we will only require that for every $T \in (0, \infty)$, equality (16) hold for all bounded $F(x.)$ that depend on x_t only for $t \in [0, T]$ (that is, for all bounded $\mathfrak{A}_T(C)$-measurable functions).

Fix $x \in D$, $S, T \in (0, \infty)$, $S \geq T$, and set $Q = (-\infty, S) \times D$, $\Gamma = \{S\} \times D$. It easily follows that

$$\mathbf{E}\left\{F(\xi.(x))|\tau(x) > S\right\} = \mathbf{E}\left\{F(\xi.(x))I_{\tau(x)\geq S}|\tau(x) > S\right\},$$

and for our Q and Γ the last expression is just the left-hand side of (3) with 0 and $F(\xi.(x))I_{\tau(x)\geq S}$ in place of t and $F(\hat{\theta}_t\xi.(x))$, respectively (the indicator is introduced here to ensure that the former function depend on $\xi_s(x)$ only for $s \in [0, \tau(0,x)]$, where $\tau(0,x) = S \wedge \tau(x)$). Thus, by what was said in Subsec. 3

$$\mathbf{E}\left\{F(\xi_{\cdot}(x)) | \tau(x) > S\right\} = \mathbf{E} F(\eta^S_{\cdot}(x)) I_{\gamma^S(x) \geq S}, \tag{18}$$

where $\gamma^S(x)$ is the first exit time from D of $\eta^S_s(x)$, $\eta^S_s(x) = \eta^S_s(0,x)$ and $\eta^S_s(t,x)$ is defined for $S \geq s \geq t \geq 0$ as a solution of equation (13), where naturally $\sigma(r,y) = \sigma(y)$, $b(r,y) = b(y)$ and β is defined as in (12) for the function $u = u^S$ defined by (4). Note that $u^S(r,y) = p(S-r,y)$ by Corollary V.4.4. We saw in Subsec. 3 that η^S leaves Q only via Γ, hence it remains in D up to time S and the last indicator in (18) can be omitted.

We will now adopt some more assumptions, to be discussed in the next subsection. Let us assume that there is a function $h(y)$ such that

$$\left(\ln u^S(r,y)\right)_y = \left(\ln p(S-r,y)\right)_y \to h(y), \tag{19}$$

as $S \to \infty$, and moreover, $\eta^S_s(x)$ tends in an appropriate sense to a process $\eta_s(x)$ which is a solution of the equation

$$\eta_s = x + \int_0^s \sigma(\eta_r)\, dw_r + \int_0^s \left[b(\eta_r) + a(\eta_r)h(\eta_r)\right] dr. \tag{20}$$

Finally, let us assume that the right-hand side of (18) has a limit as $S \to \infty$. Under these assumptions we obtain the desired equality (16), where the process $\eta_s(x)$ will never leave D since the processes $\eta^S_s(x)$ stay in D for $s \in [0, S]$.

6. A few words about condition (19). If $S \to \infty$, then also $S - r \to \infty$. Hence it is natural that h is independent of r.

Furthermore, fix a point $x_0 \in D$ and set $p(t) = p(t, x_0)$. It turns out that in all cases of practical interest there exists a function $q(r,y)$, $r \geq 0$, $y \in \bar{D}$, such that $q(r,y) > 0$ if $y \in D$, and

$$q_S(r,y) := \frac{p(S-r,y)}{p(S)} = \frac{u^S(r,y)}{u^S(0,x_0)} \to q(r,y) \tag{21}$$

as $S \to \infty$. Moreover, the convergence is uniform on every compact subset of $[0,\infty) \times \bar{D}$, and is not affected if we differentiate all the terms in (21) once with respect to (r,y) or twice with respect to y.

5. Problem. Using the hints given in Problem II.9.3, show that the case in which $d = d_1 = 1$, $D = (-1,1)$, $x_0 = 0$, $\sigma = 1$ and $b = \text{const}$ is a "case of practical interest."

A reader familiar with the theory of second-order partial differential equations of parabolic type (Harnack's inequality and estimates in $C^{2+\alpha}$) will notice that since u^S satisfies equation (V.6.5), then, as a rule, the relevant expressions in (21), their first-order derivatives with respect to (r,y), and second-order derivatives with respect to y form compact sets in the space of continuous functions on $[0,T] \times \bar{D}$ for any $T \in (0,\infty)$ and $S \geq 2T$. In this situation our assumption about (21) is justified, with the reservation that it may be worthwhile to consider a sequence $S_n \to \infty$ rather than all S. To avoid discussing subtle points and to explain as fast as possible how the first eigenfunction of L naturally arises, we will assume that the convergence in (21) holds on all $S \to \infty$.

It follows at once from (21) that the assumption (19) is justified and $h(y) = \big(\ln q(r,y)\big)_y$. In addition, since h depends only on y, it follows that $\ln q(r,y) = \ln q(0,y) + \lambda(r)$, for some function $\lambda(r)$. Consequently, $q(r,y) = \psi_1(y)\exp\lambda(r)$, where $\psi_1(y) = q(0,y)$. Furthermore, $q(r,y)$ satisfies equation (V.6.5) since that was the case for the expressions in (21) that tend to $q(r,y)$, i.e.,

$$\psi_1(y)\lambda'(r) + L(y)\psi_1(y) = 0.$$

Hence it is clear that λ' is independent of r, $\lambda(r) = \lambda r$, where λ is constant which is positive since $p(r,y)$ is obviously a decreasing function of r and $q(r,y)$ an increasing function. Thus $\lambda \geq 0$, $\psi_1 > 0$ in D,

$$\lambda\psi_1 + L\psi_1 = 0 \text{ in } D, \qquad \psi_1 = 0 \text{ on } \partial D, \tag{22}$$

$$h(y) = \big(\ln\psi_1(y)\big)_y. \tag{23}$$

In many situations there exists a unique $\lambda \geq 0$ for which we can find ψ_1 satisfying (22) such that $\psi_1 > 0$ in D. It turns out that $\lambda > 0$ and positive solutions of (22) differ at most by a multiplicative constant, so any positive solution of (22) is good for (23).

Now let us make some comments about the passage to the limit on the right of (18). As already mentioned, the indicator may be omitted. Applying Girsanov's Theorem, we easily see that (remember that $S \geq T$)

$$\mathbf{E}F\big(\eta^S_\cdot(x)\big) = \mathbf{E}F\big(\eta_\cdot(x)\big) \times \exp\left[\int_0^T \alpha_S^*\big(r,\eta_r(x)\big)\,dw_r - \frac{1}{2}\int_0^T \big|\alpha_S\big(r,\eta_r(x)\big)\big|^2 dr\right], \tag{24}$$

where

$$\alpha_S(r,y) = \big(\ln q_S(r,y)q^{-1}(r,y)\big)_y.$$

As stated above, $\alpha_S(r,y) \to 0$ as $S \to \infty$. As a rule, α_S is uniformly bounded on $[0,T]\times D$ for $S \geq 2T$, hence, the passage from (18) to (16) is elementary (and even independent of the convergence of $\eta^S_s(x)$ to $\eta_s(x)$).

7. Conditional processes are interesting not only in themselves; they may also help to investigate the original processes. For example, let us concentrate our attention on the fact that the term β in equation (11) appeared as a result of transformations of (5). Therefore, now that equation (20) and formulas (22) and (23) are known, it is natural to introduce ah in the equation for $\xi_s(x)$ in the same way, thus simultaneously obtaining another derivation of (16). In other words, as in Subsec. 4, having guessed the answer we are going to obtain it once again by another method.

If we set $u(r,y) = \psi_1(y)\exp\lambda r$, $t = 0$, $Q = (-\infty, T)\times D$ in (5), then, by (22), we can again write (6) and (7), obtaining (the first equality is obvious)

$$\begin{aligned}&\psi_1^{-1}(x)\,\mathbf{E}F\big(\xi_\cdot(x)\big)I_{\tau(x)>T}\psi_1\big(\xi_T(x)\big)e^{\lambda T}\\ &\quad= \psi_1^{-1}(x)\,\mathbf{E}F\big(\xi_\cdot(x)\big)I_{\tau(x)>T}\psi_1\big(\xi_{\tau(x)\wedge T}(x)\big)e^{\lambda(\tau(x)\wedge T)} = \mathbf{E}F\big(\eta_\cdot(x)\big)\end{aligned} \tag{25}$$

for all bounded or positive $\mathfrak{A}_T(C)$-measurable functions F. Replacing F with $F(x_\cdot)I_D(x_T)\psi_1^{-1}(x_T)$, we see that for all positive $\mathfrak{A}_T(C)$-measurable F

$$\mathbf{E}F\big(\xi_\cdot(x)\big)I_{\tau(x)>T} = \psi_1(x)\,\mathbf{E}F\big(\eta_\cdot(x)\big)\psi_1^{-1}\big(\eta_T(x)\big)e^{-\lambda T}. \tag{26}$$

In Sec. 5, a formula similar to (26) *will actually give us one of the main methods for reducing the study of functions like* (V.6.13) *for* $D \neq E_d$ *to the case* $D = E_d$.

We can obviously take $S \geq T$ instead of T in (26) for the same $\mathfrak{A}_T(C)$-measurable functions F, put $F \equiv 1$, and find $P\{\tau(x) > T\}$. We obtain

$$(27)\qquad \begin{aligned} p(t,x) = P\{\tau(x) > T\} &= \psi_1(x)e^{-\lambda T}\,\mathbf{E}\,\psi_1^{-1}(\eta_T(x)) =: \psi_1(x)e^{-\lambda T}v(T,x),\\ \mathbf{E}\Big\{F(\xi.(x))\,\Big|\,\tau(x) > S\Big\} &= v^{-1}(S,x)\,\mathbf{E}\,F(\eta.(x))\psi_1^{-1}(\eta_S(x)). \end{aligned}$$

Let us transform the right-hand side of the last expression using the Markov property of $\eta_s(x)$. We obtain

$$(28)\qquad \mathbf{E}\Big\{F(\xi.(x))\,\Big|\,\tau(x) > S\Big\} = v^{-1}(S,x)\,\mathbf{E}\,F(\eta.(x))v(S-T,\eta_T(x)).$$

For the moment, we have carried out these transformations under minimal assumptions. Comparing (28) with (24) and (18), it is easy to guess and then prove directly using Itô's formula (assuming that $p(t,x)$ is a smooth function) that the process $v(S-T,\eta_T(x)) = p(S-T,\eta_T(x))\psi_1^{-1}(\eta_T(x))\exp\lambda(S-T)$ is a martingale for $T \in [0,S]$. Rewriting it in the multiplicative form, we transform the right-hand side of (28) into that of (24). Subject to appropriate assumptions, we can now repeat part of the arguments of the previous subsection thus deducing (16) from (28).

Formula (28) suggests yet another proof of (16). In many cases, the process $\eta_s(x)$, having no possibility to exit from D, becomes very "well mixed" as $s \to \infty$ in the sense that there exists a nonnegative function $\nu(y)$ on D such that

$$(29)\qquad \lim_{s\to\infty}\mathbf{E}\,f(\eta_s(x)) = \int_D f(y)\nu(y)\,dy$$

for any $x \in D$ and any bounded continuous function f. Formula (29) shows that in a certain sense the distribution of $\eta_s(x)$ tends as $s \to \infty$ to the distribution with density ν, the latter being independent of the initial point x. Let us also assume that the quotient $p(1,x)/\psi_1(x)$ is continuous and bounded in D. Then $v(1,x)$ is bounded and continuous (see (27)). Also observe that by (27) we have

$$v(t,x) = \mathbf{E}\,\psi_1^{-1}(\eta_t(x)).$$

Consequently, by the Markov property of $\eta_s(x)$ and by (29)

$$(30)\qquad v(1+s,x) = \mathbf{E}\,v(1,\eta_s(x)),\qquad \lim_{s\to\infty} v(s,x) = \int_D v(1,y)\nu(y)\,dy.$$

We see from the first relationship that $v(s,x)$ is bounded for $s \geq 1$, $x \in D$; the second one together with the Dominated Convergence Theorem immediately deduces (16) from (28) when $S \to \infty$. This proof of (16) is based on more qualitative and more probabilistic assumptions than the proof described in Subsecs. 5, 6. On the other hand, it is trickier, because it begins from the left-hand side of (25), and the idea of introducing the first eigenfunction ψ_1 of L on the left of (25) seems artificial unless one already knows the answer. Formulas (27) and (30) also demonstrate the probabilistic meaning of λ and ψ_1 and can sometimes be used to prove that they are unique.

8. Finally, let us discuss some properties of ν. It follows from (29) with $f \equiv 1$ that ν is the density of a *probability* distribution on D. If the initial value for the process η_s is a random vector η_0 independent of $w_\cdot$, whose distribution has density ν, then by the Markov property and by (29) for any bounded continuous f and $t \geq 0$ we obtain

$$
(31)\qquad
\begin{aligned}
\mathbf{E} f(\eta_t) &= \mathbf{E}\,\mathbf{E} f(\eta_t(y))\Big|_{y=\eta_0} = \int_D \mathbf{E} f(\eta_t(y))\nu(y)\,dy \\
&= \lim_{s\to\infty} \mathbf{E}\,\mathbf{E} f(\eta_t(y))\Big|_{y=\eta_s(x)} = \lim_{s\to\infty} \mathbf{E} f(\eta_{t+s}(x)) \\
&= \int_D f(y)\nu(y)\,dy = \mathbf{E} f(\eta_0).
\end{aligned}
$$

The equality of the first and the last terms shows that η_t and η_0 have the same distribution. For that reason the distribution with density ν is called *an invariant distribution* and ν is called *an invariant density*. In addition, the equality of the third and the penultimate terms in (31), which is valid for all bounded continuous functions f, can be extended in a standard way, by using π- and λ-systems (see Theorem I.5.4), to all nonnegative Borel functions. In particular, for $t \geq 0$, $f = \psi_1^{-1}$

$$
\int_D v(t,y)\nu(y)\,dy = \int_D \psi_1^{-1}(y)\nu(y)\,dy.
$$

Together with (30) and (27), this implies that

$$
(32)\qquad \lim_{t\to\infty} e^{\lambda t} P\{\tau(x) > t\} = \psi_1(x)\int_D \psi_1^{-1}(y)\nu(y)\,dy.
$$

Thus, in addition to (26), we have expressed one more property of ξ_s in terms of the conditional process η_s.

6. Problem. The expression $\mathbf{E}\exp \nu\tau(x)$ can be written as the integral of $P\{\tau(x) > t\}\exp \nu t$ with respect to t. Use this to prove that the eigenvalue λ of the operator $(-L)$ possesses the following property: $\mathbf{E}\exp \nu\tau(x) < \infty$ for $\nu < \lambda$, $\mathbf{E}\exp \lambda\tau(x) = \infty$.

The following consideration may sometimes help to determine ν. Take an arbitrary smooth function f that vanishes on ∂D and substitute $Lf(x_T)$ for $F(x_\cdot)$ in our main formula (26), where $L = L^{a,b}$, as before, is the operator corresponding to ξ_s. Now multiply the resulting equality by $\bar\nu(x) := \nu(x)\psi_1^{-1}(x)$, integrate with respect to $x \in D$, and recall what was said about (31). Finally, integrate with respect to T from 0 to ∞ assuming that $\mathbf{E}\,\tau(x)$ is bounded in D and noting that by Itô's formula

$$
\mathbf{E} f(\xi_{\tau(x)}(x)) = f(x) + \mathbf{E}\int_0^{\tau(x)} Lf(\xi_T(x))\,dT,
$$

where the left-hand side vanishes because $f = 0$ on ∂D. Then we obtain

$$
(33)\qquad -\int_D f(x)\bar\nu(x)\,dx = \frac{1}{\lambda}\int_D \bar\nu(x)\left[\frac{1}{2}\sum_{i,j=1}^d a^{ij}(x) f_{x^i x^j}(x) + \sum_{i=1}^d b^i(x) f_{x^i}(x)\right] dx.
$$

It is proved in the theory of partial differential equations that since f in (33) is an arbitrary function, it follows in many situations that $\bar{v}$ is continuous in D, can be extended continuously to $\bar{D}$, and

$$(34)\qquad \lambda\bar{v} + \frac{1}{2}\sum_{i,j=1}^{d}\left(a^{ij}\bar{v}\right)_{x^i x^j} - \sum_{i=1}^{d}\left(b^i\bar{v}\right)_{x^i} = 0 \text{ in } D, \qquad \bar{v} = 0 \text{ on } \partial D.$$

In addition, recall that since $v = \psi_1\bar{v}$,

$$\int_D \psi_1(x)\bar{v}(x)\,dx = 1.$$

If all the positive solutions of problem (34) are proportional, and the same is true for problem (22), then it follows from (32) that for any positive solutions of problems (22) and (34) we have

$$(35)\qquad \lim_{t\to\infty} e^{\lambda t}\,P\{\tau(x) > t\} = \psi(x)\int_D \bar{v}(y)\,dy\left(\int_D \psi(y)\bar{v}(y)\,dy\right)^{-1}.$$

If, in addition, L is a formally self-adjoint operator, that is, problems (34) and (22) coincide, then $\bar{v} = c\psi_1$, where c is a constant, and (35) gives

$$(36)\qquad \lim_{t\to\infty} e^{\lambda t}\,P\{\tau(x) > t\} = \psi_1(x)\int_D \psi_1(y)\,dy\left(\int_D \psi_1^2(y)\,dy\right)^{-1}.$$

7. PROBLEM. Determining ψ_1 and $\bar{v}$ under the assumptions of Problem 5, use the *result* of Problem II.9.3 to prove the *validity* of (35) and (36) for $x = 0$.

5. A method for investigating the function (V.6.13) for $D \neq E_d$

1. In addition to the assumptions listed in the introduction to the chapter, we assume in this section that σ, b, c, f depend only on x, are twice continuously differentiable with respect to x and bounded together with their first- and second-order partial derivatives with respect to x.

Let $\psi(x)$ be a real-valued function on E_d which is continuous and bounded together with all its partial derivatives up to the fourth order. Assume that

$$D = \{x \in E_d : \psi(x) > 0\} \neq \emptyset.$$

Before stating the main result of this section, we introduce a few more objects. We need an auxiliary space $E_{d+4} = E_d \times E_4$, whose points will be viewed as the pairs $z = (x, y)$ with $x \in E_d$ and $y = (y^1, \dots, y^4) \in E_4$. Functions originally defined on E_d are naturally extended to E_{d+4} as functions independent of y. Let us assume that on the original probability space $(\Omega, \mathcal{F}, P)$, besides the d_1-dimensional Wiener process $(w_t, \mathcal{F}_t)$, we also have $4d_1$ independent one-dimensional Wiener with respect to $\{\mathcal{F}_t\}$ processes w_t^{ij}, $i = 1, \dots, 4$, $j = 1, \dots, d_1$, $t \geq 0$ (see Exercises II.2.7 and II.4.12). Denote $a = \sigma\sigma^*$,

$$L = L(x) = \frac{1}{2}\sum_{i,j=1}^{d} a^{ij}(x)\frac{\partial^2}{\partial x^i \partial x^j} + \sum_{i=1}^{d} b^i(x)\frac{\partial}{\partial x^i}, \qquad \tilde{c}(x) = -L\psi(x),$$

and consider the following system of Itô stochastic equations in $\zeta_s = (\xi_s, \eta_s)$, $\xi_s \in E_d$, $\eta_s \in E_4$, for $s \geq 0$:

$$(1)\qquad \begin{aligned} d\xi_s^i &= \sum_{k=1}^{4}\sum_{j=1}^{d_1} \chi(|\eta_s|)\eta_s^k \sigma^{ij}(\xi_s)\,dw_s^{kj} \\ &\quad + \Big(\sum_{j=1}^{d} a^{ij}(\xi_s)\psi_{x^j}(\xi_s) + \psi(\xi_s)b^i(\xi_s)\Big)ds, \qquad i = 1, \dots, d, \end{aligned}$$

$$(2)\qquad d\eta_s^k = \frac{1}{2}\sum_{i=1}^{d}\sum_{j=1}^{d_1} \psi_{x^i}(\xi_s)\sigma^{ij}(\xi_s)\,dw_s^{kj} - \frac{1}{2}\chi(|\eta_s|)\eta_s^k \bar{c}(\xi_s)\,ds, \qquad k = 1, \dots, 4,$$

where $\chi(r)$ is an infinitely differentiable function on E_1 such that $\chi(r) = 1$ if $r^2 \leq \sup|\psi|$, $\chi(r) = 0$ if $r^2 \geq 2\sup|\psi|$.

It is not difficult to verify that the coefficients of system (1), (2) satisfy Lipschitz conditions as functions of $z = (x, y)$. Moreover, they are bounded and continuous on E_{d+4} together with their first- and second-order derivatives with respect to z. Consequently, for any $z \in E_{d+4}$, system (1), (2) with initial value $(\xi_0, \eta_0) = z$ has a unique solution $\zeta_s(z) = (\xi_s(z), \eta_s(z))$ (up to indistinguishability).

1. THEOREM. *Assume that*

$$(3)\qquad -\bar{c}(x) := L\psi(x) - c(x)\psi(x) \leq -\delta$$

on E_d, where $\delta > 0$ is a constant. Then for any $x \in \bar{D}$ the expression

$$(4)\qquad v(x) := \mathbf{E}\int_0^{\tau(x)} f(\xi_s(x))e^{-\varphi_s(x)}\,ds,$$

where $\tau(x) = \inf\{s \geq 0 \colon \xi_s(x) \notin D\}$, is meaningful and finite. If in addition $y \in E_4$ and $|y|^2 = \psi(x)$, then $v(x) = \psi(x)\bar{v}(z)$, where $z = (x, y)$ and

$$\bar{v}(z) := \mathbf{E}\int_0^{\infty} f(\zeta_s(z))\exp\Big(-\int_0^{s} \bar{c}(\zeta_r(z))\,dr\Big)ds.$$

PROOF. For $T \in (0, \infty)$ and $t \in [0, T]$ define

$$\bar{v}^T(t, z) = \mathbf{E}\int_t^{T} f(\zeta_s(t, z))\exp\Big(-\int_0^{s} \bar{c}(\zeta_r(t, z))\,dr\Big)ds,$$

where $\zeta_s(t, z)$ is a solution of system (1), (2) for $s \geq t$ with initial value $\zeta_t = z$. By Theorem V.7.4, the function $\bar{v}^T$ is continuously differentiable and twice differentiable with respect to z, continuously in (t, z) in $[0, T] \times E_{d+4}$, and it satisfies in that domain the corresponding Kolmogorov equation, that is, the equation (cf. the comments after (IV.1.2))

$$
\begin{aligned}
&\frac{\partial u}{\partial t}+\frac{1}{2}\chi^2(|y|)|y|^2\sum_{i,j=1}^{d}a^{ij}(x)u_{x^ix^j}+\sum_{i,j=1}^{d}a^{ij}(x)\psi_{x^j}(x)u_{x^i}\\
&+\psi(x)\sum_{i=1}^{d}b^i(x)u_{x^i}+\frac{1}{2}\chi(|y|)\sum_{k=1}^{4}\sum_{i,j=1}^{d}a^{ij}(x)\psi_{x^j}(x)y^k u_{x^iy^k}\\
&+\frac{1}{8}(a(x)\psi_x(x),\psi_x(x))\sum_{k=1}^{4}u_{y^ky^k}\\
&-\frac{1}{2}\chi(|y|)\tilde{c}(x)\sum_{k=1}^{4}y^k u_{y^k}-\bar{c}(x)u+f(x)=0,
\end{aligned}
\tag{5}
$$

where $u=u(t,z)=u(t,x,y)$ and the column-vector ψ_x is the gradient of ψ. Moreover, it obviously follows from condition (3) and from the fact that f is bounded, that $\bar{v}^T$ is bounded and $\bar{v}^T(0,z)\to\bar{v}(z)$ uniformly on E_{d+4} as $T\to\infty$. It also follows easily from Itô's formula (cf. the discussion of Theorem IV.1.1) that equation (5) with the boundary condition $u(T,z)=0$ has only one continuous bounded solution in $[0,T]\times E_{d+4}$ with continuous derivatives with respect to t, z, zz.

Now, direct and quite elementary calculations show that, for any solution of equation (5) and any orthogonal (4×4)-matrix A, the function $u(t,x,Ay)$ also satisfies (5). Since, as pointed out above, the solution of (5) is unique, it follows that $\bar{v}^T(t,x,Ay)=\bar{v}^T(t,x,y)$ and, consequently,

$$
\bar{v}^T(t,x,y)=u^T(t,x,|y|),
\tag{6}
$$

where $u^T(t,x,r)$, its first-order derivatives with respect to (t,x,r) and its second-order derivatives with respect to (x,r) are continuous and bounded for $t\in[0,T]$, $x\in E_d$, $r\ge 0$. Consequently, the function

$$
w^T(t,x):=\psi(x)u^T\left(t,x,\sqrt{\psi(x)}\right),
\tag{7}
$$

defined on $[0,T]\times\bar{D}$, is bounded and continuous and has in $[0,T]\times D$ continuous first-order derivatives with respect to t, x and continuous second-order derivatives with respect to x.

Differentiating (6) and (7) and omitting the obvious values of arguments, we obtain

$$
\bar{v}^T_{x^i}=u^T_{x^i},\qquad \bar{v}^T_{y^k}=\frac{y^k}{|y|}u^T_r,\qquad \bar{v}^T_{x^ix^j}=u^T_{x^ix^j},\qquad \bar{v}^T_{x^iy^k}=\frac{y^k}{|y|}u^T_{rx^i},
$$

$$
\sum_{k=1}^{4}\bar{v}^T_{y^ky^k}=u^T_{rr}+\frac{3}{|y|}u^T_r,\qquad w^T_{x^i}=\psi_{x^i}u^T+\psi u^T_{x^i}+\frac{1}{2}\sqrt{\psi}\psi_{x^i}u^T_r,
$$

$$
\begin{aligned}
w^T_{x^ix^j}&=\psi_{x^ix^j}u^T+\psi_{x^i}u^T_{x^j}+\psi_{x^j}u^T_{x^i}+\frac{1}{2}\sqrt{\psi}\psi_{x^ix^j}u^T_r+\psi u^T_{x^ix^j}\\
&+\frac{1}{2}\sqrt{\psi}\left(\psi_{x^i}u^T_{rx^j}+\psi_{x^j}u^T_{rx^i}\right)+\frac{1}{4}\psi_{x^i}\psi_{x^j}\left(u^T_{rr}+\frac{3}{\sqrt{\psi}}u^T_r\right).
\end{aligned}
$$

Recalling that $\bar{v}^T$ satisfies (5) and taking $|y| = \sqrt{\psi}$ in that equation, we conclude that

$$\frac{1}{\psi}\frac{\partial w^T}{\partial t} + Lw^T - cw^T + f = 0 \quad \text{on } [0, T] \times D. \tag{8}$$

Now fix $x \in D$, take $\varepsilon > 0$ such that $x \in D(\varepsilon) := \{y: \psi(y) > \varepsilon\}$, set $\xi_s = \xi_s(x)$ and

$$\gamma(T) = \inf\Big\{s \geq 0: \int_0^s \psi^{-1}(\xi_r)\,dr \geq T\Big\},$$

$$\gamma(T, \varepsilon) = \gamma(T) \wedge \inf\{s \geq 0: \xi_s \notin D(\varepsilon)\},$$

and apply Itô's formula to

$$w^T\Big(\int_0^{s\wedge\gamma(T,\varepsilon)} \psi^{-1}(\xi_r)\,dr, \xi_{s\wedge\gamma(T,\varepsilon)}\Big) \exp\Big(-\int_0^{s\wedge\gamma(T,\varepsilon)} c(\xi_r)\,dr\Big).$$

We note, incidentally, that the Markov time $\gamma(T)$ is obviously bounded and the above determined derivatives of w^T are continuous and bounded in $[0, T] \times \bar{D}(\varepsilon)$. We obtain

$$\begin{aligned} &\psi(x)u^T(0, x, \sqrt{\psi(x)}) = w^T(0, x) \\ &= \mathbf{E}\Big[\int_0^{\gamma(T,\varepsilon)} f(\xi_r)e^{-\varphi_r}\,dr + w^T\Big(\int_0^{\gamma(T,\varepsilon)} \psi^{-1}(\xi_r)\,dr, \xi_{\gamma(T,\varepsilon)}\Big)e^{-\varphi_{\gamma(T,\varepsilon)}}\Big], \end{aligned} \tag{9}$$

where $\varphi_r = \varphi_r(x)$. We let here $\varepsilon \downarrow 0$ and we also use the fact that, by Itô's formula and (3), and since f is bounded,

$$\begin{aligned} \psi(x) &= \mathbf{E}\Big[\int_0^{s\wedge\tau(x)} (L - c)\psi(\xi_r)e^{-\varphi_r}\,dr + \psi(\xi_{s\wedge\tau(x)})e^{-\varphi_{s\wedge\tau(x)}}\Big] \\ &\geq \delta\,\mathbf{E}\int_0^{s\wedge\tau(x)} e^{-\varphi_r}\,dr, \end{aligned}$$

$$\delta\,\mathbf{E}\int_0^{\tau(x)} e^{-\varphi_r}\,dr \leq \psi(x), \qquad \mathbf{E}\int_0^{\tau(x)} |f(\xi_r)|e^{-\varphi_r}\,dr < \infty.$$

By the way, these formulas imply that the right-hand side of (4) is defined and finite. In addition, the second term under the expectation sign in (9) vanishes on the set $\{\omega: \gamma(T, \varepsilon) = \gamma(T)\}$ since $w^T(T, y) = 0$ for $y \in D$. On the set $\{\omega: \gamma(T, \varepsilon) < \gamma(T)\}$ we have $\xi_{\gamma(T,\varepsilon)} \in \partial D(\varepsilon)$, and since w^T is bounded and continuous in $[0, T] \times \bar{D}$ and $w^T(t, y) = 0$ for $t \in [0, T]$, $y \in \partial D$, the term in question vanishes by virtue of the Dominated Convergence Theorem when $\varepsilon \downarrow 0$. Finally, we see that

$$\psi(x)u^T(0, x, \sqrt{\psi(x)}) = \mathbf{E}\int_0^{\gamma(T)\wedge\tau(x)} f(\xi_r)e^{-\varphi_r}\,dr.$$

Again, by the Dominated Convergence Theorem, letting $T \to \infty$ in this equality we obtain $\psi(x)\bar{v}(z) = v(x)$, where $z = (x, y)$, $|y| = \sqrt{\psi(x)}$. We have obtained the desired equality for $x \in D$. Since $\tau(x) = v(x) = \psi(x) = 0$ for $x \in \partial D$, our assertion holds for all $x \in \bar{D}$. □

2. REMARK. The equality $\psi(x)\bar{v}(z) = v(x)$ reduces the investigation of properties of v to the same question for $\bar{v}$. Since the definition of $\bar{v}$ does not involve first exit times, this opens up the possibility to investigate it by the methods of Secs. 1–3. Incidentally, going back to Remark 3.1, we observe that in the situation in which the function $\bar{v}$ appears, one can always introduce additional parameters related to the group of all rotations in the space of the variables y. The simplest consequence of the equality $\psi\bar{v} = v$ is that since the continuous (smooth) functions $\bar{v}^T(0, z)$ converge to $\bar{v}(z)$ uniformly in z, the function $\bar{v}$ is continuous, and hence v is continuous in $\bar{D}$.

3. REMARK. The function (4) differs from (V.6.13) in the term involving g. If g is twice continuously differentiable and bounded together with its first and second derivatives, we can apply Itô's formula and bring this term to an integral form.

2. The proof of Theorem 1 is perhaps too technical; in particular, it may seem to the reader that the relation between equations (5) and (8) is more or less coincidental. The remainder of the section will therefore be devoted to explaining *why the assertion of Theorem* 1 *is natural.* We will continue certain arguments of Sec. 4, again with no pretensions to absolute rigor. To avoid misunderstanding, we note that the notation used below is not that of Subsec. 1 of this section, but follows that of Subsecs. 5–7 in Sec. 4.

In the central formulas (4.25) and (4.26) of Sec. 4.7, replace ψ_1 by any other function ψ that is smooth in E_d, positive in D and equal to zero on ∂D. For $x \in D$, $\xi_s = \xi_s(x)$, $s < \tau = \tau(x)$ we have by Itô's formula (cf. (4.6)–(4.8))

$$d \ln \psi(\xi_s) = \alpha^*(\xi_s)\, dw_s - \frac{1}{2}|\alpha(\xi_s)|^2\, ds + \frac{L\psi}{\psi}(\xi_s)\, ds,$$

$$\psi(\xi_{\tau\wedge T}) \exp\Big(- \int_0^{\tau\wedge T} \frac{L\psi}{\psi}(\xi_s)\, ds\Big)$$

$$= \psi(x) \exp\Big(- \int_0^{\tau\wedge T} \alpha^*(\xi_s)\, dw_s - \frac{1}{2}\int_0^{\tau\wedge T} |\alpha(\xi_s)|^2\, ds\Big), \tag{10}$$

where $\alpha = \sigma(\ln \psi)_x$. Repeating the corresponding arguments of Sec. 4.3 with obvious modifications, we obtain the following generalization of formula (4.25)

$$\mathbf{E} F(\xi_\cdot(x))\, I_{\tau(x)>T}\psi(\xi_T(x)) \exp\Big(- \int_0^T \frac{L\psi}{\psi}(\xi_s(x))\, ds\Big)$$

$$= \mathbf{E} F(\xi_\cdot(x))\, I_{\tau(x)>T}\psi(\xi_{\tau(x)\wedge T}(x)) \exp\Big(- \int_0^{\tau(x)\wedge T} \frac{L\psi}{\psi}(\xi_s(x))\, ds\Big) \tag{11}$$

$$= \psi(x)\mathbf{E} F(\zeta_\cdot(x)) I_{\tau(\zeta,x)>T},$$

where $F(x)$ is any $\mathfrak{A}_T(C)$-measurable function such that at least one of expressions in (11) is meaningful, $\tau(\zeta, x)$ is the first exit time of $\zeta.(x)$ from D, $\zeta.(x)$ being defined as a solution of (4.20) with $h = (\ln \psi)_x$. Unlike (4.25), we have retained the indicator in the last expression in (11), since the meaning of our operations no longer implies directly, as it did in Sec. 4.5, that $\tau(\zeta, x) = \infty$ (a.s). Nevertheless, we have not gone too far from the notion of conditional process, and the behavior of ζ is closely connected with that of the process η from Subsec. 4.5. Indeed,

$$a(\ln \psi)_x = a(\ln \psi_1)_x + \sigma\left(\sigma^*\left(\ln \frac{\psi}{\psi_1}\right)_x\right),$$

whence it follows by Girsanov's Theorem, that if $\psi_1\psi^{-1}$ and the first derivatives of $\psi_1\psi^{-1}$ are bounded in D, then ζ can be obtained from η by an appropriate change of measure, that is,

$$\mathbf{E} F(\zeta.(x)) I_{\tau(\zeta,x)>T} = \mathbf{E}\rho_T F(\eta.(x)) I_{\tau(\eta,x)>T}, \tag{12}$$

where as we know, the indicator on the right can be omitted, and

$$\rho_T = \exp\left\{\int_0^T \left(\ln \frac{\psi}{\psi_1}\right)_x^* \sigma(\eta_s)\, dw_s - \frac{1}{2}\int_0^T \left|\sigma^*\left(\ln \frac{\psi}{\psi_1}\right)_x\right|^2 (\eta_s)\, ds\right\}.$$

(Incidentally, formula (12) is easily applied to the derivation of (11) from (4.25).) It follows from (12) with $F \equiv 1$ that $P\{\tau(\zeta, x) > T\} = 1$ and $\tau(\zeta, x) = \infty$ (a.s.). Thus the last indicator in (11) can be omitted.

4. **Problem.** In general it is not necessarily true that $\tau(\zeta, x) = \infty$ (a.s). Prove, for example, that if $d_1 = d \geq 2$, $\sigma(x) = |x|^2(\delta^{ij}) - xx^*$ in $D = \{|x| < 1\}$, $b = 0$, $\psi(x) = 1 - |x|^2$, then $\tau(x) = \tau(\zeta, x) = (d-1)^{-1}(|x|^{-2} - 1)$ (a.s). In order to understand how this case relates to the above argument try to find ψ_1 and λ.

Notice by the way that formula (4.25) obviously follows from (11) with $\psi = \psi_1$. Further, we obtain from (11), as in the case of (4.26),

$$\mathbf{E} F(\xi.(x)) I_{\tau(x)>T} = \psi(x)\,\mathbf{E} F(\zeta.(x))\psi^{-1}(\zeta_T(x)) \exp \int_0^T \frac{L\psi}{\psi}(\zeta_s(x))\, ds.$$

Substituting

$$F(x.) = f(x_T) \exp\left(-\int_0^T c(x_r)\, dr\right),$$

using the notation (3), (4) and integrating with respect to T from 0 to ∞, we conclude that

$$v(x) = \psi(x)\,\mathbf{E}\int_0^\infty \psi^{-1} f(\zeta_s(x)) \exp\left(-\int_0^s \frac{\bar{c}}{\psi}(\zeta_r(x))\, dr\right) ds. \tag{13}$$

3. We have thus reduced the investigation of $v(x)$ to that of a similar function which does not involve the first exit time. But there is still a considerable difference between the present situation and that of Sec. 1: neither $\psi^{-1}f$ nor the coefficient $a(\ln\psi)_x$ in the equation for ζ_s are bounded in D, since $\psi(x)\to 0$ as x approaches ∂D.

This difficulty can be overcome by a random time change, as was done in a similar situation in Example V.2.2. Indeed, if we set

$$\delta_s(x)=\int_0^s \psi^{-1}(\zeta_r(x))\,dr,\qquad \gamma_s(x)=\inf\{t\geq 0\colon \delta_t(x)\geq s\},$$

then, by the results of Sec. IV.2, the process $\bar{\xi}_s(x):=\zeta_{\gamma_s(x)}(x)$ satisfies the equation

$$\bar{\xi}_s=x+\int_0^s \sqrt{\psi}\sigma(\bar{\xi}_r)\,d\bar{w}_r+\int_0^s(\psi b+a\psi_x)(\bar{\xi}_r)\,dr, \tag{14}$$

where $\bar{w}_r$ is a d_1-dimensional Wiener process and the second factor in (13) equals

$$\mathbf{E}\int_0^\infty f(\bar{\xi}_s(x))\exp\left(-\int_0^s\bar{c}(\bar{\xi}_r(x))\,dr\right)ds. \tag{15}$$

We can now explain why we have replaced ψ_1 by ψ. First, in more or less complicated cases of degenerate operators L, it is hardly likely that we will be able to establish the smoothness of ψ_1 and, consequently, of the coefficients of (14) by applying the theory of differential equations to problem (4.22). Moreover, in the situation of Problem 4 the necessary function ψ_1 does not exist at all. Second, in a great many cases we can ensure the validity of inequality (3) by a suitable choice of ψ. This is necessary if we want to apply Theorem 1.1 to (15), and it can never be done if $\psi=\psi_1$, since then $\bar{c}=(\lambda+c)\psi_1=0$ on D. Thus we have every reason to forget about the eigenvalues and eigenfunctions of $(-L)$.

5. Problem. The time change has actually enabled us to overcome one more difficulty. Prove that the first exit time of $\bar{\xi}_s(x)$ from D (for $x\in D$) in Problem 4 is infinite (a.s.). Prove also that if $c\geq 0$ in D, then $\bar{c}>0$ in D.

4. We have reached the second crucial point in our project to reduce the investigation of v in D to that of a similar function but with $\tau=\infty$. Equation (14) is not satisfactory in that the conditions $\bar{c}=c\psi-L\psi\geq\delta$ in D, $\psi=0$ on ∂D, usually imply that ψ behave like the distance between x and ∂D as x approaches ∂D. For such functions the coefficient $\sqrt{\psi}\sigma$ in (14) does not have bounded derivatives in D. This disadvantage could be avoided by one more random time change in (14), as a result of which, say, $\sqrt{\psi}\sigma$ might be multiplied by $\sqrt{\psi}$ and $(\psi b+a\psi_x)$ by ψ. But at the same time f and $\bar{c}$ in (15) would then be multiplied by ψ, and we would not only have to drop conditions like the second one in (1.1), but conditions of type (1.2) would almost always break down for x close to ∂D, even for $\delta=0$.

This approach, therefore, is obviously useless, and that is why one turns to the behavior of the process $(\psi(\bar{\xi}_s))^{1/2}$. If we raise (10) to the power (-1), transfer the exponential function from the right to the left, recall that one of the models for the process $\zeta.(x)$ may be taken as $\xi.(x)$ after appropriate change of measure, and finally apply the second part of Theorem IV.3.4, then we see that the process

$$\frac{1}{\psi(\zeta_s(x))} \exp \int_0^s \frac{L\psi}{\psi}(\zeta_r(x))\,dr$$

is a local martingale. It follows from the results of Sec IV.2 that a random time change does not violate this property, and so, if we define

$$\pi_s(x) := \sqrt{\psi(\bar{\xi}_s(x))} \exp\left(-\frac{1}{2}\int_0^s L\psi(\bar{\xi}_r(x))\,dr\right),$$

then the process $\pi_s^{-2}(x)$ is a local martingale. (It should be mentioned, of course, that this result can be established without referring to (10) but instead using (14) to evaluate $d\pi_s^{-2}(x)$. However, this would not reveal the arguments that led to the expression $\pi_s^{-2}(x)$.)

Let us find another process with the same property. It turns out that we have already met such process in the proof of Theorem II.10.4, namely, $|y + B_s|$, where B_s is a four-dimensional Wiener process and y any nonzero point of E_4. The process $|y + B_s|^{-2}$ clearly remains a local martingale if we make any random time change. Hence, by Theorem IV.2.3 or simply by Itô's formula for any random *real-valued* function v_r satisfying natural conditions of measurability and integrability, the process

$$\left|y + \int_0^s v_r\,dB_r\right|^{-2} \tag{16}$$

is a local martingale. It should be mentioned that in this sense the process

$$\left|y + \int_0^s v_r\,dB_r\right| \tag{17}$$

not only "looks like" $\pi_s(x)$ but, since $\pi_s^{-2}(x)$ is a local martingale, there always exist y, v, B such that $\pi_s(x)$ coincides with (17). This assertion, which we shall not discuss here, is very similar to Theorem III.10.8 and Problem III.10.10.

From the computational point of view, it is easier to find y and v by comparing, not $\pi_s(x)$ and (17), but $\pi_s^{-2}(x)$ and (16). A necessary condition for these two local martingales to coincide is that they coincide for $s = 0$ and have the same quadratic variation, which may be found by Itô's formula. This yields

$$|y|^2 = \psi(x), \qquad 4v_s^2 = (a\psi_x, \psi_x)(\bar{\xi}_s) \exp \int_0^s \tilde{c}(\bar{\xi}_r)\,dr,$$

where $\bar{\xi}_s = \bar{\xi}_s(x)$ and $\tilde{c} = -L\psi$. To find v_s we have to extract the square root of $(a\psi_x, \psi_x)$, which may not be sufficiently smooth at the points where $\psi_x = 0$. We can overcome this difficulty by using another model of the stochastic integrals in (16) and

(17). Namely, let $w_s^{(i)}$, $i = 1, \dots, 4$, be independent d_1-dimensional Wiener processes. Consider

$$y^i + \int_0^s \frac{1}{2}\psi_x^*\sigma(\bar\xi_t)\exp\frac{1}{2}\int_0^t \tilde c(\bar\xi_r)\,dr\,dw_t^{(i)}. \tag{18}$$

By Theorem III.10.8 (or, more precisely, Problem III.10.10), there exist independent one-dimensional Wiener processes B_s^i such that the processes (18) coincide with

$$y^i + \int_0^s v_r\,dB_r^i.$$

Note that we can now write the equality of (17) and $\pi_s(x)$ as

$$\psi(\bar\xi_s) = |\bar\eta_s|^2, \tag{19}$$

where $\bar\eta_s = (\bar\eta_s^1, \dots, \bar\eta_s^4)$ and $\bar\eta_s^i$ are defined as the products of (18) with

$$\exp\left(-\frac{1}{2}\int_0^s \tilde c(\bar\xi_r)\,dr\right). \tag{20}$$

It immediately follows from Itô's formula that the processes $\bar\eta_s^i$ satisfy the equation

$$\bar\eta_s^i = y^i + \frac{1}{2}\int_0^s \psi_x^*\sigma(\bar\xi_r)\,dw_r^{(i)} - \frac{1}{2}\int_0^s \bar\eta_r^i\tilde c(\bar\xi_r)\,dr. \tag{21}$$

Finally, the processes (17) and $\pi_s(x)$ may coincide only provided there is some definite relation between $\bar w$ and B (cf. the proof of Theorem III.10.8). Similarly, equality (19) is possible only provided some definite relation holds between $\bar w$ and w^i. Raising both sides of (19) to the power (-1), multiplying the result by the (-2)nd power of (20) and applying Itô's formula, we conclude that for (19) to hold it is necessary that

$$\psi^{-2}\psi_x^*\sqrt{\psi}\sigma(\bar\xi_s)\,d\bar w_s = |\bar\eta_s|^{-4}\psi_x^*\sigma(\bar\xi_s)\sum_{i=1}^4 \bar\eta_s^i\,dw_s^{(i)},$$

$$\psi_x^*\sigma(\bar\xi_s)\left[\sqrt{\psi}(\bar\xi_s)\,d\bar w_s - \sum_{i=1}^4 \bar\eta_s^i\,dw_s^{(i)}\right] = 0.$$

It is natural here to assume that the expression in the square brackets vanishes, so we can rewrite (14) as

$$\bar\xi_s = x + \sum_{i=1}^4 \int_0^s \bar\eta_r^i\sigma(\bar\xi_r)\,dw_r^{(i)} + \int_0^s (\psi b + a\psi_x)(\bar\xi_r)\,dr. \tag{22}$$

System (21), (22) differs from (1), (2) only in the notation and in the appearance of χ in the latter. This function was introduced to ensure that the derivatives of the coefficients of (1), (2) with respect to z be bounded. It can be introduced into system (21), (22) as well since by (19) we have $\chi(|\bar\eta_s|) \equiv 1$ (for $|y|^2 = \psi(x)$).

This ends our discussion of Theorem 1. We have shown that many properties of the solutions of (1), (2) are natural, and in this connection we formulate our last problem.

6. PROBLEM. Define $\bar{\psi}(z) = \psi(x) - |y|^2$, where $z = (x, y) \in E_{d+4}$. Letting $\zeta_s(z)$ be a solution of system (1), (2) and evaluating the stochastic differential of $\bar{\psi}(\zeta_s(z))$, prove that $\bar{\psi}(\zeta_s(z)) = 0$ for all s at once (a.s.), if $\bar{\psi}(z) = 0$, that is, the process $\zeta_s(z)$ is always located on the surface defined in E_{d+4} by the equation $\bar{\psi}(z) = 0$. Also prove that if $x \in D$, $|y|^2 < \psi(x)$, then $\xi_s(z)$ never leaves D (for $|y|^2 < \psi(x)$ it is helpful to recall the discussion of Theorem II.10.4). Finally, prove that the derivatives in probability of $\zeta_s(z)$ with respect to z are always tangent to the above-mentioned surface, provided that was the case at $s = 0$ and $\bar{\psi}(z) = 0$.

APPENDIX A

Proof of Lemma II.2.4

First, we assume that f is continuously differentiable on $[0, \pi]$. We can obviously assume that the integral in (II.2.2), which we denote by I, is finite and that f is not a constant. For any fixed $y \in [0, \pi]$ the integrand is equivalent to $|x - y|^{p - \alpha p - 1} |f'(y)|^p$ as $x \to y$. The integral of this expression with respect to x diverges if $f'(y) \neq 0$ and $\alpha \geq 1$. Hence the assumption that I is finite yields $\alpha < 1$. Therefore,

$$K := \sup_{t,s \in [0,\pi]} \big|f(t) - f(s)\big| \cdot |t - s|^{\frac{1}{p} - \alpha} < \infty.$$

Now let $0 \leq s < t \leq \pi$. Integrating the inequality

$$\big|f(t) - f(s)\big| \leq \big|f(t) - f(u)\big| + \big|f(u) - f(v)\big| + \big|f(v) - f(s)\big|$$

with respect to $u \in [t - \varepsilon, t]$, $v \in [s, s + \varepsilon]$, where $0 < \varepsilon \leq t - s$, we obtain

$$\begin{aligned} \big|f(t) - f(s)\big| \leq \frac{1}{\varepsilon} \int_{t-\varepsilon}^{t} \big|f(t) - f(u)\big|\, du + \frac{1}{\varepsilon} \int_{s}^{s+\varepsilon} \big|f(v) - f(s)\big|\, dv \\ + \frac{1}{\varepsilon^2} \int_{s}^{t} \int_{s}^{t} \frac{\big|f(u) - f(v)\big|}{|u - v|^{\frac{1}{p} + \alpha}} |u - v|^{\frac{1}{p} + \alpha}\, du dv. \end{aligned} \tag{1}$$

To estimate the last integral we use the Hölder inequality ($p > \alpha^{-1} > 1$) and the inequalities $|u - v| \leq |t - s|$ and

$$\left(\int_{s}^{t} \int_{s}^{t} |u - v|^{(\frac{1}{p} + \alpha)\frac{p}{p-1}}\, du dv \right)^{\frac{p-1}{p}} \leq \left(|t - s|^{(\frac{1}{p} + \alpha)\frac{p}{p-1} + 2} \right)^{\frac{p-1}{p}}.$$

To estimate the first and second terms on the right in (1), note that, for example,

$$\big|f(t) - f(u)\big| \leq K |t - u|^{\alpha - \frac{1}{p}}, \qquad \int_{t-\varepsilon}^{t} \big|f(t) - f(u)\big|\, du \leq K \varepsilon^{\alpha - \frac{1}{p} + 1}.$$

Now we see that

$$\big|f(t) - f(s)\big| \leq 2K \varepsilon^{\alpha - \frac{1}{p}} + \varepsilon^{-2} |t - s|^{2 + \alpha - \frac{1}{p}} I^{\frac{1}{p}}.$$

If $\delta \in (0, 1)$ is a constant, t, s arbitrary points of $[0, \pi]$, and $\varepsilon = \delta |t - s|$, then by the definition of K this inequality yields

$$K \leq 2\delta^{\alpha - \frac{1}{p}} K + \delta^{-2} I^{\frac{1}{p}}.$$

Choosing δ so that the coefficient of K on the right equals $1/2$, we complete the proof of (II.2.2) for smooth f. Now the length of the interval $[0, \pi]$ was immaterial to our arguments, so that inequality (II.2.2) remains valid with the same constant N even when arbitrary a, b are substituted for $0, \pi$, provided only that $t, s \in [a, b]$ and f is smooth on $[a, b]$. In particular, if f is continuously differentiable on $[s, t]$, then

$$|f(t) - f(s)|^p \leq N|t - s|^{p\alpha - 1} \int_s^t \int_s^t \frac{|f(x) - f(y)|^p}{|x - y|^{1+p\alpha}} dxdy. \tag{2}$$

Passing to the general case, we note that it will suffice to prove (II.2.2) for $t, s \in (0, \pi)$. Fix such t, s, $t \geq s$, and let ξ be a random variable such that $|\xi| \leq 1$ and the density $p(x)$ of the distribution of ξ is infinitely differentiable. Denote

$$f_\varepsilon(x) = \mathbf{E} f(x + \varepsilon\xi)$$

where $0 < \varepsilon < s \wedge (\pi - t)$, $x \in [s, t]$.

It follows from the Dominate Convergence Theorem that $f_\varepsilon \to f$ as $\varepsilon \downarrow 0$ on $[s, t]$. In addition, since on $[s, t]$

$$\mathbf{E} f(x + \varepsilon\xi) = \int_{-1}^{1} f(x + \varepsilon y) p(y)\, dy = \frac{1}{\varepsilon} \int_0^\pi f(y) p\Big(\frac{y - x}{\varepsilon}\Big) dy,$$

f_ε is infinitely differentiable on $[s, t]$. Finally, by Hölder's inequality and Fubini's Theorem,

$$\int_s^t \int_s^t \frac{|\mathbf{E}[f(x + \varepsilon\xi) - f(y + \varepsilon\xi)]|^p}{|x - y|^{1+p\alpha}} dxdy$$

$$\leq \mathbf{E} \int_s^t \int_s^t \frac{|f(x + \varepsilon\xi) - f(y + \varepsilon\xi)|^p}{|(x + \varepsilon\xi) - (y + \varepsilon\xi)|^{1+p\alpha}} dxdy.$$

An obvious substitution shows that for the ε's under consideration the last expression is less than the integral on the right of (II.2.2). To complete the proof of (II.2.2) it remains to substitute f_ε for f in (2) and let $\varepsilon \downarrow 0$. □

PROBLEM. Prove that if the integral in (II.2.2) is finite for $\alpha = (p + 1)/p$, then f is a constant. (Hint: see (2).)

APPENDIX B

Proof of Theorem II.8.1

Observe that $h^{1/p} \le (T+\varepsilon)^{(p-1)d/(2p)}h$. Therefore, $|g(x)| \le N_0 h(T, x)$, where N_0 is a constant. Now if $t_n \in [0, T]$, $x_n \in E_d$, $t_n \to t$, $x_n \to x$, then by Exercise II.6.8 and Fatou's Lemma we obtain,

$$\begin{aligned}\varliminf_{n\to\infty} [N_0 h \pm u](t_n, x_n) &= \varliminf_{n\to\infty} \mathbf{E}\big[N_0 h(T, x_n + w_{T-t_n}) \pm g(x_n + w_{T-t_n})\big] \\ &\ge \mathbf{E} \varliminf_{n\to\infty} \big[N_0 h(T, x_n + w_{T-t_n}) \pm g(x_n + w_{T-t_n})\big] \\ &= \mathbf{E}\big[N_0 h(T, x + w_{T-t}) \pm g(x + w_{T-t})\big] \\ &= N_0 h(t, x) \pm u(t, x).\end{aligned}$$

It obviously follows that u is continuous and $|u| \le N_0 h$. Set

$$p(x, t) = (2\pi t)^{-d/2} e^{-|x|^2/(2t)}, \qquad t > 0,\ x \in E_d.$$

Then for $t \in [0, T)$

$$u(t, x) = \int_{E_d} g(x+y)p(T-t, y)\, dy = \int_{E_d} g(y)p(T-t, y-x)\, dy. \tag{1}$$

Moreover, it is not difficult to prove that if $t \le t_0 < t_1 < T$, then, denoting by $p'(T-t, y-x)$ any derivative of *any order* of $p(T-t, y-x)$ with respect to (t, x), there exists a constant N_1 such that

$$\big|p'(T-t, y-x)\big| \le N_1 p(T-t_1, y-x). \tag{2}$$

It follows almost immediately from the inequality $|y-x|^2 \ge (1-\delta)|y|^2 + (1-\delta^{-1})|x|^2$ that, if $t_1 < t_2 < T$, $R \in (0, \infty)$, $|x| \le R$, then there exists a constant N_2 such that the right-hand side of (2) is bounded from above by $N_2 p(T-t_2, y)$. Estimate $|g| \le N_0 h(T, x)$, together with Exercise II.6.8 and Weierstrass's Theorem now proves that if we substitute p' for p in (1), we obtain an integral, which converges uniformly with respect to $(t, x) \in [0, t_0] \times \{|x| \le R\}$. This implies that $u(t, x)$ is infinitely differentiable with respect to (t, x) in Q and its derivatives can be evaluated by differentiating the last term in (1) under the integral sign. In particular, since $\partial p/\partial t - (1/2)\Delta p = 0$, this proves (II.8.2) in Q.

The inequality $|u|^p \le Kh$ follows from Exercise II.6.8 and from Hölder's inequality, according to which

$$\big|u(t, x)\big|^p = \big|\mathbf{E} g(x + w_{T-t})\big|^p \le \mathbf{E}\big|g(x + w_{T-t})\big|^p.$$

To prove the remaining assertions it suffices to find u_x by differentiating p in (1), integrate by parts, verify that $u_x(t, x) = \mathbf{E}\, g_x(x + w_{T-t})$, and repeat the first and last parts of the previous discussion of $u(t, x)$.

List of Notations

The notation X, $\mathfrak{A}$ and μ, introduced on p. 84, is used later without alteration.

Comments

Chapter I

Sections 1–3, 5. The basic results of measure theory were established at the end of the 19th and the beginning of the 20th centuries. Nevertheless, Scheffé's theorem, for example (Problem 2.18), which is essentially a criterion for compactness in $\mathcal{L}_1$, was proved considerably later, in 1947. Incidentally, Problem 2.18 can be solved immediately by noting that $(\xi - \xi_n)_+ \leq \xi$ and using the formula $|\xi_n - \xi| = |\xi - \xi_n| = 2(\xi - \xi_n)_+ - (\xi - \xi_n)$.

The Monotone and Dominated Convergence Theorems are often called Lebesgue's theorems, although the first was proved earlier by B. Levi. The reader may find more detailed historical information in Saks [44] (1937) and Dunford and Schwartz [10] (1958).

Section 4. Characteristic functions were first used in probability theory by A. M. Lyapunov. For their properties and applications see, for example, Shiryaev [45] (1980) and Loève [38] (1977-78).

Chapter II

Section 1. Objections are sometimes raised as to the interpretation of the probability of an event in an experiment as the limit of the frequency of its occurrence in an infinite sequence of repeated experiments. One might argue that until one performs an infinite number of the experiments, the very notion of the probability of an event is senseless. But a similar problem arises in connection with applications of any mathematical theory. For example, in real life we use geometry even without verifying one of its basic axioms that, say, any two points in Moscow, can be connected by a straight line.

Sections 2–3. Wiener [49] (1923) constructed the Daniel's integral on the space C of continuous functions. The measure corresponding to this integral is known as Wiener measure on C, and a random element with values in C whose distribution is Wiener measure is known as the Wiener process. The Fourier sine expansion of $w_t - tw_\pi/\pi$ for $t \in [0, \pi]$ may be found in Paley, Wiener, and Zygmund [41] (1933). The use of embedding theorems (Lemma 2.4) to construct a Wiener process is not the first application of these theorems in probability theory. Lemma 2.4 can easily be derived from more advanced results of Campanato [5] (1963). One should also mention the Lévy's [34] (1965) construction of a Wiener process by using series in Haar functions. The first part of Theorem 3.2 and Theorem 3.6 can be found in Wiener [49] (1923).

Sections 4–10. The notion of the strong Markov property was introduced by Dynkin and Yushkevich [13] (1956). Many of the properties of Wiener processes described in this book, as well as their connection with the theory of differential equations, may be generalized to a broad range of Markov processes (see, for example, Dynkin [12] (1963)).

Within this theory formula (8.11) is a particular case of the general *Dynkin formula*. The reader can find an exposition of the general theory of martingales in Liptser and Shiryaev [37] (1986). Remark 7.4 belongs in essence to Bismut and Yor [3] (1983). For the Feynman-Kac formula, see Kac [25] (1951). Theorem 9.7, in a slightly different form, was proved by Petrovskiĭ [Petrowsky] [42] (1934–35). It can be generalized to processes with independent increments (see Skorokhod [46] (1986)). The Cauchy process introduced in Problem 10.8 is left-continuous in t. The term "Cauchy process" is usually reserved for its right-continuous version.

Chapter III

Sections 1–2. Paley, Wiener, and Zygmund [41] (1933) considered the stochastic integral of a nonrandom function with respect to a Wiener process, starting with formula (2.12) for smooth functions and then taking closures in $\mathcal{L}_2$. Formula (2.14) may be found in Itô [21–23] (1951). Our definitions are most close to those used in Krylov [29] (1986). The Clark's theorem [6] (1970) is sometimes called Itô's theorem, referring to Itô [23] (1951).

Sections 3–8. The material of these sections may be found in any book on the theory of stochastic integration, such as Liptser and Shiryaev [37] (1986). The different kinds of convergence used in the theory of probability are discussed in Shiryaev [45] (1980) and Loève [38] (1977-78). The generalized Itô inequality is sometimes called Lenglart's inequality bearing in mind his work of 1977.

Sections 9–10. These sections mostly follow Kunita and Watanabe [33] (1967) and Stroock and Varadhan [47] (1979).

Section 11. In addition to predictable processes, pseudopredictable functions (which need not be random processes), well-measurable processes, and regularly measurable processes, one encounters few other notions, such as progressively measurable processes (see, for example, Liptser and Shiryaev [37] (1986)).

Chapter IV

Section 2. Dubins and Schwarz [9] (1965) obtained a large class of martingales by using random time change in Wiener processes (see also Kunita and Watanabe [33] (1967)). Before that random time change was used in investigations of Markov processes; see Dynkin [12] (1963).

Sections 3–4. For generalizations of these results, and also those of Sec. 2 see Liptser and Shiryaev [37] (1986), Ikeda and Watanabe [20] (1989). Theorem 3.5 (c) is due to Portenko [43] (1982).

Chapter V

Section 1. The proof of Theorem 1 is due to Krylov [32] (1990), where the reader may find references to earlier works. We mention only Gyöngy and Krylov [17] (1980), where a version of Theorem 5 may be found. Many other existence theorems for stochastic equations may be found in the references at the end of the book. This also relates to other questions considered in Chapter V.

Section 2. The result of Example 1 was previously established by Krylov [28] (1985) by translating the arguments presented here into the language of the theory of differential equations. Example 2 arose as a counterexample to a certain incorrect result in the theory of differential equations, which stated, in particular, that the only bounded solution of equation (5) is the trivial solution.

The connection between explosion times and the solutions of homogeneous differential equations is discussed by Ikeda and Watanabe [20] (1989).

Sections 4–8. Gikhman [18] (1951) was the first to use Euler's method to solve stochastic differential equation and to prove the existence of solutions of degenerate differential equations by probabilistic methods using Kolmogorov's equation. His definition of a stochastic differential equation differs from the one accepted now. Partly because of this, Gikhman's 1951 paper was not noticed at the time, and his results were proved again by Blagoveshchenskiĭ and Freĭdlin [4] (1961), who used a different method. Degenerate elliptic equations were considered later by Freĭdlin [15] (1968). The results of Secs. 7, 8 admit a generalization to the case of controlled diffusion processes and nonlinear differential equations (see Krylov [26] (1977)).

Sections 9–10. The results presented in these sections follow from the Feller property of solutions of stochastic differential equations. Therefore, they remain valid for many Feller processes (see Dynkin [12] (1963)).

Chapter VI

Sections 1–3. The smoothness conditions given here for probabilistic solutions of differential equations slightly generalize those of Krylov [30] (1988).

Section 4. The contents of this section belong to the folklore of the subject, although there also exist rigorous results in the theory of conditional processes, which may be found starting from the book by Dynkin [12] (1963). In modern notation, it is only formula (15) that shows that $\eta_t(z)$ is a conditional Wiener process. Formula (36) was proved for the Wiener process by Kac [25] (1951). The definition of conditional processes by means of change of measure remind Cramér's transformation in the theory of large deviations.

Section 5. A generalization of Theorem 1 may be found in Krylov [27] (1981). An application of Theorem 1 to the derivation of almost necessary conditions for the smoothness of probabilistic solutions of differential equations in a domain may be found in Krylov [31] (1989).

References

1. M. T. Barlow, *One-dimensional stochastic differential equations with no strong solution*, J. London Math. Soc. (2) **26** (1982), 335–347.
2. S. N. Bernstein, *The theory of probability*, 4th ed., OGIZ, Moscow and Leningrad, 1946. (Russian)
3. J. M. Bismut and M. Yor, *An inequality for processes which satisfy Kolmogorov's continuity criterion. Application to continuous martingales*, J. Funct. Anal. **51** (1983), 166–173.
4. Yu. N. Blagoveshchenskiĭ and M. I. Freĭdlin, *Some properties of diffusion processes depending on a parameter*, Dokl. Akad. Nauk SSSR **138** (1961), 508–511; English transl. in Soviet Math. Dokl. **2** (1961).
5. S. Campanato, *Proprieta di holderianita di alcune classi di funzioni*, Ann. Scuola Norm. Sup. Pisa (3) **17** (1963), 175–188.
6. J. M. C. Clark, *The representation of functionals of Brownian motion by stochastic integrals*, Ann. Math. Statist. **41** (1970), 1282–1295.
7. M. G. Crandall and P.-L. Lions, *Viscosity solutions of Hamilton-Jacobi equations*, Trans. Amer. Math. Soc. **277** (1983), 1–42.
8. J. L. Doob, *Stochastic processes*, Wiley, New York, 1953.
9. L. E. Dubins and G. Schwarz, *On continuous martingales*, Proc. Nat. Acad. Sci. U.S.A. **53** (1965), 913–916.
10. N. Dunford and J. T. Schwartz, *Linear operators. Part* II: *Spectral theory. Selfadjoint operators in Hilbert space*, Interscience, New York, 1963.
11. R. Durrett, *Brownian motion and martingales in analysis*, Wadsworth, Belmont, CA, 1984.
12. E. B. Dynkin, *Markov processes. Vols.* I, II, Fizmatgiz, Moscow, 1963; English transl., Grundlehren Math. Wiss., vols. 121, 122, Springer-Verlag, Berlin, 1965.
13. E. B. Dynkin and A. A. Yushkevich, *Strong Markov processes*, Theory Probab. Appl. **1** (1956), 134–139.
14. I. Gyöngy and N. V. Krylov, *On stochastic equations with respect to semimartingales*. I, Stochastics **4** (1980), 1–21.
15. M. I. Freĭdlin, *On the smoothness of solutions of degenerate elliptic equations*, Izv. Akad. Nauk SSSR Ser. Mat. **32** (1968), 1391–1413; English transl. in Math. USSR-Izv. **2** (1968).
16. ———, *Functional integration and partial differentiation equations*, Ann. of Math. Stud., vol. 109, Princeton Univ. Press, Princeton, NJ, 1985.
17. A. Friedman, *Stochastic differential equations and applications. Vols.* 1, 2, Probability and Math. Statist., vol. 28, Academic Press, New York, 1975; 1976.
18. I. I. Gikhman, *On the theory of differential equations of random processes*. II, Ukrain. Mat. Zh. **3** (1951), 317–339; English transl. in Amer. Math. Soc. Transl. Ser. 2 **1** (1955).
19. I. I. Gikhman and A. V. Skorokhod, *Theory of random processes. Vol.* III, "Nauka", Moscow, 1975; English transl., Springer-Verlag, Berlin, 1979.
20. N. Ikeda and S. Watanabe, *Stochastic differential equations and diffusion processes*, 2nd ed., North-Holland, Amsterdam, 1989.
21. K. Itô, *On stochastic differential equations*, Mem. Amer. Math. Soc **4** (1951), no. 4.
22. ———, *On a formula concerning stochastic differentials*, Nagoya Math. J. **3** (1951), 55–65.
23. ———, *Multiple Wiener integral*, J. Math. Soc. Japan **3** (1951), 157–169.
24. K. Itô and H. P. McKean, *Diffusion processes and their sample paths*, Grundlehren Math. Wiss., vol. 125, Springer-Verlag, Berlin, 1965.
25. M. Kac, *On some connections between probability theory and differential and integral equations*, Proceedings of the Second Berkeley Symposium on Mathematical Statistics and Probability, Univ. California Press, Berkeley and Los Angeles, 1951, pp. 189–215.

26. N. V. Krylov, *Controlled diffusion processes*, "Nauka", Moscow, 1977; English transl., *Appl. Math.*, vol. 14, Springer-Verlag, New York, 1980.
27. ———, *On the control of a diffusion process until the moment of the first exit from the domain*, Izv. Akad. Nauk SSSR Ser. Mat. **45** (1981), no. 5, 1029–1048; English transl. in Math. USSR-Izv. **19** (1982).
28. ———, *Nonlinear elliptic and parabolic equations of the second order*, "Nauka", Moscow, 1985; English transl., Reidel, Dordrecht and Boston, 1987.
29. ———, *Introduction to the theory of random processes.* I, MGU, Moscow, 1986. (Russian)
30. ———, *Moment estimates for the quasiderivatives, with respect to the initial data, of solutions of stochastic equations and their applications*, Mat. Sb. **136** (1988), no. 4, 510–529; English transl. in Math. USSR-Sb. **64** (1989).
31. ———, *Smoothness of the payoff function for a controllable diffusion process in a domain*, Izv. Akad. Nauk SSSR Ser. Mat. **53** (1989), no. 1, 66–96; English transl. in Math. USSR-Izv. **34** (1990).
32. ———, *Simple proof of the existence of a solution to the Itô equation with monotone coefficients*, Teor. Veroyatnost. i Primenen. **35** (1990), no. 3, 576–580; English transl. in Theory Probab. Appl. **35** (1990).
33. H. Kunita and S. Watanabe, *On square integrable martingales*, Nagoya Math. J. **30** (1967), 209–245.
34. P. Lévy, *Processus stochastiques et mouvement brownien*, Gauthier-Villars, Paris, 1965.
35. P.-L. Lions, *Optimal control of diffusion processes and Hamilton-Jacobi-Bellman equations.* II. *Viscosity solutions and uniqueness*, Comm. Partial Differerential Equations **8** (1983), 1229–1276.
36. R. S. Liptser and A. N. Shiryaev, *Statistics of random processes.* I, "Nauka", Moscow, 1974; English transl., Springer-Verlag, Berlin and New York, 1977.
37. ———, *Theory of martingales*, "Nauka", Moscow, 1986; English transl., Kluwer, Dordrecht, 1989.
38. M. Loève, *Probability theory*, 3rd ed., Van Nostrand, Princeton, NJ, 1963.
39. M. Motoo and S. Watanabe, *On a class of additive functionals of Markov processes*, J. Math. Kyoto Univ. **4** (1965), 429–469.
40. S. M. Nikol′skiĭ, *Approximation of functions of several variables and imbedding theorems*, "Nauka", Moscow, 1969; *Grundlehren Math. Wiss.*, vol. 205, Springer-Verlag, Berlin, 1975.
41. R. K. A. C. Paley, N. Wiener, and A. Zygmund, *Note on random functions*, Math. Z. **37** (1933), 647–668.
42. I. Petrowsky, *Zur ersten Randwertaufgabe der Wärmelteinungsgleichung*, Compositio Math. **1** (1935), 383–419.
43. N. I. Portenko, *Generalized diffusion processes*, "Naukova Dumka", Kiev, 1982; English transl., Amer. Math. Soc., Providence, RI, 1990.
44. S. Saks, *Theory of the integral*, Hafner, New York, 1937.
45. A. N. Shiryaev, *Probability*, "Nauka", Moscow, 1980; English transl., Springer-Verlag, Berlin and New York, 1984.
46. A. V. Skorokhod, *Random processes with independent variations*, "Nauka", Moscow, 1986; English transl., Reidel, Dordrecht, 1991.
47. D. W. Stroock and S. R. S. Varadhan, *Multidimensional diffusion processes*, Grundlehren Math. Wiss., vol. 233, Springer-Verlag, Berlin and New York, 1979.
48. A. Yu. Veretennikov, *Strong solutions and explicit formulas for solutions of stochastic integral equations*, Mat. Sb. **111** (1980), no. 3, 434–452; English transl. in Math. USSR-Sb. **39** (1981).
49. N. Wiener, *Differential space*, J. Math. Phys. **2** (1923), 131–174.

Index

COPYING AND REPRINTING. Individual readers of this publication, and non-profit libraries acting for them, are permitted to make fair use of the material, such as to copy a chapter for use in teaching or research. Permission is granted to quote brief passages from this publication in reviews, provided the customary acknowledgment of the source is given.

Republication, systematic copying, or multiple reproduction of any material in this publication (including abstracts) is permitted only under license from the American Mathematical Society. Requests for such permission should be addressed to the Assistant to the Publisher, American Mathematical Society, P.O. Box 6248, Providence, Rhode Island 02940-6248. Requests can also be made by e-mail to reprint-permission@ams.org.

Other Titles in This Series

(*Continued from the front of this publication*)

106 **G.-C. Wen,** Conformal mappings and boundary value problems, 1992
105 **D. R. Yafaev,** Mathematical scattering theory: General theory, 1992
104 **R. L. Dobrushin, R. Kotecký, and S. Shlosman,** Wulff construction: A global shape from local interaction, 1992
103 **A. K. Tsikh,** Multidimensional residues and their applications, 1992
102 **A. M. Il'in,** Matching of asymptotic expansions of solutions of boundary value problems, 1992
101 **Zhang Zhi-fen, Ding Tong-ren, Huang Wen-zao, and Dong Zhen-xi,** Qualitative theory of differential equations, 1992
100 **V. L. Popov,** Groups, generators, syzygies, and orbits in invariant theory, 1992
99 **Norio Shimakura,** Partial differential operators of elliptic type, 1992
98 **V. A. Vassiliev,** Complements of discriminants of smooth maps: Topology and applications, 1992 (revised edition, 1994)
97 **Itiro Tamura,** Topology of foliations: An introduction, 1992
96 **A. I. Markushevich,** Introduction to the classical theory of Abelian functions, 1992
95 **Guangchang Dong,** Nonlinear partial differential equations of second order, 1991
94 **Yu. S. Il'yashenko,** Finiteness theorems for limit cycles, 1991
93 **A. T. Fomenko and A. A. Tuzhilin,** Elements of the geometry and topology of minimal surfaces in three-dimensional space, 1991
92 **E. M. Nikishin and V. N. Sorokin,** Rational approximations and orthogonality, 1991
91 **Mamoru Mimura and Hirosi Toda,** Topology of Lie groups, I and II, 1991
90 **S. L. Sobolev,** Some applications of functional analysis in mathematical physics, third edition, 1991
89 **Valeriĭ V. Kozlov and Dmitriĭ V. Treshchëv,** Billiards: A genetic introduction to the dynamics of systems with impacts, 1991
88 **A. G. Khovanskiĭ,** Fewnomials, 1991
87 **Aleksandr Robertovich Kemer,** Ideals of identities of associative algebras, 1991
86 **V. M. Kadets and M. I. Kadets,** Rearrangements of series in Banach spaces, 1991
85 **Mikio Ise and Masaru Takeuchi,** Lie groups I, II, 1991
84 **Ðào Trọng Thi and A. T. Fomenko,** Minimal surfaces, stratified multivarifolds, and the Plateau problem, 1991
83 **N. I. Portenko,** Generalized diffusion processes, 1990
82 **Yasutaka Sibuya,** Linear differential equations in the complex domain: Problems of analytic continuation, 1990
81 **I. M. Gelfand and S. G. Gindikin, Editors,** Mathematical problems of tomography, 1990
80 **Junjiro Noguchi and Takushiro Ochiai,** Geometric function theory in several complex variables, 1990
79 **N. I. Akhiezer,** Elements of the theory of elliptic functions, 1990
78 **A. V. Skorokhod,** Asymptotic methods of the theory of stochastic differential equations, 1989
77 **V. M. Filippov,** Variational principles for nonpotential operators, 1989
76 **Phillip A. Griffiths,** Introduction to algebraic curves, 1989
75 **B. S. Kashin and A. A. Saakyan,** Orthogonal series, 1989
74 **V. I. Yudovich,** The linearization method in hydrodynamical stability theory, 1989
73 **Yu. G. Reshetnyak,** Space mappings with bounded distortion, 1989
72 **A. V. Pogorelev,** Bendings of surfaces and stability of shells, 1988
71 **A. S. Markus,** Introduction to the spectral theory of polynomial operator pencils, 1988
70 **N. I. Akhiezer,** Lectures on integral transforms, 1988
69 **V. N. Saliĭ,** Lattices with unique complements, 1988
68 **A. G. Postnikov,** Introduction to analytic number theory, 1988
67 **A. G. Dragalin,** Mathematical intuitionism: Introduction to proof theory, 1988

(See the AMS catalog for earlier titles)